International Conference on Aviation Engineering and Transportation (AviaEnT 2020)

IOP Conference Series: Materials Science and Engineering
Volume 1061

Irkutsk, Russia
21 - 26 September 2020

Editors:

Igor Bobarika
Belomestnykh Aleksandr Sergeevich

ISBN: 978-1-7138-3340-6
ISSN: 1757-8981

Printed from e-media with permission by:

Curran Associates, Inc.
57 Morehouse Lane
Red Hook, NY 12571

Some format issues inherent in the e-media version may also appear in this print version.

Institute of Physics
Dirac House, Temple Back
Bristol BS1 6BE UK

Phone: 44 1 17 929 7481
Fax: 44 1 17 920 0979

techtracking@iop.org

Additional copies of this publication are available from:

Curran Associates, Inc.
57 Morehouse Lane
Red Hook, NY 12571 USA
Phone: 845-758-0400
Fax: 845-758-2633
Email: curran@proceedings.com
Web: www.proceedings.com

TABLE OF CONTENTS

IOP Conf. Series: Materials Science and Engineering 1061 (2021) 011001 doi:10.1088/1757-899X/1061/1/011001

PREFACE

The international conference Aviation Engineering and Transportation (AviaEnT 2020) was successfully held at Irkutsk National Research Technical University, Irkutsk, Russia, near the Baikal Lake, on 21-26 September 2020. The main organizers of AviaEnT 2020 were Irkutsk National Research Technical University (Irkutsk, Russia) and the Belarusian-Russian University (Mogilev, Republic of Belarus). The conference was originally scheduled on May, but due to the CoViD 2019 pandemic, it was postponed.

AviaEnT 2020 aims to develop the scientific cooperation between engineers and industry researchers and provides a forum for them to present and discuss the latest innovations and developments in the fields of Aircraft dynamics and control, Mechanical analysis of alloys and composites, Reaction engines design and physics, Aircraft structural mechanics, Antennas and RF systems, Positioning and aeronavigation, Data acquisition and signal processing, Control engineering and regulators, FMS software, CAD and CALS, Risk and reliability assessment, and Air traffic management models.

The participants presented their research on topical issues of aviamechanical engineering, machine-building technologies, road-building and mining machinery, transport management and logistics. After careful reviewing, the most valuable and significant research papers were selected for publication. We sincerely hope that the AviaEnT 2020 proceedings will show the readers a broad overview of the latest research results in related fields. For the best convenience of non-local participants, organizing committee introduced an opportunity of virtual participation via Zoom. Six days of the conference has certainly enriched all the participants with new ideas for their future research.

We would like to express our sincere gratitude to all the participants and members of the organizing committee for their hard work when preparing for the conference. We are grateful to Irkutsk National Research Technical University for hosting the AviaEnT 2020. We also would like to thank all the reviewers and editors for their great contribution to preparing the conference proceedings.

On behalf of the Organizing Committee of AviaENT 2020,
Igor O. Bobarika

Editors

Igor Bobarika, Irkutsk National Research Technical University, 83 Lermontova str., Irkutsk, 665462, Russia, e-mail: bobarika_io@ex.istu.edu

Belomestnykh Aleksandr Sergeevich, Irkutsk National Research Technical University, Irkutsk, 665462, Russia, e-mail: berkut1@mail.ru

Program Chair:

Kornyakov Mikhail Viktorovich – Dr., Rector of the Irkutsk National Research Technical University (Irkutsk, Russia);

Pashkevich Victor Mikhaylovich – Dr., Vice Rector of the Belarusian-Russian University (Republic of Belarus, Mogilev);

Gromyko Pyotr Nikolaevich – Dr., prof., Belarusian-Russian University (Republic of Belarus, Mogilev);

Yakubovich Dmitry Ivanovich – PhD, prof., Belarusian-Russian university (Republic of Belarus, Mogilev) ;

Malozyomov Boris Vitalyevich – PhD, Novosibirsk State Technical University (Novosibirsk, Russia);

Vilberger Mikhail Evgenyevich – Novosibirsk State Technical University (Novosibirsk, Russia);

Pavel Stibor – Executive Director of the "Domist" company (Czech Republic);

Saidaminov Isokhon – Dr., prof., Tajik university of technologies (Tajikistan);

Nguyen Huu Locke – prof., University of technologies of Ho Chi Minh (Vietnam);

WU Xiaogang – professor Harbin University of Science and Technology (Harbin, China).

Organizing Committee:

Organizing Chair:

Semyonov E.Yu. - PhD, Vice Rector of Irkutsk National Research Technical University (Irkutsk, Russia)

Govorkov A. S. – PhD, Head of department of scientific activity of Irkutsk National Research Technical University (Irkutsk, Russia)

IOP Conf. Series: Materials Science and Engineering 1061 (2021) 011001 doi:10.1088/1757-899X/1061/1/011001

Bobarika I. O. – PhD, Irkutsk National Research Technical University (Irkutsk, Russia)

Akhatov R.H. – PhD, Irkutsk National Research Technical University (Irkutsk, Russia)

Lytkina A.A. – Irkutsk National Research Technical University (Irkutsk, Russia)

Martyushev N.V. – PhD, Tomsk Polytechnic University (Tomsk, Russia).

IOP Conf. Series: Materials Science and Engineering 1061 (2021) 011002 doi:10.1088/1757-899X/1061/1/011002

Peer review declaration

All papers published in this volume of *IOP Conference Series: Materials Science and Engineering* have been peer reviewed through processes administered by the Editors. Reviews were conducted by expert referees to the professional and scientific standards expected of a proceedings journal published by IOP Publishing.

- **Type of peer review:** Single-blind peer review (by two committee members – session head and volume editor).
- **Conference submission management system:** submissions were received and handled via e-mail (aviaent@mail.ru)
- **Number of submissions received:** 98
- **Number of submissions sent for review:** 62
- **Number of submissions accepted:** 46
- **Acceptance Rate (Number of Submissions Accepted / Number of Submissions Received X 100):** 47%
- **Average number of reviews per paper:** 2
- **Total number of reviewers involved:** 8
- **Any additional info on review process:** the review pipeline was organized as follows
  1. secretary of the committee handled the correspondence via email aviaent@mail.ru and formally checked submissions;
  2. all qualified submissions were forwarded to one of the editors;
  3. editors reviewed papers by themselves and assigned an extra referee to each paper;
  4. at the final stage style and spell checks of each accepted paper were performed by the language editor.
- **Contact person for queries:** Igor O. Bobarika, editor-in chief, head of the department of Aircraft Construction and Maintenance, Irkutsk National Research Technical University, bobarika_io@ex.istu.edu

Passenger trips analysis determined by processing validation data of the electronic tickets in public transport

A I Fadeev, S Alhusseini

Siberian Federal University, 79, Svobodny Avenue, Krasnoyarsk, 660041, Russia

E-mail: fai@ak1967.ru

Abstract. Based on the set of trip-legs obtained by the algorithm described in our earlier work- trip-legs were sorted by their start or end time for each ticket- we managed to identify the set of passenger correspondence which consists of one or several trip-leg based on the analysis of the transfer conditions. Advanced and in-depth study of passenger correspondence is pivotal for solving transport-planning tasks and avoiding any mistake happened at some point at trip identification. In this work, we will process two matters: distribution transfers by time and identify short activities trips. This work shows that for a two-hour period of free transfer, the share of free trips will be about 17% of the total number of trips. In details, 12% of passengers per day will take one free trip and the share of passengers who will perform two or more free transfers is about 5% of trips. For a one-hour free transfer period the share of free transfer will be about 10%. Passengers with one free transfer will make about 8% of trips. Also, after identifying the short activities trips and dividing these trips into separate trips, the share of trip without transfer increased by 7%. The transfer factor decreased by 0.04.

1. Introduction

According to electronic tickets validation data it is possible to determine transit demand and form a set of passenger trips when it is necessary. The passenger trip (transit rider) from the origin stop to the destination stop is determined by analysing the sequence of the electronic tickets validation data, which were recorded and stored by automated fare collection system. Passenger trip is formed from passenger trip segments (trip-leg), the passenger's correspondence can consist of one or several trip-legs.

Based on the set of trip-legs received from the algorithm described in our earlier work [1, 2], which were sorted by their start or end time for each ticket, we managed to identify set of passenger trips consisting of one or several trip-leg based on the analysis of the transfer conditions [4-8] (figure 1):
-Walk access link;
-Transfer penalty.

In our previous work, we used a transfer conditions for discrimination between trip segment and trip, where exceeding one of these two limits is sufficient to consider trip-leg as trip.
The first condition was walk access link [3] where we assumed that if the distance between alighting stop and the next boarding stop exceed walk access link boundary, then the alighting stop would consider as destination stop for the previous trip and the next boarding as origins for next trip.
Advanced and in-depth study of passenger trips determined by processing validation data of the electronic tickets in public transport is pivotal for solving transport-planning tasks and avoiding any

mistake happened at some point at trip identification. In this work, we will process two matters: identify short activities trips and distribution transfers by time.

2. Identify short activities trips

We observed some passenger trips, the purpose of which is to distinguish short actions at transfer points (often the duration of such actions falls within the transfer penalty) so our algorithm fails to identify these trips and combines them in one trip. We will call such trips as short activities trips. An example of short activities trips is given in figure 2 and table 1. With a considerable length of trip consisting of several trip-legs, the origin and destination stops are located within walk access link.

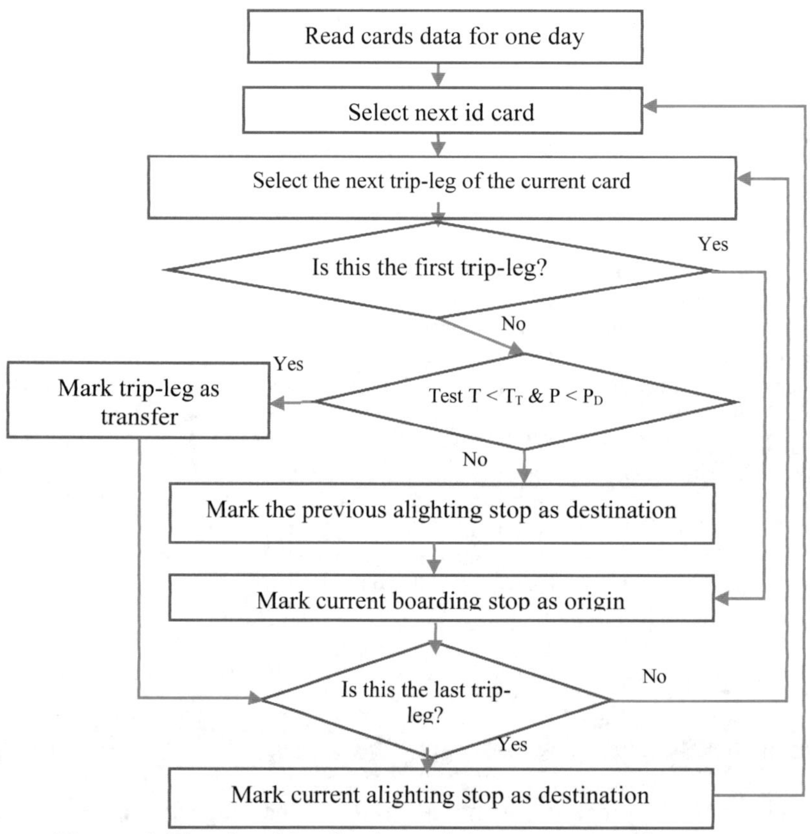

Figure 1. The algorithm for determining passenger trips

The short activities trips are required to be divided into separate trips in order to correctly determine transit demand. In work [4], the following criteria are proposed for identifying short activities trips:

- The time between the alighting for previous trip segment and the boarding for the current trip segment (transfer penalty) exceeds 30 minutes;
- two successive trip segments are made in the same route;
- the length of the trip (compound trip segments) is more than two times the Euclidean distance between the origin and destination stops.

The first criterion (transplantation time) is accepted and already was included in our algorithm; the second criterion proposed in [4] do not fully reflect the real situation. In many cases, the passenger's trip cannot consider as short activities trip, even if its compound trip-legs length is significantly greater than the Euclidean distance between the origin and destination stops. An example of such trip is given in figure 3: the trip was made from stop 75 to stop 103 with a transfer at stop 66 to stop 95. A significant excess of the length of the trip over the Euclidean distance between the origin and destination stops was due to the configuration of the route network of the city.

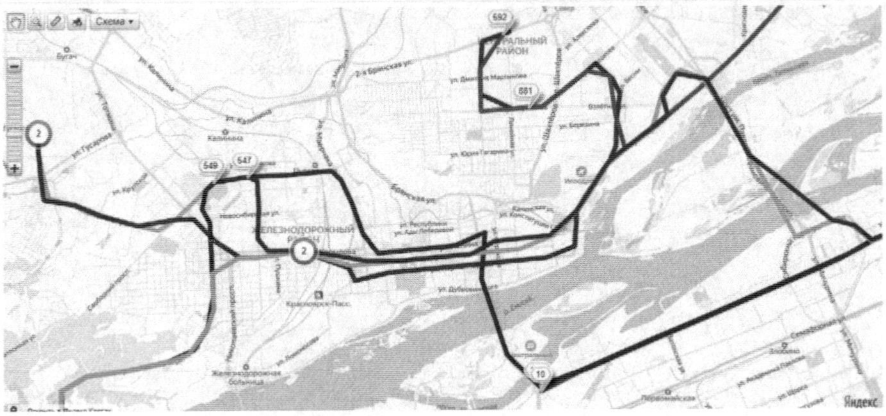

Figure 2. Scheme of short activities trip: start at point 549 and end point 547 which are within walk access link (Schemes of trip and routes obtained using the Yandex.Maps)

Table 1. Trips-legs list for short activities trip

Id trip-leg	Rout and dir.	Id boarding stop	Boarding time (hh:mm)	Id alighting stop	Alighting time (hh:mm)	Length, (km)	Transfer penalty, (h)
13141301	53B	549	11:59	615	12:11	4,4	0,00
13141300	49B	616	12:20	881	13:24	19,4	0,14
13141318	20B	592	13:42	418	14:46	23,7	0,31
13141299	2B	418	14:53	175	15:28	10,0	0,11
13141332	2A	10	15:58	212	16:21	8,9	0,50
13141319	2B	213	16:33	641	17:47	24,7	0,20
13141283	83B	640	17:52	1006	18:12	6,9	0,09
13141298	85A	1006	18:14	384	19:29	26,2	0,04
13141331	85B	384	19:38	547	20:57	27,7	0,15

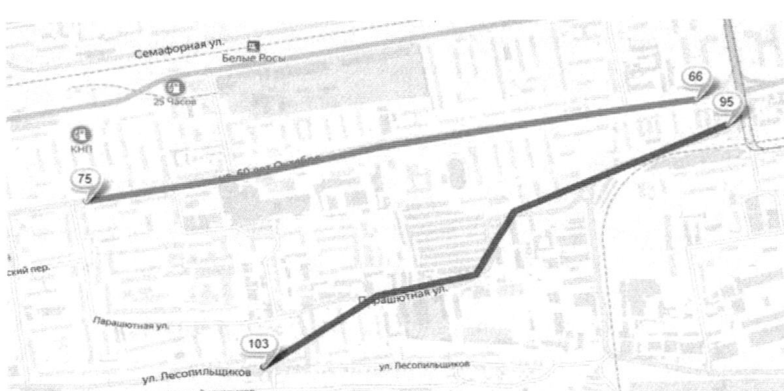

Figure 3. Illustration of trip, which its length 4.9 times longer than the Euclidean distance between the origin and destination stops

In addition, the third criterion, which based on the identification of successive trips along the same route, is not cover all cases. In figure 4 an illustration of successive trip segments made in different routes but on the same network section. Obviously, these trip segments should be considered as two trips, despite the fact that the transfer penalty (at stops 198, 199) was less than 30 minutes (14.4 minutes) and formally they were made in different routes.

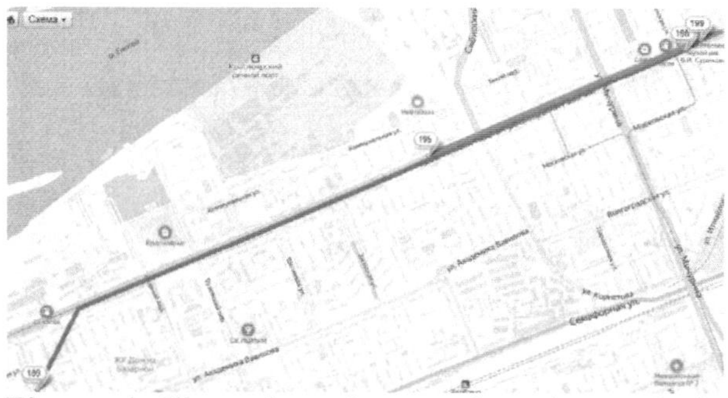

Figure 4. Illustration of successive trips made in different routes on the same network section

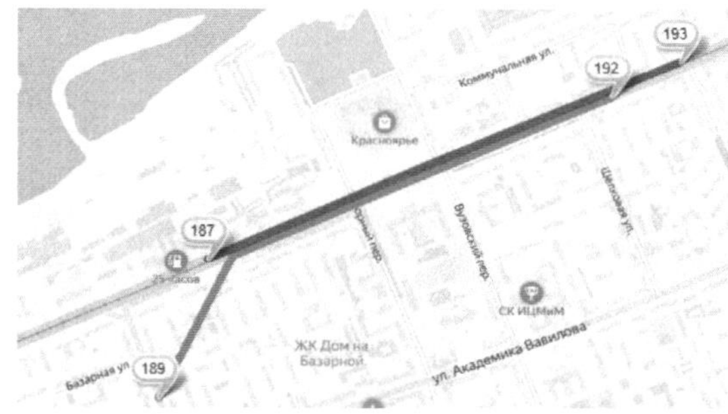

Figure 5. Illustration of successive trip segments, which should be considered as separate trips according to the criterion of walk access link between the origin and destination stops

In this paper, the following criteria are proposed for identifying short activities trips. The passenger trip is considered as a short activities trip, which is subjected to fragmentation if at least one of the following conditions is met:

- Transfer penalty exceeds 30 minutes;
- the origin and destination stops of trip are located within walk access link (see figure 5);
- there are more than three trip-legs in the passenger trip (2 transfers);
- the length of trip is more than 2 times the shortest length along the route network between the origin and destination stops of the trip, this coefficient value was established on the basis of a selective analysis of trips with transfers along the route network of the city of Krasnoyarsk for April 2019;
- trip segments were made on the same network section in different directions (see figure 4): the final (alighting) stop of the later segment is within walk access link from the stopping-point of the previous segment or the starting (boarding) stop of the previous segment is within walk access link from the stopping-point of the later segment (see figure 6).

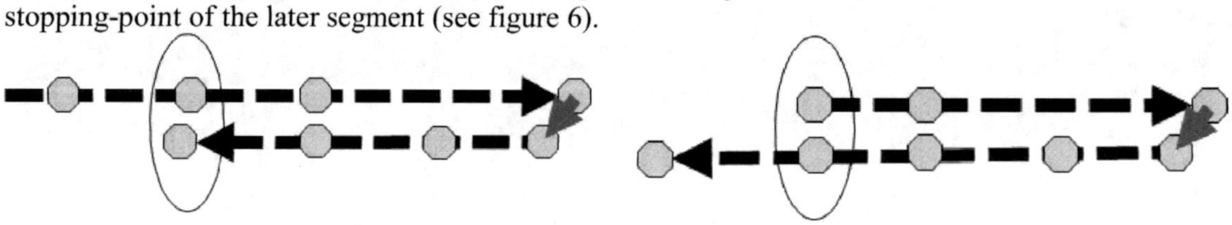

A) B)

Figure 6. Illustration of passenger trip over a network section in different directions
A) the alighting stop of the later segment is within walk access link from the intermediate stop of the previous segment
B) the boarding stop of the previous segment is within walk access link from the intermediate stop of the later segment

In table 2 shows the distribution of short activities trips determined in accordance with the proposed indicators (criteria). The table shows that 91% of passenger trips is without transfer (consists of one segment), 8% of trips is made with one transfer and the remaining about 1% of trips with two or more transfers. The largest number of passenger trip segments is 8 (7 transfers), in April 2019 there were 2 such trips.

The criterion of transfer penalty (0.5 hours) was used in the algorithm of formation passenger trips (see figure 1), for this reason, in table 2 data on the criterion of transfer penalty is not given.

According to the criterion of walk access link, about 14% of trips with transfers (consisting of 2 or more segments) were revealed as short activities trips. For these trips, the origin and destination stops were located within walk access link (less than 500 m), while the total length of trips is a few kilometers (tens of kilometers).

Table 2. Short activities trips structure identified by proposed Criteria

Segment	Trips		Short activities trips (criteria)							
	TOTAL		Walk access link		Trans. over the same net. sec.		actual to shortest possible length		TOTAL	
	Amount	Weight (%)	Amount	Weight (%)	Amount	Weight (%)	Amount	Weight (%)	Amount	Weight (%)
1	2154165	91,4								
2	189363	8,0	25557	13,5	71897	38,0	54487	28,8	79716	42,1
3	10226	0,4	1669	16,3	8203	80,2	5715	55,9	8577	83,9
4	1689	0,1	615	36,4	1642	97,2	1446	85,6	1689	100,0
5	146	0,0	54	37,0	146	100,0	135	92,5	146	100,0
6	27	0,0	4	14,8	27	100,0	26	96,3	27	100,0
7	5	0,0	1	20,0	5	100,0	5	100,0	5	100,0
8	2	0,0			2	100,0	2	100,0	2	100,0
*	201458		27900	13,8	81922	40,7	61816	30,7	90162	44,8
**	2355623	8,6	27900	1,2	81922	3,5	61816	2,6	90162	3,8

Note. * - trips with transfers; ** all trips.

More than 40% of the total trips with transfers were classified as short activities trips by the criterion of transference over the same network section and more than 30% by the criterion of the actual length of trip to the shortest possible length (in accordance with the routes network configuration) ratio. When determining the possible trip length, trips with one segment (without transfer) or with two segments (with one transfer) was considered. Trips with a large number of segments are classified as short activities trips according to the criterion of the number of transfers.

3. Distribution transfers by time

We will study the distribution of transfers and determine their preparation during a specific period by using validation data of the electronic tickets. This study can be used to impose a reduced fare on the second trip segment, which occur after transfer during a specific time period or even provide a period of time in which the transfers are free, which gives our study the importance as it determines the number of trips that will be discounted or even free.

In table 3 is given the distribution of electronic ticket holders by the number of carried out trips per day.

Table 3. Distribution of electronic ticket holders by the number of carried out trips per day

Num. of trips	Passenger-day	Weight (%)	Num. of trips	Passenger-day	Weight (%)
1	876131	34,5	11	54	0,0
2	1134825	44,6	12	25	0,0
3	315165	12,4	13	11	0,0
4	154731	6,1	14	9	0,0
5	40250	1,6	15	3	0,0
6	14041	0,6	17	1	0,0
7	4204	0,2	19	2	0,0
8	1544	0,1	24	1	0,0
9	478	0,0	31	1	0,0
10	145	0,0	Total	2541621	100,0

The table shows that 34% of passengers using an electronic ticket make one trip a day, about 47% make two trips, 12% make three trips, etc. There is a small amount of passengers, which are carried out more than 5 trips per day. The proportion of such passengers is about 0.8%.

Consider the distribution of free or discounted fare trips under various conditions of free transfer's period. We will consider two options for the condition of free transfers: within two hours (a) and within one-hour (b). We highlighted the trips with transfers and denote trips, for which fare should be debited or free of charge based on the conditions:

$$\begin{cases} t^0 = \min(t_i), i = \overline{1,n} \\ t^0 = \min(t_i), if\ t_i > \max\left(t_j^0\right) + \tau, i = \overline{1,n}; j = \overline{1,n'} \end{cases} \quad (3.1)$$

Where: n - is the number of trips for ticket per day;
τ - Period of free transfer, hour.

In accordance to (3.1), each first trip of the day is considered paid and then each trip, the start time of which exceeds the interval allotted for a free transfer also is considered paid.

4. Conclusion

Accordance to table 2 on the public network of the Krasnoyarsk city, more than 8% of passenger trips currently are tagged as trips with transfers. Of these, about half (3.8%) are short activities trips, i.e. trips carried out for specific (short) purposes at transfer points. Moreover, in many cases, the conditions of match trip segments in time and walk access link are fulfilled. From table 2 shows that the sum of shares for each criteria exceeds the total share of short activities trips. This is because some trips satisfy more than one indicator, for example, the criterion of walk access link and the criterion of transference over the same network section are linked.

In table 4 a set of trips and trip segment (trip-leg) are given with and without taking into account the existence of short activities trips in the structure of transit demand. The table shows that initially in the structure of passenger trips, the share of trips without transfer amounted to about 84% of the total number of trips. The transfer factor (i.e., the ratio of the all trip segments to trips number) amounted to 1.09 (about 9% of passengers carry out transport needs with transfers).

After identifying the short activities trips and dividing these trips into separate trips, the share of trip without transfer increased to 91%. The transfer factor decreased to 1.05, i.e. only about 5% of passengers' trips consist of two or more segments.

In table 5 provides a set of free trips for two interval options: two hours and one hour. The table shows that for a two-hour period of free transfer, the share of free trips will be about 17% of the total number of trips. In details, 12% of passengers per day will take one free trip and the share of passengers who will perform two or more free transfers is about 5% of trips.

For a one-hour period of free transfer, the share of free transfer will be about 10%. Passengers with one free transfer will make about 8% of trips.

Obviously, the number of free transfer can be limited, for example, at the level of two free transfer per day. A smaller number of free transfer is illogical, as the passenger performs two work trips per day. However, the restriction to two free transfers per day is inefficient, since in this case the number of free transfers is reduced by only one percent.

Table 4. Set of trips and trip segment (April 2019, Krasnoyarsk city)

Segment num.	Total trips			Trips after identifying the short activities trips		
	Trips	Trip segments	Share (%)	Trips	Trip segments	Share (%)
1	2154165	2154165	83,8	2347027	2347027	91,3
2	189363	378726	14,7	109647	219294	8,5
3	10226	30678	1,2	1649	4947	0,2
4	1689	6756	0,3	0	0	0,0
5	146	730	0,0	0	0	0,0
6	27	162	0,0	0	0	0,0
7	5	35	0,0	0	0	0,0
8	2	16	0,0	0	0	0,0
Total	2355623	2571268	1,09	2458323	2571268	1,05

Table 5. Distribution of trips with free transfers

Transfer number	Two-hour period of free transfer			One-hour period of free transfer		
	Passenger-day	Free trips number	Share (%)	Passenger-day	Free trips number	Share (%)
1	342639	342639	11,84	220660	220660	7,63
2	56682	113364	3,92	30241	60482	2,09
3	8698	26094	0,90	2308	6924	0,24
4	1183	4732	0,16	232	928	0,03
5	195	975	0,03	24	120	0,00
6	35	210	0,01	4	24	0,00
7	7	49	0,00	1	7	0,00
9	1	9	0,00			
Total		488072	16,87		289145	9,99
All trips		2893013			2893013	

5. Acknowledgments

The authors are grateful to Municipal Public Institution of the Krasnoyarsk City (MSE) "Krasnoyarskcitytrans" for providing the data for this research.

References

[1] Fadeev A, Alhusseini S and Belova E N, 2018, Monitoring Public Transport Demand Using Data From Automated Fare Collection System *Advances in Engineering Research*, vol. **158**, pp. 5–12

[2] Fadeev A and Alhusseini S 2019 Using Automated Fare Collection System Data To Determine Transport Demand *Advances in Engineering Research*, vol. **188**

[3] Munizaga, M.A. and Palma, C., 2012. Estimation of a disaggregate multimodal public transport origin-destination matrix from passive Smart card data from Santiago, Chile. *Transportation Research* **24**, 9–18.

[4] Barry J J, Freimer R and Slavin H L 2009 Use of entry-only automatic fare collection data to estimate linked transit trips in New York City *Transp. Res. Rec. J. Transp. Res. Board* **2112** 53–61

[5] Barry J J, Newhouser R, Rahbee A and Sayeda S 2002 Origin and destination estimation in New York City with automated fare system data *Transportation Research* **1817** 183–7

[6] Zhao J, Rahbee A and Wilson N 2007 Estimating a rail passenger trip origin- destination matrix using automatic data collection systems *Computer Aided Civil and Infrastructure Engineering* **22** 376–87

[7] Trépanier M, Tranchant N and Chapleau R 2007 Individual trip destination estimation in a transit smart card automated fare collection system *Journal of Intelligent Transportation Systems* **11** 1–14

[8] Devillaine F, Munizaga M A and Trépanier M 2012 Detection activities of public transport users by analyzing smart card data *Transportation Research Record* **2276** 48–55

[9] Munizaga M A, Devillaine F, Navarrete C and Silva D 2014 , Validating travel behavior estimated from smartcard data *Transportation Research Part C Emerging Technologies* **44**:70–79

Research of recognition accuracy of dangerous and safe x-ray baggage images using neural network transfer learning

N A Andriyanov[1,2], Al K Volkov[3], An K Volkov[3], A A Gladkikh[2,4]

[1] JSC "RPC "Istok" named after Shokin", Vokzalnaya Str., 2A, 141190, Moscow Region, Russia
[2] Ulyanovsk State Technical University, Severnyj venec, 32, 432027, Ulyanovsk, Russia
[3] Ulyanovsk Civil Aviation Institute, Mozhaisky Street, 8/8, 432071, Ulyanovsk, Russia

[4] E-mail: a_gladkikh@mail.ru

Abstract. The article considers the use of neural networks to solve the problem of recognizing dangerous and safe objects carried in the luggage of airport passengers. A comparative analysis is performed to define the accuracy achieved on the test sample for different convolutional neural networks. It also explores the influence of various regularizations on the accuracy of a two-class classification. The increased probability of correct recognition is achieved due to augmentation, reset weights and saturation of the network. The method of transfer training is used to increase the efficiency of the recognizer. In this case, a study was carried out for the transfer of various neural networks.

1. Introduction

Among relevant tasks on computer vision, one can single out the task of recognizing dangerous objects that can be used to commit an act of unlawful interference, according to the results of an X-ray baggage or hand luggage scan by the aviation security service [1-3]. Moreover, automation of such a process is possible using convolutional neural networks (CNN) [4,5], which are currently replacing algorithms focused on mathematical image modeling [6, 7]. Now there are many approaches to the learning process of neural networks, and their application to solve this problem is advisable. At the same time, the special interest in solving a narrow specialized problem, such as transfer learning and the use of regularizations for X-ray data, has not been fully studied. Therefore, in this paper, special attention is paid specifically to improving recognition accuracy due to transfer learning and regularization.

2. A brief analysis of the current state of the problem

Currently, screening at airport checkpoints is carried out by security officers who visually check x-ray images of baggage and hand luggage of passengers in order to determine whether they are harmless or contain prohibited items and, therefore, ask additional screening using explosives detection or manual inspection of baggage [2]. Examples of prohibited items are pistols, knives, improvised explosive devices, self-defense gas cans or devices for electric shock [8]. Inspection of x-ray images by airport security officers (inspectors) includes visual search and decision-making [9,10]. The problems of visual search include low target prevalence, changing target visibility, finding an unknown set of goals, and the possible presence of multiple targets [11,12].

However, in the considered works [2,8–12], computer vision methods are considered only as secondary. The main attention is paid to aspects of the operator's perception of the visual flow in order to classify the recorded images as accurately as possible. In [13], a neural network was developed that could effectively separate dangerous and safe objects in X-ray images. However, with the training used from scratch, a sufficiently large error remains.

At the same time, one of the popular techniques for preliminary adjustment of neural networks to solve a narrow task in computer vision is to use the experience of a neural network that has proven itself in a wider class of problems, or, in other words, transfer learning [14]. With this approach, a trained neural network is used as a layer (in fact, many layers) of feature extraction. At the same time, a narrow task leaning occurs only in the output fully connected layer, since the coefficients of the internal network do not change. It provides quite high results in accuracy with much lower computational work.

In addition, it is necessary to develop neural networks in such a way that they have the greatest generalizing ability for a specific narrow task. To do this, the widest range of options needs to be considered. Augmentation algorithms are often used due to the lack of data [15]. Finally, great versatility can be achieved by using a weight reset [16], since this avoids neural network overfitting.

In this article, transfer learning will be considered for the following neural networks:

- VGG-19: is a convolutional neural network that contains 19 layers in depth. It is possible to download a pre-trained version of the network, trained on more than a million images from the ImageNet database. A pre-trained network can classify images into 1,000 categories of objects, such as a keyboard, mouse, pencil, and many animals. The network has an input image size of 224x224.
- U-Net:is a convolutional neural network that was created in 2015 to segment biomedical images in the Computer Science division of the University of Freiburg. The network architecture is a fully-connected convolutional network modified so that it can work with fewer examples (images of learning) and make more accurate segmentation.
- GoogleNet: is a convolutional neural network that contains 22 layers in depth. It is possible to download a pre-trained version of the network, trained on either ImageNet or Places365 databases. An ImageNet network categorizes images into 1,000 categories of objects, such as keyboards, mice, pencils, and many animals. A Places365 network is similar to ImageNet network, but classifies images into 365 different place categories, such as field, park, runway and lobby. An ImageNet network will be used. Pre-trained networks have an input image size of 224x224.

3. Description of the original data sample

The database of x-ray images included both dangerous and safe objects. The following groups belong to dangerous objects: improvised explosive devices, cold weapons and firearms, as well as other objects (stun guns, gas sprays, etc.). Safe items include various household items, clothing, food, etc. It should be noted that the database of x-ray images of safe objects will objectively always surpass the similar base with dangerous objects due to their low target prevalence. The task of binary classification is also complicated by the following facts: inside the classes "dangerous" and "safe", which are the outputs of the network, x-ray images of several object groups are presented with their uneven distribution, moreover, a significant area in the images can have different sizes. Fig. 1 shows examples of x-ray images. The first row of X-ray images includes dangerous objects, the second row includes safe objects.

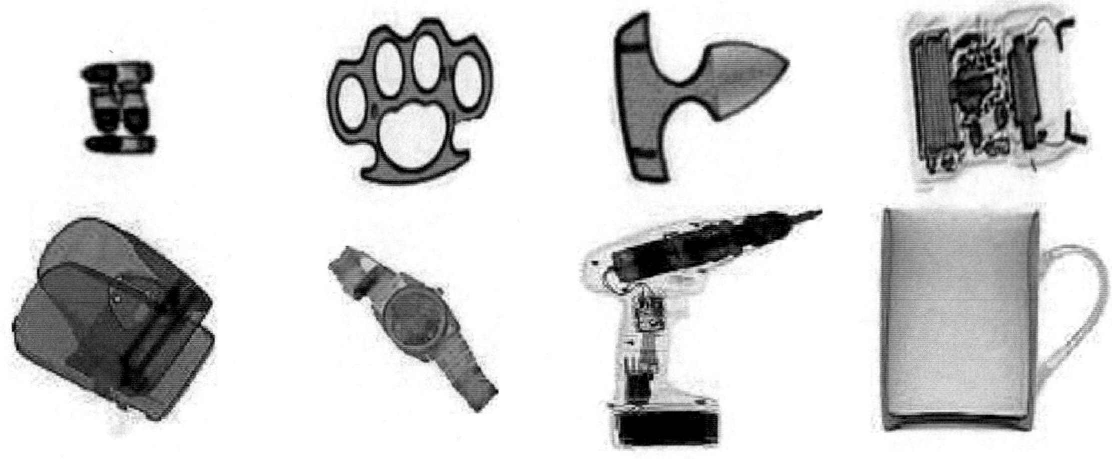

Figure 1. X-ray image examples

Before studying recognition efficiency under various training methods, all available images were divided as follows: 1894 safe and 855 dangerous in the learning sample, 106 safe and 52 dangerous in the test sample

4. The study of weight regularization and augmentation impact on a neural network

First, we will perform learning and inference of a neural network trained on the basis of a conventional convolutional architecture with zero coefficients. We'll set up 50 learning epochs with ADAM optimization, and each layer will contain 64 neurons. The fully connected layer contains 128 neurons. All image sizes are reduced to 224x224.

Table 1 presents the recognition results on a test sample for neural networks with different depth (number of layers). Table 2 presents the recognition characteristics of a 7-layer network for various weight loss probability, that is a change in the weight coefficient for a neuron with a certain probability set to 0 at each epoch. Finally, Table 3 shows results when applying augmentation to a learning image in a given epoch with some given probability for each image. It should be noted that the number of augmentations considered shifts, turns, scaling and adding white noise, as well as blurring images. In this case, the data augmentation algorithm works so that it is possible to overlay different augmentations to each other.

Increased network depth and augmentation results in better recognition accuracy.

Table 1. Accuracy of the layer number

Number of layers	Recognized, %
1	67.722
3	74.684
5	77.848
7	82.278

Table 2. Accuracy from weight loss

Probability of weight loss	Recognized, %
0.1	79.114
0.2	77.215
0.3	62.025
0.4	33.544
0.5	32.911

Table 3. Accuracy from data augmentation

Augmentation probability	Recognized, %
0.1	82.911
0.2	86.076
0.3	84.177
0.4	82.911
0.5	81.013

Obviously, the accuracy in this case is directly proportional to the depth of the network. However, over time, a further increase in the number of layers will lead to a much larger increase in computational work than a gain in recognition accuracy. It is interesting that weight loss in this problem led to a significant decrease in recognition accuracy, therefore, the parameter characterizing the probability of weight loss was chosen to be 0.1 for future studies. At the same time, for augmentations, the maximum accuracy was achieved with a probability of 0.2. At the same time, for the considered example, resetting the scales did not lead to an improvement in the quality of the network, and a slight augmentation made it possible to increase the accuracy by about 3.8%.

5. The study of transfer learning impact on a neural network

Now consider the results of the neural network, which learning took place with pre-trained weights. There were 128 neurons in the fully connected learning layer. Given the transfer learning, a fully connected layer has been trained for 10 epochs. When submitting test images, there were used augmentations with probability $p = 0.2$ and the probability of weight regularization on a fully connected layer was set at $p = 0.05$.

Table 4 presents the results for various neural networks.

Table 4. Transfer learning

Neural network	Accuracy, %
Trained from scratch (2nd line of the 3rd table)	86.076
U-Net	91.244
GoogleNet	91.365
VGG-19	93.018

The results show that transfer learning is extremely advisable. The accuracy was increased by 7% using VGG-19 network weights compared to the best network trained from scratch.

6. Conclusion

Thus, with a relatively small initial image sample, sufficiently high recognition percentage was achieved when augmentation and transfer learning were used. In particular, the best results were obtained for the VGG-19 network, which can be recommended for further research in the field of X-ray image processing.

7. Acknowledgments
The reported study was funded by RFBR, project number 19-47-730011.

Reference
[1] Chavaillaz A, Schwaninger A, Michel S, Sauer J 2019 Expertise, Automation and Trust in X-Ray Screening of Cabin Baggage *Frontiers in Psychology* **10(256)** pp 1-11 DOI: 10.3389/fpsyg.2019.00256

[2] Sterchi Y, Schwaninger A 2015 A first simulation on optimizing EDS for cabin baggage screening regarding throughput *International Carnahan Conference on Security Technology (ICCST)* pp 55–60 DOI: 10.1109/CCST.2015.7389657

[3] D'Arcy G, Márquez-Grant N, Lane D W 2020 Baggage scanners and their use as an imaging resource in mass fatality incidents *International Journal of Legal Medicine* **134** pp 1419–1429 DOI: https://doi.org/10.1007/s00414-019-02132-y

[4] Indoliaa S, Kumar G A, Mishrab S P, Asopaa P 2018 Conceptual Understanding of Convolutional Neural Network – A Deep Learning Approach *Procedia Computer Science* **132** pp 679–688

[5] Boukaye Boubacar Traore, Bernard Kamsu-Foguem, Fana Tangara 2018 Deep convolution neural network for image recognition *Ecological Informatics* **48** pp 257–268

[6] Andriyanov N A, Dementiev V E 2019 Developing and studying the algorithm for segmentation of simple images using detectors based on doubly stochastic random fields *Pattern Recognition and Image Analysis* **29(1)** pp 1–9 DOI: 10.1134/S105466181901005X

[7] Smith J, Best E, Sum E, Guzel Y, Saville M A, LoMonte L, Wicks M 2015 Physical-model-based image processing for feature aided analysis *2015 International Conference on Electromagnetics in Advanced Applications (ICEAA)* pp 565–568 DOI: 10.1109/ICEAA.2015.7297179

[8] Schwaninger A 2005 Increasing efficiency in airport security screening *WIT Transactions on the Built Environment* **82** pp 407–416

[9] Koller S M, Drury C G, Schwaninger A 2009 Change of search time and non-search time in X-ray baggage screening due to training. *Ergonomics* **52(6)** pp 644–56

[10] Wolfe J M, Van Wert M J 2010 Varying target prevalence reveals two dissociable decision criteria in visual search *Current Biology* **20(2)** pp 121–124 DOI 10.1016/j.cub.2009.11.066

[11] Mitroff S R, Biggs A T, Cain M S 2015 Multiple-target visual search errors *Policy Insights the Behavioral Brain Sciences* **2** pp 121-128 DOI: 10.1177/2372732215601111

[12] Biggs A T, Kramer M R, Mitroff S R 2018 Using cognitive psychology research to inform professional visual search operations *Journal of applied research in memory and Cognition* **7** pp 189–198 DOI: 10.1016/j.jarmac.2018.04.001

[13] Andriyanov N A, Volkov Al K, Volkov An K, Gladkikh A A, Danilov S D 2020 Automatic x-ray image analysis for aviation security within limited computing resources *IOP Conference Series: Materials Science and Engineering* **862(1)** pp 1–5 DOI:10.1088/1757-899X/862/5/052009

[14] Renu Khandelwal 2019 Deep Learning using Transfer Learning *Towards Data Science* https://towardsdatascience.com/deep-learning-using-transfer-learning-cfbce1578659

[15] Buslaev A, Parinov A, Khvedchenya E, Iglovikov V, Kalinin A 2018 Albumentations: fast and flexible image augmentations https://arxiv.org/abs/1809.06839

[16] Srivastava N, Hinton G, Krizhevsky A, Sutskever I, Salakhutdinov R 2014 Dropout: A simple way to prevent neural networks from overfitting *Journal of Machine Learning Research* **15** pp 1929–1958

IOP Conf. Series: Materials Science and Engineering 1061 (2021) 012003 doi:10.1088/1757-899X/1061/1/012003

The internal structure of microprocessors for industrial control and data processing systems

A A Antonov[1], A A Krasnyuk[1,2]

[1] Scientific Research Institute of System Analysis of the Russian Academy of Sciences, Nahinovskij Prospect 36/1, Moscow, 117218, Russia
[2] NRNU «MEPHI» 115409, Russia, Moscow, Kashirskoe hwy, 31, Russia

E-mail: antonov@niisi.msk.ru

Abstract. The fundamental objective of the problem is a comprehensive study of the internal structure of microprocessors for industrial microelectronics, design technology to ensure VLSI efficiency at extremely high temperature levels with external destructive factors. Based on the data we have studied, the main problem of choosing optimal structural solutions for industrial systems based on them, providing the best resistance and fault tolerance of memory and microprocessors under difficult operating conditions, requires a preliminary comparative analysis of various schemes and functional solutions used for IP blocks.

1. Introduction

The industrial equipment system for each application and for each unique object requires the creation of a new computer system. Currently, there is a significant increase in the complexity of objects using high performance computing systems. At the same time, the cost of industrial equipment is growing many times over, many tasks, due to their high complexity and cost, become international and, as a result, require the necessary standardization of electronic equipment. The basis of industrial computing systems is currently not only a multichannel device for collecting, storing and processing information, but also a decision-making system based on the concept of artificial intelligence (AI) and the industrial Internet of things (IoT). System unification can take place both at the level of microprocessor cores, external interfaces, communication media, RAM and external storage devices, data transmission and processing circuits. Unification should also take place at the system software level. The article discusses microcircuits developed in SRISA for computing and controlling industrial systems.

2. Microprocessor architecture

One of the main directions of SRISA for the last 10 years is the design of electronic computer components for space and industrial applications. To create new design rules and experimental studies, it was necessary to determine the desired composition of VLSI based on known industry and new technologies [1]. It is assumed that the developed microprocessors will have the main negative external factor primarily high temperature (more than 150C) and failures caused by cosmic particles with high energies. Thus, the internal structure of the microprocessor must be selected based on high

IOP Conf. Series: Materials Science and Engineering 1061 (2021) 012003 doi:10.1088/1757-899X/1061/1/012003

performance and fault tolerance. This is in line with industry design rules and IP blocks for low-power (LP), ultra-low-power (ULP), and fault-tolerant systems.

The industrial microelectronics is one of the most popular directions of development of the modern electronics [2]. The fundamental objective of the prodlem is a comprehensive study of the physical and technological and schematic solutions, design techniques to ensure efficiency VLSI at extremely high levels of temperature and external destructive factors. This part of the work investigated the fundamental features of the existing CMOS and SOI CMOS microprocessors. There is a tendency to reduce the topological rules, for example from 0.35 μm to 0.065 μm. Also the maximum operating temperature for industrial μP is reducing from 250-300°C до 175-225°C. Studies have shown that problem of high temperature VLSI may be solved by using not only most important parameters for memory circuits, new temperature models and methods of investigations, but also the development recommendations and design considerations in SOI CMOS Integrated Circuits. Solving these problems will create a fundamental basis for the development of high temperature microprocessors, original topological and structural solutions, libraries, VLSI topology, improving main VLSI characteristics.

Basic MIPS-like microprocessor architecture KOMDIV32 and KOMDIV64 include various blocks composition, such as: controller of exchange with external memory implements the fault-tolerant exchange protocol; controller GPIO; controller interface MIL-STD-1553B (MKIO) with a redundant channel; SpaceWire interface controller with a backup channel; controller RS-232 interface; controller serial interface SPI and so on – Figure 1 [3].

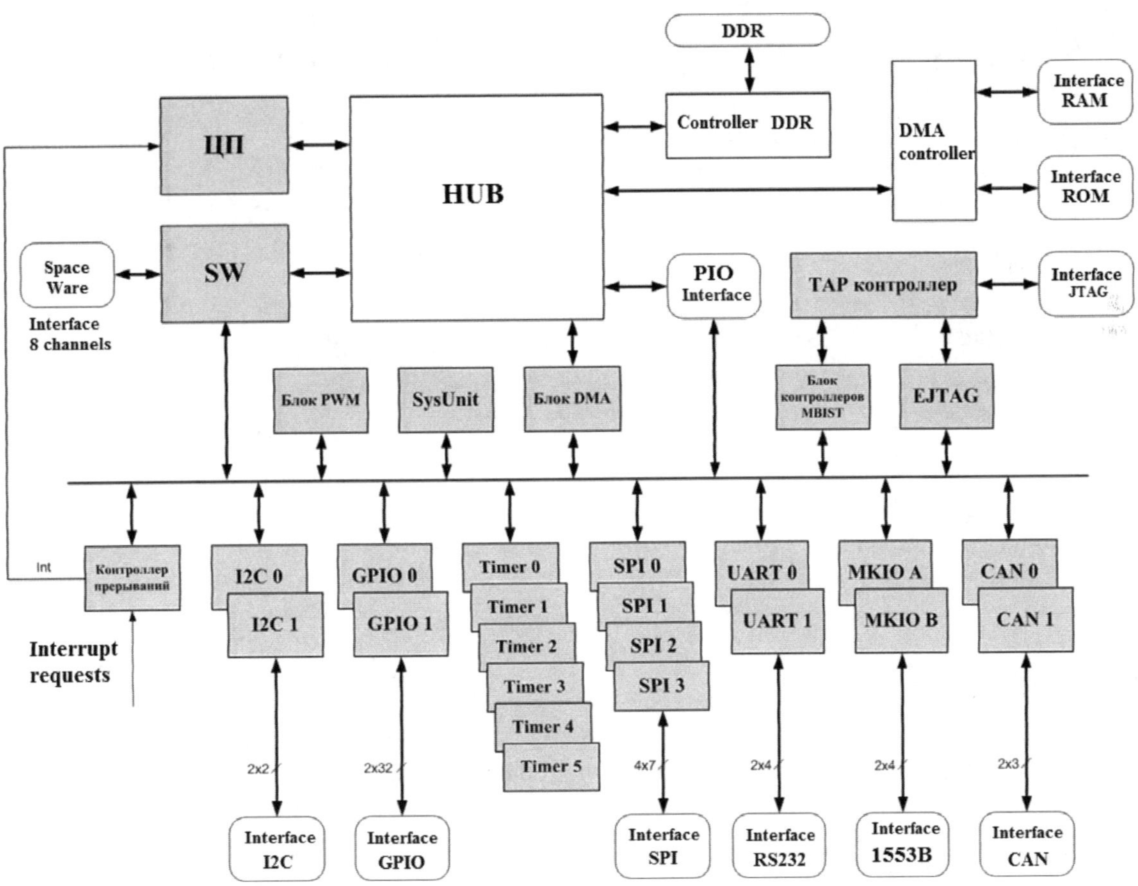

Figure 1. The MIPS-like μP structure KOMDIV32 and KOMDIV64

The proposed architecture is based on the use of IP solutions for a complex of blocks, which allows you to form various versions of microprocessors from industrial microcontrollers to high-performance Space class processors. The list industrial SRISA chips are presented in table 1.

Table 1. The list industrial SRISA chips

Parameters and characteristics	1907VM014	1907VM044	1907VM056	Current project
Microprocessor architecture	KOMDIV32	KOMDIV32	KOMDIV32	KOMDIV32
Maximum core frequency (МГц)	100	66	100	150
Fault tolerance	Parity and Hamming ECC	Parity and Hamming ECC TMR for blocks	Parity and Hamming ECC blocks	Parity and Hamming Total TMR
Using SEC/DED for external memory	+	+	+	+
Integrated system controller	+	+	+	+
Internal PLL	+	—	+	+
FPU coprocessor	+	+	+	+
CPU cache	8K + 8K	4K + 4K	8K + 8K	16K + 16K
Built-in memory	—	—	—	128КБ + 32КБ
Interface MIL-STD-1553B (MKIO)	2	2	2	2
Number of SPI interfaces	1	1	2	4
Number of CAN interfaces	—	—	1	2
SpaceWire channels	2	2	8	8
SpaceWire signal levels	2 CMOS	2 CMOS	8 LVDS or 6 LVDS+2 CMOS	8 LVDS or 6 LVDS+2 CMOS
Universal timer	3	3	3	6

To increase the stability of chips to the effects of temperature and single failures used special circuits of functional elements, and to ensure the protection of the control process in microprocessors of errors applies triple modular redundancy (TMR). As cores were used established at the SRISA of core MIPS-like architecture KOMDIV32 and KOMDIV64. Thanks to the unification of processor cores and peripheral interface blocks for high-performance systems, and systems for space applications held in-depth verification of the chip for space applications and software for them in a wide range of equipment: various control computer of the "Baget" family, high-performance processing devices for signal processing, etc. For various applications of the systems based on microprocessors can be used operating systems (OS) real-time: operating system real-time (RTOS) "Baget 3.0"; RTOS "Baget 2.0"; QNX. RTOS "Baget 3.0" developed in SRISA, and is intended for software development for single and multiprocessor computer systems working in real time, and to ensure their functioning. "Baget 3.0" is a secure and technologically independent domestic development. These qualities are confirmed by the long operation experience (more than 10 years with the RTOS "Baget 2.0") and support equipment developed in SRISA [2].

For use in flight control systems, space-based with high demands on reliability of the designed system-on-chip 1907VM044 [3]. Chip 1907VM044 is a system-on-chip with fault-tolerant 32-bit RISC processor architecture the KOMDIV and the following set of interfaces: controller of exchange with external memory implements the fault-tolerant exchange protocol; controller GPIO; controller interface MIL-STD-1553B (MKIO) with a redundant channel; SpaceWire interface controller with a backup channel;two controller RS-232 interface; controller serial interface SPI. The main feature of the chip is the implementation of TMR the entire logical structure of the VLSI. All the blocks of memory have a special circuit and topological solutions for protecting data against failures. The controller exchanges with an external storage device implements the fault-tolerant exchange protocol, allows major booting from the PROM. Power consumption of the chip is about 7W on the frequency 66MHz. Ceramic case contains 256 pins.

The computing system consists of specialized computing, control microprocessors, controllers, executive devices is combined through the SpaceWire interface by means of chip multi-port switch 1907VM056. Chip 1907VM056 radiation-resistant multi-port switch with embedded SpaceWire processor architecture division KOMDIV32 is designed for building compact on-board multiprocessor data processing systems for space applications and missile technology. Chip 1907VM056 consist of: controller SpaceWire interface with 8 channels of built-in switch interface SpaceWire; two controllers interface according to MIL-STD-1553B; JTAG interface according to standard IEEE-1149.1 support boundary scan; controller external memory; controller general purpose input-output signals (GPIO); two controllers RS-232 interface; three real time timerds; two controllers of SPI interface; controller CAN interface.

Embedded memory is protected by parity codes have a dedicated circuit and topological solutions to protect stored data against failures. The controller exchanges with an external memory implements the fault-tolerant exchange protocol, allows fault-tolerant booting from the PROM. The supply voltage of the IC is 3.3 V, the maximum power consumption is 6 W at a clock frequency of 100 MHz; the chip has a ceramic case PGA with 407 pins.

3. Technological and architectural techniques and trends

Special requirements on the composition of the internal RAM blocks is not shown, except for a general recommendation to use the most conservative design rules memory arrays and compliance LP and ULP approaches. To ensure the reliability of the systems operating in conditions of high temperature, requires both technological and architectural techniques. To include all technological methods related to chip manufacturing technology, libraries and items, etc. One method is to finalize the radiation-resistant technology such as SOI CMOS to decrease leakage current [4]. Technology includes all methods related to the production technology of chips used original models, libraries, layouts etc. For example, according to Honywell and Vorago SOI CMOS Technologies they have managed to reduce the leakage currents to the level of 1/10 of the original, but in both cases the property of resistance to radiation were reduced (Table 2) [5]. Architectural methods to ensure reliability include the actions associated with changes in functional model: the decrease of heat dissipation; software fault tolerance; monitoring loss of efficiency; the use of LP and ULP-(IP) blocks, i.e. elements with reduced leakage currents. Reducing heat dissipation from the blocks of the chip can be achieved in the following ways: reducing the frequency if disable sync on the unit. Disable blocks that are not used at the moment will reduce the heat dissipation and reduce the temperature.

IOP Conf. Series: Materials Science and Engineering 1061 (2021) 012003 doi:10.1088/1757-899X/1061/1/012003

Table 2. HTE SOI CMOS processes

Options	Honywell Deep Trek	SST(Vorago Technologies), HardSIL	X-FAB XI10	Fraunhofer H035
Size	800 nm	130 nm	1 μm	350 nm
Technology	SOI CMOS	CMOS	SOI CMOS	SOI CMOS
Confirmed time	8000 hours (250°C)	2500 hours (250°C)	>=2250°C	More than 10000 hours (250°C)

4. Conclusion

Based on MIPS-like μP structure, unified microprocessor elements for industrial applications have been developed, allowing you to build various μP versions from the minimum complexity of microcontrollers to high-performance "Space" class μP. To reduce power consumption and increase reliability, the architecture is made hierarchical. The upper layer is a switching medium SpaceWire with a transmission rate of up to 200 Mbps and a lower level of the low-speed multiplex channel over MIL-STD-1553B and RS232/RS485. To create high-level computing systems, it is proposed to use two types of standards: VPX (Euromechanics 6U) for building high-performance computing systems and PCI/104-Express or PCIe/104 for small systems.

The operating system involved the use of the domestic RTOS "Baget 3.0." The operating system is made in accordance with international standards, be portable to various platforms and tested in several systems (several hundred).

References

[1] Bobkov S G *et al.* 2015 *Physics Procedia* **74** 224

[2] Godunov A N *et al.* 2014 *Programming and Computer Software* **40(5)** 259

[3] Bobkov S G *et al.* 2013 *Software & Systems* **4** 49

[4] Bobkov S G 2016 *Journal of Physics: Conference Series* **675** 032014

[5] Volkov S I *et al.* 2018 *Nanoindustry* **9** 220-28

IOP Conf. Series: Materials Science and Engineering 1061 (2021) 012004 doi:10.1088/1757-899X/1061/1/012004

The elastic, strength and thermophysical characteristics modeling of mono- and polycrystalline aluminum Al

I Kh Badamshin[1]

[1] Ufa State Aviation Technical University, 12, Karl Marks ave. Ufa, 450000, Russia

E-mail: adbadamshin@ugatu.ac.ru

Abstract. Temperature dependences of mechanical and thermophysical properties of the material are the initial data for estimating the stress-strain state of structural elements. However, when designing, the amount of this data is usually not enough. Under these conditions, the problem of modeling these properties arises. This problem is solved by taking into account the forces of interatomic interaction in the crystal structure on the example of a single crystal and polycrystal of aluminum. The results of calculating the elastic modulus, elastic limit and yield strength, Poisson's ratio, coefficients of thermal conductivity, thermal expansion and material density are presented. A satisfactory convergence with the available reference data is obtained. The results obtained can be used as input data for the construction of compositions and alloys based on aluminum.

1. Introduction

To assess the strength of the part material, it is necessary to know its temperature dependences of mechanical and thermophysical properties. In the case when there is not enough reference data, it becomes necessary to model them. In this paper the problem of modeling material characteristics is solved by taking into account the forces of interatomic interaction in the crystal structure.

2. Assumptions and calculation formulas

Assumptions

1. An elementary atomic cell of a defect-free single crystal with a lattice period of a_0 is considered.
2. The most closely packed planes of a single atomic cell of a single crystal and the crystallographic directions corresponding to these planes are considered. In particular, for a cubic lattice, planes and directions are shown in Figure 1. Thus, the anisotropy of the properties of a elementary atomic cell of a single crystal is taken into account.
3. Due to the symmetry of the crystal structure, the properties of an elementary atomic cell are transmitted to the macro volume of a single crystal.
4. The stress-strain diagram of a defect-free single crystal is characterized by a so-called "yield tooth» (figure 2, a). The stress-strain diagram model is shown in figure 2, b.

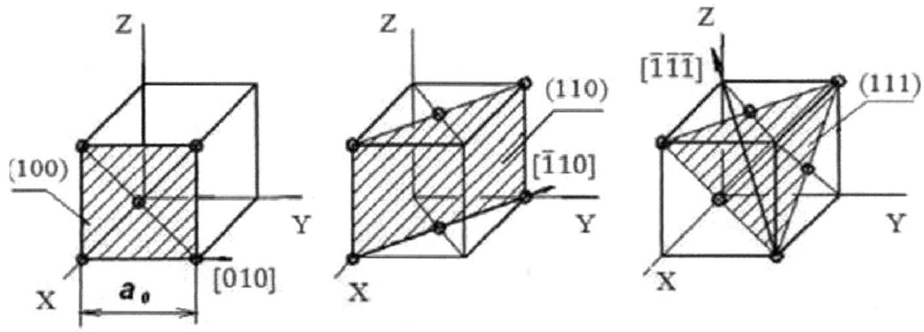

Figure 1. Directions and planes of the crystal lattice.

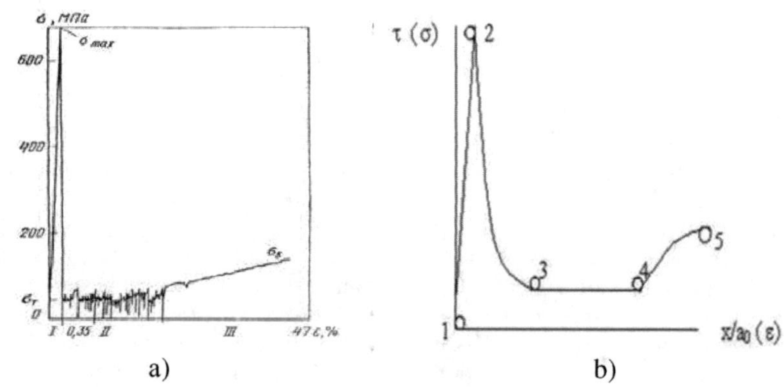

a) experiment [1]; b) calculation model; 1-2-3-single crystal; 3-4-5-polycrystal
Figure 2. Stress-strain Diagram of filamentous single crystal

In particular, the elastic characteristics of a single crystal are determined by the elastic modulus and the elastic limit (point 2 in figure 2, b). At that, the elastic limit exceeds the yield strength. Therefore, to calculate the yield strength of a single crystal, it is necessary to know the value of the elastic limit.

Due to the limited volume of the article, only some basic formulas are given here. The rest of the formulas can be found in the work [2].

The tangential component of the elastic limit for shear in a single atomic cell of a single crystal is calculated by the theoretical formula [1]

$$\tau_{max} = \frac{e^2}{\left(a_0^{\;2}\pi\varepsilon_0 A\right)}\left(1-\cos^2\gamma_{max}\right)\cos\left(\frac{\pi}{2}-\gamma_{max}\right) \qquad (1)$$

where γ - maximum shear angle corresponding to the elastic limit; A - shear area; e = 1,6 · 10^{-19}, C - the value of interacting charges equal to the electron charge; π = 3,14; ε_0 = 8,85 · 10^{-12}, $C^2/N \cdot m^2$ - electric constant; a_0 - the period of the crystal lattice.

The tensile yield strength of a single crystal is calculated by the theoretical formula [1]. It is based on the mechanism of displacement edge dislocation and the electrostatic nature of interatomic interaction

$$\tau = \frac{c}{\left(\frac{x}{a_0}\right)^2}, \qquad (2)$$

where c - coefficient determined at the point [τ_{max}; x / a_0]; x - the amount of displacement in an elementary atomic cell.

The relationship between the normal and tangential projections of the shear stress of a defect-free single crystal is determined by the formula [1]

$$\sigma = \frac{\tau}{(\cos\alpha\cos\beta)} \quad (3)$$

where α - is the angle between the horizontal plane and the shift plane; β - the angle between the shear plane and the direction of normal stress.

The change in the period of the crystal lattice due to an increase in temperature is calculated based on the change in the binding energy of two opposite charges in the crystal lattice. The dependence of the change in the period of the crystal lattice on the temperature is calculated using the theoretical formula [1]

$$\Delta a_0 = 0.5\left(\frac{e^2}{\frac{e^2}{a_0} - c^1_v \Delta T 2\pi\varepsilon_0} - a_0\right), \quad (4)$$

where Δa_0 - change in the period of the crystal lattice due to thermal action; ΔT – temperature change; $c'v$ – the heat capacity of the atom at a constant volume.

The formula for calculating the density of a single crystal of a simple substance □ consisting of atoms of a single chemical element has the form [2]

$$\rho = \frac{m_A \cdot M \cdot k_p}{a_0^3}, \quad (5)$$

where m_A is the atomic mass of the element; $M = 1.66 \cdot 10^{-27}$ is the atomic unit of mass; k_p is the reticular density coefficient for an elementary atomic cell.

3. The results of calculation and comparative analysis

Figures 3-7 and tables 1-5 show the results of calculations and comparison with reference data of the following characteristics of mono-and polycrystal aluminum: elastic modulus, elastic limit and yield strength, Poisson's ratio, thermal conductivity ratio, thermal expansion ratio, density of the material.

Figure 3, a shows the temperature dependences of the elastic modulus, taking into account the anisotropy of the properties of the *Al* single crystal, in the crystallographic directions ⟨100⟩, ⟨110⟩, ⟨111⟩. In figure 3, b - for a *Al* polycrystal in comparison with reference data (points on the graph) [3].

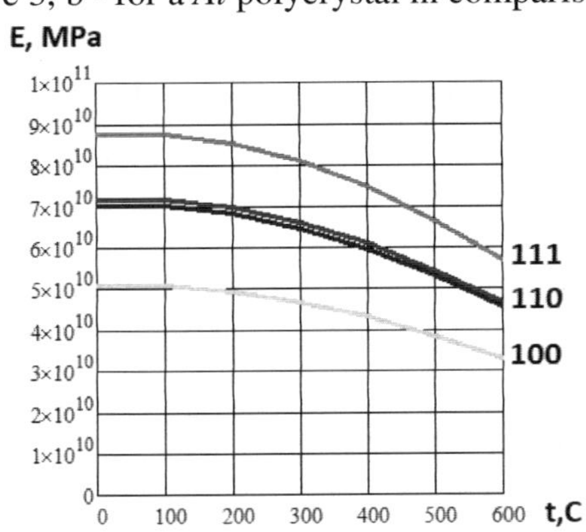

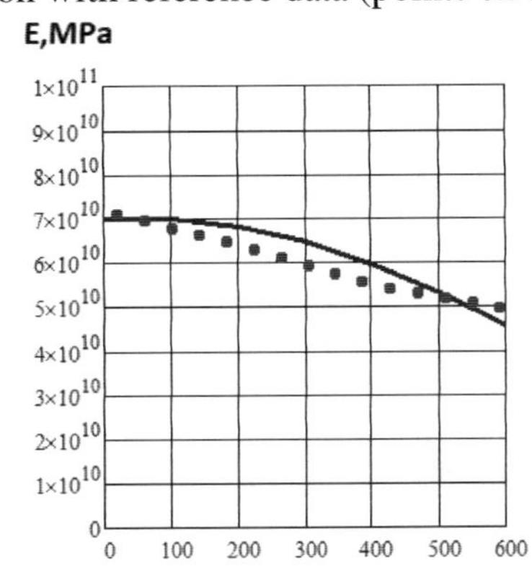

a) directions⟨100⟩, ⟨110⟩, ⟨111⟩; b) polycrystal.

Figure 3. Temperature dependences of the elastic modulus

The coefficients of transverse strain (Poisson's ratio) ν_{xy}, ν_{xz}, are determined for uniaxial tension along the x axis (figure 1). If in an elementary atomic cell the x coordinate axis is combined with the direction ⟨100⟩, then there is an "orthogonal isotropy", in other words, the elastic characteristics along the x, y, and z coordinate axes are the same. In this case, the Poisson's ratio is determined by the work formula [2]. If in an elementary atomic cell the x coordinate axis is combined with the direction ⟨110⟩ or ⟨111⟩, then anisotropy occurs. The calculation results are shown in table 1.

Table 1. Poisson's Ratio

Parameter designation	The value of the parameter when the X axis coincides with the direction		
	⟨100⟩	⟨110⟩	⟨111⟩
ν_{xy}	0.352	0.192	0.477
ν_{xy}	0.352	0.462	0.149

Figure 4, a shows the temperature dependences of the thermal expansion coefficient of the al single crystal in the directions⟨100⟩, ⟨110⟩, ⟨111⟩. In figure 4, b - for a polycrystal in comparison with reference data (points on the graph) [3].

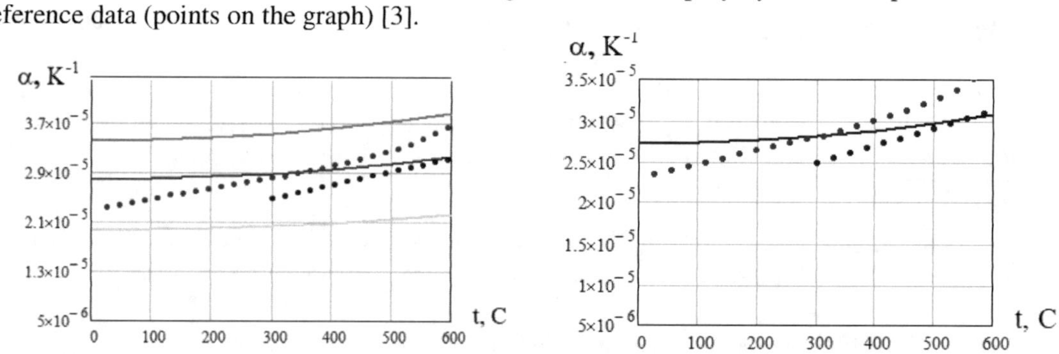

a) directions⟨100⟩, ⟨110⟩, ⟨111⟩; b) polycrystal.

Figure 4. Temperature dependences of the coefficient of thermal expansion

The temperature dependences of the elastic limit (upper notation 100,110,111) and yield strength (lower notation 100,110,111) are shown in figure 5,a. In figure 5, b is the yield strength for a polycrystal in comparison with reference data (points on the graph) [4].

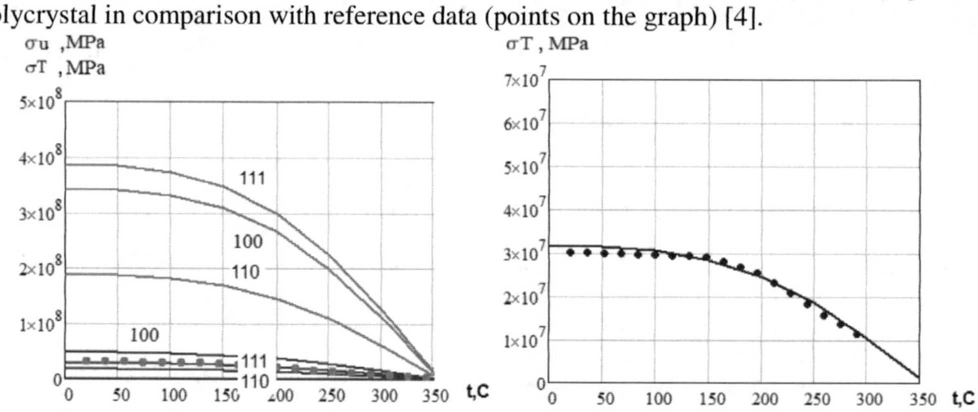

a) directions ⟨100⟩, ⟨110⟩, ⟨111⟩; b) yield strength of polycrystal.

Figure 5. Elastic limit and yield strength of *Al*.

Figure 5, a shows that the maximum value of the elastic limit corresponds to the crystallographic direction ⟨111⟩.In turn, the maximum value of the yield strength point corresponds to the crystallographic direction ⟨100⟩. In particular, since the "stress-strain" diagram of a single crystal has a "yield tooth" (figure 2,a), the maximum value of the yield point is determined by the elastic limit. This information may be necessary for the development (construction) of single-crystal aluminum alloys. Table 2 shows the yield strength at 20 C in comparison with the reference data.

Table 2. The yield strength

Element designation	$\sigma_{T\ 100}$ *(MPa)*	$\sigma_{T\ 110}$ *(MPa)*	$\sigma_{T\ 111}$ *(MPa)*	$\sigma_{T\ middle}$ *(MPa)*	$\sigma_{T\ ref}$ *(MPa)*	Δ,%
Al	48	17.6	28.8	31.5	30 [4]; 32 [5]	≈ ±1.5

To determine the thermal conductivity coefficient, it is first necessary to calculate the oscillations propagation velocity [2]. This velocity is determined in the crystallographic directions ⟨100⟩, ⟨110⟩, ⟨111⟩ and is denoted υ_{100}, υ_{110}, υ_{111}., respectively. Its values are compared with reference data [5] for a polycrystal, respectively: υ_3 – transverse velocity; υ_l - velocity in the rod; υ_2 – longitudinal velocity. The results of the calculation of υ are shown in table 3.

Table 3. The propagation velocity of oscillations

Element designation	υ_{100} (ms^{-1})	υ_3 (ms^{-1})	υ_{110} (ms^{-1})	υ_l (ms^{-1})	υ_{111} (ms^{-1})	υ_2 (ms^{-1})
Al	3534	3080	4997	5080	6123	6260

Table 3 shows that there is a satisfactory convergence between the values: υ_{100} and υ_3, υ_{110} and υ_l, υ_{111} and υ_2 In particular, for aluminum, υ_{110} = 4997 m/s and υ_l = 5080 m/s, υ_{111} = 6123 m/s and υ_2 = 6260 m/s.

The results of calculating the thermal conductivity coefficient are summarized in table 4.

Table 4. The thermal conductivity.

Element designation	λ_{100} $(Wm^{-1}K^{-1})$	λ_{110} $(Wm^{-1}K^{-1})$	λ_{111} $(Wm^{-1}K^{-1})$	λ_{middle} $(Wm^{-1}K^{-1})$	λ_{ref} $(Wm^{-1}K^{-1})$	Δ,%
Al	368.6	488.3	359.3	405.4	401	1.1

The average value values for a single crystal $\lambda_{middle} = (\lambda_{100} + \lambda_{110} + \lambda_{111})/3$. is used as the thermal conductivity coefficient for a polycrystalline state. Table 4 shows that the calculation results satisfactorily converge with the experimental values for a polycrystal from the lspr reference literature [5] with a discrepancy of 1.3 %. These results also indirectly confirm the adequacy of the model for determining the rate of oscillations propagation velocity (table 4).

The results of calculating the temperature dependence of the crystal lattice period and the density of the Al polycrystal are shown in figures 6.7 in comparison with the reference data (points on the graph) [3,4].

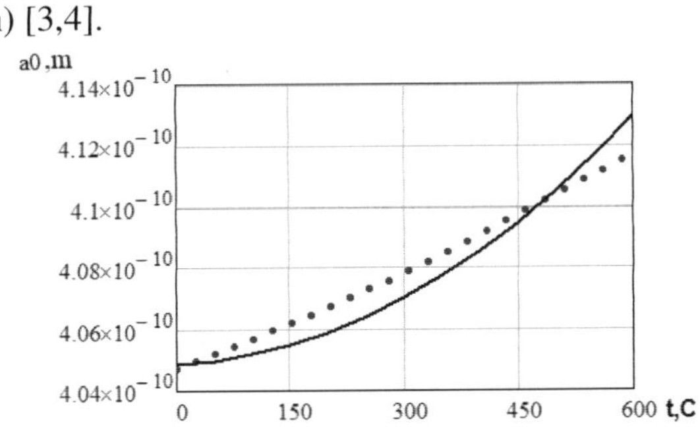

Figure 6. Temperature dependence of the crystal lattice period

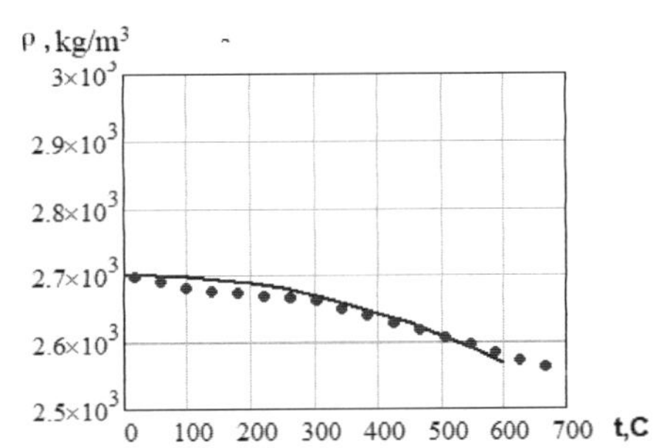

Figure 7. Temperature dependence of density

The density of Al as a function of temperature is also shown in table 5.

Table 5. The density of aluminum

Element designation	Density ρ, kg / m3 at temperature, C						
	20	100	200	300	400	500	600
Al	2699	2697	2687	2669	2643	2609	2586

Thus, the satisfactory convergence of the calculation results with reference data confirms the adequacy of the model, and also allows you to fill in the missing experimental data with calculated values.

4. Conclusion

1. The Developed model allows to Supplement the missing volume of reference data on elastic, strength, thermal and physical characteristics of the material.

2.The Model allows you to get the necessary information in conditions when the experiment is difficult to conduct for technical reasons, for example, at high temperatures.

3. The calculation results satisfactorily agree with the reference data for polycrystalline condition of the aluminum. In the case where reference information about the properties of a single crystal is available, there is also a satisfactory convergence, in particular for the elastic modulus of aluminum.

4. Calculations make up for the lack of reference information for the single-crystal state of aluminum on its elastic, strength and thermal characteristics, taking into account anisotropy, as well as depending on temperature.

5. The Results obtained are the initial data for the construction of compositions and alloys based on aluminum.

References

[1] Svetlov I 1964 *Plant laboratory* Machine for mechanical testing of filamentous crystals (whiskers) No 9 1133-1136.

[2] Badamshin I Kh 2016 From four to one. Forces of interatomic interaction and strength of materials (Moscow, Publishing House of the Academy of Natural Sciences) 134

[3] Samsonov G V 1976 Elements properties: Reference Book, In 2 hours 1 (Moscow, Metallurgy, 1976) 600

[4] Dritz M E 1985 Elements properties: Reference Book, (Moscow, Metallurgy, 1985) 672

[5] Grigorjev I S, Mejlihov E Z 1991 Physical Values: Reference Book (Moscow Energoatomizdat) 1232

IOP Conf. Series: Materials Science and Engineering 1061 (2021) 012005 doi:10.1088/1757-899X/1061/1/012005

Identification of factors affecting accidents on the intercity road network

К V Baklanova, E S Voevodin, E V Fomin, A S Kashura, E P Cheban

Siberian Federal University, 30, Svobodnyy prospect 79, Krasnoyarsk, 660041, Russia

E-mail: pulkristina@yandex.ru

Abstract. This article examines the influence of various factors that affect the accident rate on the intercity road network. Ten road sections that differ in the level of accidents, traffic characteristics and geometric parameters were selected for the research. The experiment conducted with the help of specialized equipment showed that these road sections with similar parameters of the roadbed being in compliance with all standards have a strikingly different accident rate. The lack of correlation between the measured standardized indicators of road sections and the accident rate on them indicates the influence of unaccounted factors on the latter indicator, which are not considered by approved methods.

1. Introduction

The President of the Russian Federation Vladimir Putin has appointed the government up to the 1st of December to develop and approve the third state program of the Russian Federation on road safety for the period up to 2030 aimed to reduce the amount of deaths and injuries caused by road accidents, as well as to improve the effectiveness of the management system in the field of road safety. At the moment the 2nd Federal target program for improving road safety aimed to reduce the number of road accidents (hereinafter referred to as road vehicle accidents) by 30% by 2020 compared to 2010 is being completed [1,2]. This confirms the relevance of issues related to reducing the risk of road vehicle accidents.

2. Methods and equipment

In the research carried out earlier, various methods of road safety assessment were analyzed [3]. A comparative analysis of existing road safety assessment methods has shown that with help of the accident rate method [4], which is used in practice, it is possible to assess road safety more accurately because the assessment takes the influence of several elements of the VADS system into account.

Taking into account the inevitable changes in traffic streams and local conditions it is necessary to clarify and enhance the accident rates based on the generalization of regional accident statistics. Thus, the given list of partial accident rates cannot be considered exhaustive and final. Therefore, we have conducted an investigation on the territory of the Krasnoyarsk region on ten sections of federal roads R255 and R257 (711 km, 714 km, 763 km, 770 km, 796 km, 799 km, 875 km, 879 km, 27 km, 30 km) which differ in the level of road vehicle accidents, traffic characteristics and geometric parameters.

The full-scale experiment was performed on Toyota Caldina passenger car. The selected sections were passed at a low speed (10-20 km / h) therefore the condition of the roadway was examined (figure 1) by driving. A video camera was installed at the driver's eye level in the car in order to capture the roadway, roadsides, slopes and surrounding environment on video.

Figure 1. Inspection of the roadway.

To conduct this experiment we used a universal rail, a KP-203 type curvimeter (road wheel), a steel band, a rangefinder, a portable PPK-f and a Nikon D3100 camera (figure 2). To process and analyze the collected data on a computer, Windows Movie Maker, ASKON "COMPASS 3D" and Windows Microsoft Excel were required.

a b c

Figure 2. Linear dimension measurements: a – plot length measurement; b – curb width measurement; c – barrier height measurement.

The measuring instruments were verified and prepared in accordance with their operating instructions. Linear dimensions were measured in accordance with GOST (National State Standard) 26433.0 - GOST (National State Standard) 26433.2 and in accordance with the instructions. The measurement sites were fenced with portable barriers and warning signs to ensure safety.

The parameters affecting road safety such as vehicle density, lane width, shoulder width, longitudinal gradient, radius of curve in the plan, visibility in the plan, visibility in the profile, the number of main lanes on the roadway to direct traffic, the width of the separation strip, the distance from the edge of the roadway to the cliff deeper than 5 m, the coefficient of friction, width of the separation strip, the composition of the stream and speed difference have been studied [5-8].

When determining the speed and intensity of traffic a 50-100 m section was selected depending on the direct visibility on the roadway. Then the passage time of this section by single cars was measured followed by the calculation of the speed of movement and measurement data processing [5].

The speed of a single car was determined by the formula:

$$V_n = l/t_n \quad (1)$$

l – the length of the section;
t_n – the time of passing the section by a car n.

The average time interval was determined by the formula:

$$\bar{t} = (t_1 + t_2 + \dots + t_n)/n \quad (2)$$

The average speed on the section was calculated by the formula:

$$V = l/\bar{t} \quad (3)$$

Based on the video recording and the obtained speed data, the speed difference was determined by the formula:

$$V_p = V_2 - V_1 \quad (4)$$

V_2 – the speed of the overtaking car;
V_1 – speed of the car being overtaken

Based on the video recording, the entire traffic stream was divided into two groups according to the exterior: passenger vehicles and multifunctional vehicles (cargo vehicles, buses, tractors and other self-propelled vehicles).

3. Research results

With help of the universal rail the height of the fences and the slope were determined. The range finder determined visibility according to the National State Standard GOST 32963. The width of the roadway and the length of straight sections were determined by the curvimeter KP-203 (road wheel). The width of the dividing strip, the width of the shoulder, and the distance from the edge of the roadway to the cliff with a depth more than 5 m were determined by means of a steel tape. The coefficient of adhesion was determined by the portable PPK-f.

The research showed that the busiest areas are those that are located in the boundaries of the Krasnoyarsk agglomeration (Figure 3).

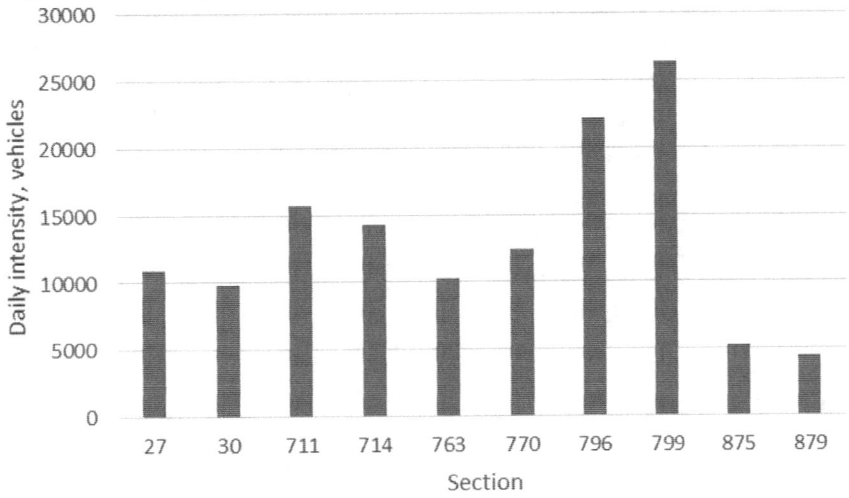

Figure 3. Daily intensity in the considered areas

It was also found out, that the hourly traffic intensity of trucks on various sections of the Federal highway ranges from 6% to 55% of the total number of recorded vehicles. The average value on sections not included in the territory of the Krasnoyarsk agglomeration is about 35% and this figure increases annually. Bus intensity is about 2% of the total number of vehicles (Figure 4).

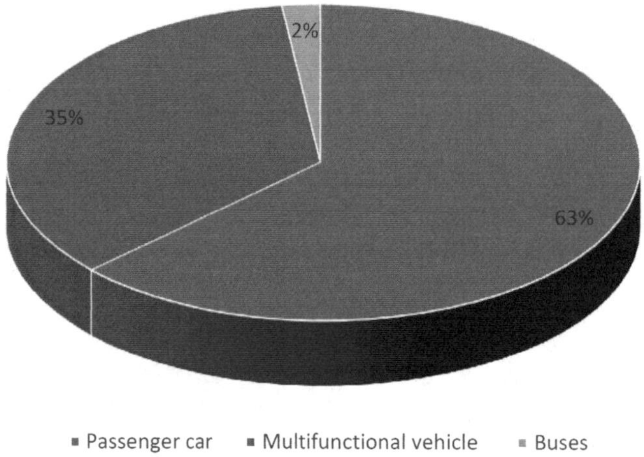

Figure 4. Traffic stream mix

The research showed that the stream rate is not uniform and 50% of the vehicles exceed the safety speed limit on the specified sections. Figure 5 shows the speed division on 799 km of the road.

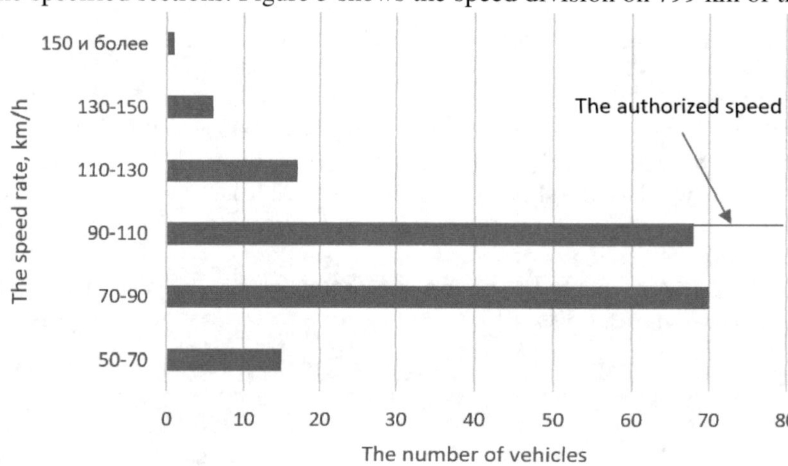

Figure 5. Distribution of speeds of vehicle on the section of 799km

The analysis of the speed difference between the overtaking vehicle and the vehicle being overtaken showed that the difference reaches up to 70 km/h. (Figure 6). In 70% of cases the speed difference in the stream occurred when there was a multifunctional vehicle (a cargo vehicle or a bus) in the stream.

All the data obtained for each section were includede in the table.

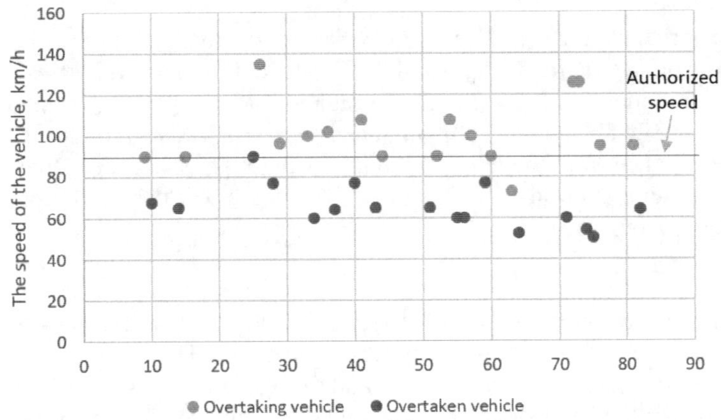

Figure 6. The Graph of the speed difference on the section of 799 km

Table 1. Results of the research

Parameter	Research section									
	27	30	711	714	763	770	796	799	875	879
Traffic intensity, Vehicle\day	10896	9809	15755	14328	10281	12382	22197	26367	5267	4445
Lane width, m	3,5	3,5	3,5	3,5	3,5	3,5	3,5	3,75	3,5	3,5
Shoulder width, m	3,5	3,5	3,5	3,5	3,5	3,5	3,5	3,75	3,5	3,5
Longitudinal slope, %	3	3	3	5	0	0	0	0,03	0	0
The radius of the curve in the plan, m	656	695	322	30000	2000	426	12900	1042	1850	1850
Visibility in the plan, m	600	600	300	820	820	400	820	820	700	700
Visibility in the profile, m	70	70	50	100	100	60	1000	80	100	100
Number of main lanes on the roadway	2	2	2	2	2	2	4	4	2	2
Width of the dividing strip, m	0,2	0,2	0,2	0,2	0,2	0,2	0,2	2,5	0,2	0,2
Distance from the edge of the roadway to a cliff more than 5 m deep, m	-	-	-	-	0,5	0,5	0,5	0,5	0,5	0,5
The coefficient of friction	0,75	0,75	0,75	0,75	0,75	0,75	0,75	0,75	0,75	0,75
The stream composition, % of cargo vehicles	8	8	33	33	27	26	5	5	17	17
The average speed difference, km/h	16	17	42	45	39	38	23	25	32	32

Thus, the characteristics of 10 road sections with various configurations and different number of road vehicle accidents on federal roads P255 and P257 were determined and series of data such as traffic volume, lane width, shoulder width, longitudinal gradient, radius of curve in the plan, visibility in the plan, visibility in the profile, the number of main lanes on the roadway for direct traffic, the width of the separation strip, the distance from the edge of the roadway to the cliff deeper than 5 m, the coefficient of friction, width of the separation strip, the length of the straight sections, the composition of the stream, speed difference were systemised.

4. Findings

The research found that the analysed sections have similar parameters of the roadbed meeting all the standards, but the accident rate on them is strikingly different. The lack of correlation between the measured standardized indicators of road sections and the accident rate on them indicates the influence of unaccounted factors on the latter indicator, which are not considered by approved methods.

The analysis of the systematized data showed that there is also a big speed difference in the sections where there is a significant number of multifunctional vehicles in the traffic stream.

In areas with a high intensity (more than 20%) of multifunctional vehicles and with a significant speed difference there is a big number of accidents over the past 5 years.

It is necessary to quantify the impact of each of the identified factors on road safety.

5. Conclusion

The qualitative composition of the traffic stream is not uniform and significantly differs on different sections of federal roads. Thus, to improve road safety it is advisable to improve the methodology of accident rates, taking into account the composition of the traffic stream and the speed difference that occurs due to the nonuniformity of the traffic composition.

References

[1] Kremlin website [Electronic resource] List of instructions based on the results of the meeting of the State Council held on June 26, 2019. Available at: http://www.kremlin.ru/acts/assignments/orders/61040

[2] Ministry of the Interior of the Russian Federation [Electronic resource] Federal Target Program Improving Road Safety in 2013–2020 Available at: https://мвд.рф/Deljatelnost/gosprogram/

[3] Pulyanova K, Voevodin E, Fadeev A, Fomin E, Askhabov A and Kashura A 2019 Analysis of road safety assessment methods *IOP Conf. Series: Materials Science and Engineering* **632** pp 1–8

[4] Babkov V 1993 *Road conditions and traffic safety* (Moscow: Transport) p 272

[5] Korchagin V, Pogodaev A, Klyavin V and Suvorov V 2016 *Bulletin of the Moscow Automobile and Road State Technical University (MADI)* **2 (45)** pp 88–94

[6] Tesic M, Hermansb E, Lipovac K and Pesic D 2018 Identifying the most significant indicators of the total road safety performance index *Accident Analysis and Prevention* **113** pp 263–78

[7] Kanugantia S, Agarwalab R, Duttac B and Bhanegaonkard P 2017 Road safety analysis using multi criteria approach: A case study in India *World Conf. on Transport Research 25* pp 4649–61

[8] Sarabaev V, Yeremin V and Korolev P 2013 Assessment of road safety by mathematical modeling *Science and technology in the road industry* **3** pp 9–11

Determination of geometric characteristics of friction areas in contact of the chassis wheel with a solid support

E V Balakina[1], D S Sarbaev[1], I V Sergienko[1]

[1] Volgograd State Technical University, 28, Lenin avenue, Volgograd, 400005, Russia

E-mail: fahrgestell2011@yandex.ru

Abstract. Elastic wheels contain the chassis of airplanes, automobiles and other vehicles. Their tires interact with a solid supporting surface – roads and airfields. As a result of this interaction, only static friction, only sliding friction or a combination of them, depending on the mode of the wheel movement, can be present in the contact patch. The geometric dimensions and relative position of the different friction areas determine the stable, unstable or boundary movement of the elastic wheel. Purpose of work: calculation and measurement of geometric characteristics of static and sliding friction areas in the wheel-to-road contact. Determined that: the slip area appears on the longitudinal axis of the contact patch, at a distance from the center of the contact patch equal to 1/4 of the length of the contact patch, then it increases due to a decrease in the static friction area; the center of the static friction area in the contact patch moves toward the current moment by an amount proportional to the moment; the maximum displacement of the center of the static friction area corresponds to the moment maximum in terms of adhesion, and is 1/3 of the length of the contact patch.

1. Introduction

Elastic wheels contain the chassis of airplanes, automobiles and other vehicles [1-6]. Their tires interact with a solid support surface – roads and airfields [7-10]. Since the phenomena occurring in contact of a moving tire with the support surface determine the movement stability properties, controllability and braking dynamics of the vehicle, when creating tires, their manufacturers strive to simultaneously provide a number of their important properties: friction, elastic at different coordinates, strength, weight, and reliability [11-15].

At the same time, most modern vehicles, including unmanned vehicles, are equipped with motion control and control systems, whose functioning algorithms provide for executive actions on the wheels and tires. Despite the fact that these systems are constantly being improved, their presence does not always contribute to maintaining safety, i.e., movement stability, controllability and brake dynamics [2, 3, 5, 7-9]. Moreover, it is known that in some driving modes, the activation of these systems leads to a loss of stability and controllability of the vehicle. These are modes associated with the presence of a side force of even a small value, amounting to 15...20% of the vehicle weight, while the presence of a brake mode further aggravates the situation [2, 3, 5]. The described phenomena are not related to malfunctions or defects in the design of the motion control and control system, which corresponds to the modern ideas of specialists-designers about the interaction of the wheel with the road [11, 16-20]. These modern concepts are not perfect due to the lack of knowledge of the phenomena that occur in the wheel-to-road contact patch, especially in the presence of lateral force. Accordingly, this lack of knowledge is manifested in the problem of incorrect modeling of phenomena in the wheel-to-road contact patch (at the design stage of the vehicle), which determine the stability and controllability of the vehicle, and does not provide the basic possibility of creating motion control and control systems that can adequately perform their functions in all driving modes, including specific ones.

In this regard, despite the existence of significant basic research [11, 21-24], there is a need for further research of the interaction of an elastic wheel with a solid support, especially in the presence of lateral force. A number of problems appear, one of them is to calculate the changing geometric characteristics of areas with static and sliding friction in the contact patch of an elastic wheel with a solid support, which perceive external forces differently and create reactions to the wheel in different ways.

2. Aim of work

Calculation and measurement of geometric characteristics of areas with static and sliding friction in the contact patch of an elastic wheel with a solid support.

3. Current state of the problem

Previously, experiments were carried out on the experimental installation created at the Volgograd State Technical University to study the phenomena in the contact patch of an elastic wheel with a solid support surface, which allowed us to qualitatively identify the mutual location of areas with static and sliding friction in the contact patch [10]. They showed that:

1. In the contact patch of an elastic wheel with a solid support in the wheel braking mode, the static friction area is shifted back relative to the wheel speed vector, towards the rear of the contact patch;
2. In the contact patch of an elastic wheel with a solid support in the driving mode of the wheel, the static friction area is shifted forward relative to the wheel speed vector, towards the front of the contact patch;
3. When a side force is applied to the wheel, the static friction area is shifted in the direction opposite to the side force of the wheel (co-directed to the side reaction vector of the wheel).

These were qualitative studies, but not quantitative. At the same time, the result described in point 3 was new, and the results described in points 1 and 2 were confirmation of earlier studies by Knoroz V.I. and Balabin I.V.

4. Methods and approaches

As is known, the interaction of an elastic wheel with a solid support in the contact patch is characterized by diagrams of normal and tangential stresses. Using a plot of the normal stresses, it is possible to define a plot of permissible tangential stresses in the contact patch. Since the sliding area in the contact patch originates not where the tangent stresses are maximum, but where they begin to exceed the permissible ones, the condition for the appearance of sliding at the i-th point of the contact patch can be written as follows:

$$\tau_{xi} \geq [\tau_{xi}] = f_{st} \cdot \sigma_{zi} \tag{1}$$

where σ_{zi} – normal stresses at the i-th point of the wheel contact patch with the support; τ_{xi} – tangential stresses at the i-th point of the wheel contact patch with the support; f_{st} – static friction coefficient between the wheel and the support surface.

This method was implemented. The plots of normal and tangent stresses in the contact patch were analytically determined [3]:

$$\sigma_{zi} \approx \left(\frac{k \cdot R_d \cdot X_i}{l_c \cdot r^2} \cdot \varphi_{\max} + \frac{1}{l_c} \right) \cdot P \Big/ 2 \cdot \left((B_c/2)^2 \cdot \left(1 - \frac{X_i^2}{(l_c/2)^2} \right) \right)^{1/2} \tag{2}$$

$$\tau_x = \tau_{x0} - \Delta\tau_x \tag{3}$$

$$\tau_{x0} \approx E \cdot \left(\frac{2 \cdot R_0}{l_c} \cdot \arcsin\left(\frac{l_c}{2 \cdot R_0} \right) - 1 \right) \cdot \sin\left(\frac{X_i}{l_c/2} \cdot \pi \right) \tag{4}$$

$$\Delta\tau_x \approx \left(k \cdot P_z \cdot \varphi_{\max} \cdot R_d \cdot \frac{\sqrt{r^2 - X_i^2}}{r^2} \right) \Big/ S_c \tag{5}$$

where R_0 – free wheel radius; R_d – dynamic wheel radius; r – tire landing radius; l_c – length of the wheel contact patch with the support surface; B_c – width of the wheel contact patch with the support surface; S_c – the area of the wheel contact patch with the support surface; $\varphi_{\max}$ – maximum friction coefficient of the tire to dry asphalt; P_z – normal load on the wheel axle; X_i – longitudinal coordinate i-th point of the contact patch; k – dimensionless coefficient equal to the ratio of braking moments ($k = M_b / M_{b\max}$); E – elastic modulus of the tire material along the contact patch.

Then we compared the tangential stresses acting in the contact patch with the permissible ones and found the positions and sizes of static and sliding friction areas in the contact patch of an elastic wheel with a solid support at different values of the moment on the wheel.

Figure 1 shows the final schemes of the appearance and growth of the sliding area in the contact patch when the brake torque increases.

Thus, the plots of normal and tangential stresses in the contact patch of an elastic wheel with a solid support are obtained by calculation for different types of wheel loading, since it is possible to identify the place of sliding in the contact patch of an elastic wheel with a non-deformable flat base only on the basis of comparing the plots of tangential and normal stresses. The tangent stresses acting in the contact patch are compared with the permissible ones, and thus the geometric arrangement of the areas with static and sliding friction in the contact patch under different modes of wheel movement is analytically found.

For experimental confirmation of the obtained results, an experimental setup was created for quantitative detection of the relative location of areas with static and sliding friction in the contact patch, shown in figure 2.

The tire 4.10/3.50-5 (for an airfield cart) is diagonal, manufactured by Omega Company (Taiwan) has a pressure according to the passport – 2.5 ATM; the load – 1100 N. The pressure was created by a DC 12V 300 PSI compressor, designed for a maximum pressure of up to 3 ATM, and was controlled by a manometer of the type MTI GOST 2405-63.

The experimental setup consists of an elastic wheel pivotally mounted in a horizontal position, interacting with a solid flat support made of optically transparent material. It is a model of the road and provides the ability to visualize phenomena in the contact patch between the wheel and the road. The experimental setup allows the wheel to be loaded from the flat support side with a normal load and a longitudinal load until a complete sliding occurs in the wheel contact patch. For video shooting of the contact patch, a high-speed video camera with a shooting speed of 8 thousand FPS was used. The image was taken during longitudinal loading until complete sliding in the contact patch occurred, using an optical lens with a focal length of 0.2 m to visually increase the size of the contact patch.

The video recording was processed using image analysis of the contact patch on a computer with visualization on a large monitor, as shown in figure 3. A large monitor and lens provide a hundredfold increase in the size of the contact patch. This eliminates the comparability of the measured values of the length parameters with the values of measurement error.

We considered those tire tread blocks that were completely located within the boundaries of the contact patch. Their characteristic parts were counted as points with numbers. Then, on the screen of the monitor, measurements were made of the geometric position of each characteristic point at each moment of time. From this data, the movements of each characteristic contact point at each moment of the process flow were calculated. The analysis of the forms of the obtained dependencies allows us to judge the location of each contact point at each moment of time in the area with static or sliding friction.

Preliminary processing of the experiment results confirmed the results of the calculations. The experimental research of pre-determined quantitative characteristics of static and sliding friction areas in the contact patch, namely, the coordinates of the location of the sliding phase relative to the center of the contact patch, the coordinates of the current location area with the sliding friction and end coordinates of this area at the time of full sliding. The coordinates are expressed in arbitrary units relative to the length of the contact patch. This allows to determine the location of the application of the lateral reaction of the support surface, implemented in the contact patch in response to external lateral force, since it is implemented only by the area with static friction in the contact patch and has the greatest effect on the deviation of the wheel from the specified trajectory.

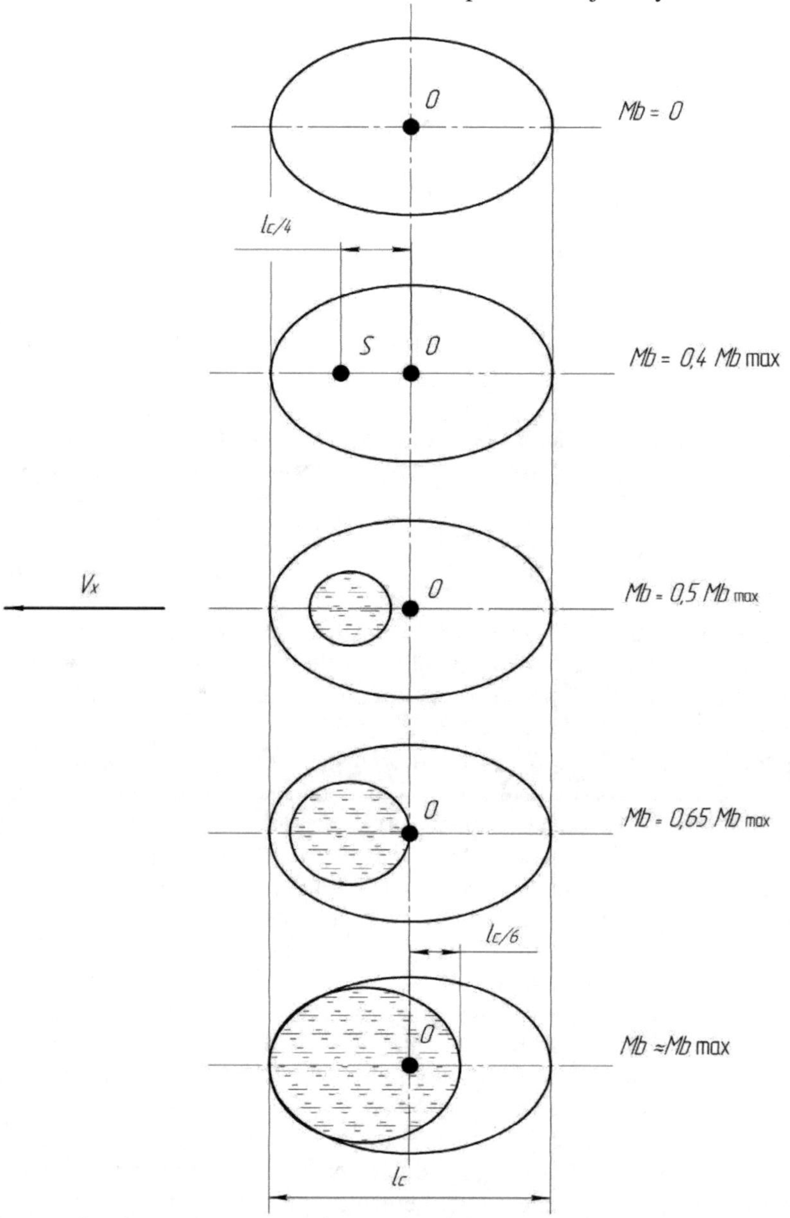

Figure 1. Schemes for the appearance and growth of the sliding area in the contact patch of an elastic wheel with a solid support when the brake torque increases: O – the center of the contact patch; S – the place of occurrence of sliding area

a b

Figure 2. Photo of the experimental setup: a – general view; b – contact patch view of the tire 4.10/3.50-5 (for an airfield cart)

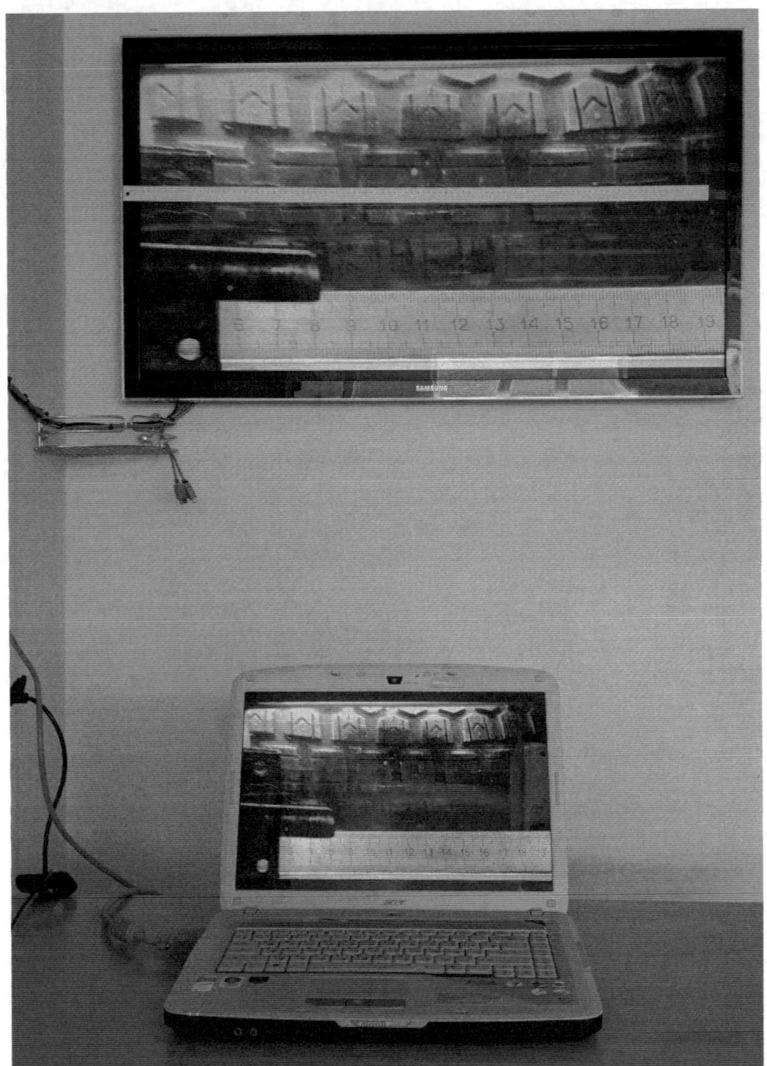

Figure 3. Processing photo

5. Conclusion

1. In the contact patch of an elastic wheel with a solid support in the driving mode of the wheel, the static friction area is shifted forward relative to the direction of the translational speed vector of the wheel, towards the front of the contact patch.
2. In the contact patch of an elastic wheel with a solid support in the braking mode of the wheel, the static friction area is shifted backward relative to the direction of the translational speed vector of the wheel, towards the rear of the contact patch.
3. In the contact patch of an elastic wheel with a solid support, the sliding area appears on the longitudinal axis of the contact patch, at a distance from the center of the contact patch equal to 1/4 of the length of the contact patch. Then it increases by reducing the static friction area.
4. The center of the static friction area in the contact patch moves in the direction of the acting moment by an amount proportional to the moment.
5. The maximum displacement of the center of the static friction area corresponds to the moment that is maximum under the friction conditions, and is 1/3 of the contact patch length for all types and states of solid support. When this movement is achieved, there is a full sliding in the contact patch, when the sliding friction area occupies its entire space.

6. Acknowledgments

The reported study was funded by RFBR according to the research project № 19-08-00011.

References

[1] Balabin I V, Putin V A and Chabunin I S 2012 *Automobile and tractor wheels and tires* (Moscow: MSTU «MAMI») p 920
[2] Balakina E V and Zotov N M 2015 Determination of the Mutual Arrangement of Forces, Reactions, and Friction Zones in the Contact Zone of an Elastic Wheel with a Solid Surface *Journal of Friction and Wear* **36(1)** pp 29–32
[3] Balakina E V 2017 Calculation of the Geometric Position and the Sizes of the Static Friction and Sliding Friction Zones at the Point of Contact between an Elastic Wheel and a Firm Surface *Journal of Friction and Wear* **38(2)** pp 144–149
[4] Zotov V M 2018 Physical evaluation method of longitudinal slip ratio of a vehicle wheel on hard surfaces *Bulletin of the Lower Volga Agro-University Complex: science and higher professional education* **3(51)** pp 338–349
[5] Kristalny S R, Toporkov M A, Fomichev V A and Popov N V 2015 Measuring complex for determining the effectiveness of electronic stability control systems for automobiles *Motor Transport Company* **6** pp 37–41
[6] Knoroz V I 1976 *Car tire operation* (Moscow: Transport) p 283
[7] Fedotov A I 2015 *Dynamic diagnostic method of pneumatic brake drive of automobiles* (Irkutsk: INRTU) p 514
[8] Fedotov A I and Gromalova V O 2020 Mathematical model for studying the braking path of an automobile with ABS on winter roads *Automobile industry* **3** pp 15–21
[9] Fedotov A I and Boyko A V 2014 Interaction mechanics of an elastic tire with a cylindrical surface of a running drum of a diagnostic stand *Automobile industry* **10** pp 33–36
[10] Balakina E V and Sarbaev D S 2019 Qualitative Research of Combined Friction in Contact of Elastic Wheel with Solid Surface *Proc. of the 4th Int. Conf. on Industrial Eng.* (Moscow, Russia) pp 773–779
[11] Pacejka H B 2012 *Tire and Vehicle Dynamics* (Published by Elsevier Ltd, USA)
[12] Kubba A I, Hall G J, Varghese S, Olatunbosun O A and Anthony C J 2018 Modeling of contact patch in dual-chamber pneumatic tires *Tire Science and Technology* **46(2)** pp 78–92
[13] Emilio Jimenez and Corina Sandu 2020 Experimental Investigation of the Tractive Performance of Pneumatic Tires on Ice *Tire Science and Technology* **48(1)** pp 22–45
[14] Flavio Farroni, Riccardo Russo and Francesco Timpone 2017 Experimental Investigations on

Rubber Friction Coefficient Dependence on Visco-Elastic Characteristics, Track Roughness, Contact Force, and Slide Velocity *Tire Science and Technology* **45(1)** pp 3–24

[15] Hojong Lee, Min Tae Kim, and Saied Taheri 2018 Estimation of Tire-Road Contact Features Using Strain-Based Intelligent Tire *Tire Science and Technology* **46(4)** pp 276–293

[16] Mehran Motamedi, Saied Taheri, and Corina Sandu 2016 Rubber-Road Contact: Comparison of Physics-Based Theory and Indoor Experiments *Tire Science and Technology* **44(3)** pp 150–173

[17] Morris de Beer, Colin Fisher and Louw Kannemeyer 2012 Tyre – pavement interface contact stresses on flexible pavements – quo vadis? *8th Conf. on asphalt pavements for Southern Africa*

[18] Pavel Sarkisov, Günther Prokop, Jan Kubenz and Sergey Popov 2019 Physical Understanding of Transient Generation of Tire Lateral Force and Aligning Torque *Tire Science and Technology* **47(4)** pp 308–333

[19] Reza N. Jazar 2008 *Vehicle Dynamics. Theory and Application* p 1015

[20] Shao L, Jin C, Lex C, Eichberger A 2019 Robust road friction estimation during vehicle steering *Vehicle System Dynamics* **57(4)** pp 493–519

[21] Seyedmeysam KHALEGHIAN 2017 A technical survey on tire-road friction estimation *Friction* **5(2)** pp 123–146

[22] Svendenius Jakob 2007 *Tire Modeling and Friction Estimation* (Department of Automatic Control Lund University, Lund, Sweden) p 194

[23] Seyedmeysam Khaleghian, Omid Ghasemalizadeh, and Saied Taheri 2016 Estimation of the Tire Contact Patch Length and Normal Load Using Intelligent Tires and Its Application in Small Ground Robot to Estimate the Tire-Road Friction *Tire Science and Technology* **44(4)** pp 248–261

[24] Wang C, Ayalew B, Rhyne T, Cron S and Adcox J 2017 A laminated ring on elastic foundation model and a feedback contact algorithm for tire-road contact analysis *Tire Science and Technology* **45(4)** pp 307–334

IOP Conf. Series: Materials Science and Engineering 1061 (2021) 012007 doi:10.1088/1757-899X/1061/1/012007

The validation system for reliability and survivability of unique mechanical systems

A F Berman[1], O A Nikolaichuk[1], A Yu Yurin[1,2]

[1]Matrosov Institute for System Dynamics and Control Theory SB RAS (ISDCT SB RAS), 664033, Irkutsk, :134 Lermontova St.
[2]Irkutsk National Research Technical Universit, 664074, Irkutsk, 83 Lermontova St.

E-mail: bafbam@mail.ru

Abstract. We propose a concept of the validation system for the properties of reliability and survivability of unique mechanical systems, the failures of which can result in accidents. The system is based on the principles of self-organization and provides the opportunity to consider off-design impacts of any origin, errors in design and operation based on the use of appropriate rules formulated by experts. The principles of self-organization are implemented by a mechanism based on the rules that step-by-step formulate the necessary tasks, the required expert systems and computing units to solve them, as well as on the rules directly linking negative factors and neutralizing their properties.

1. Introduction
The design automation of unique mechanical systems, the reliability, survivability and safety of which is ensured only by the limits of strength and durability, requires special approaches due to the lack of analytical dependencies between the main criteria.

It is important to highlight that when considering such systems, we cannot eliminate the possible exposure to off-design loads of any origin, errors in design and operation. Then, decision-making enters the area of high uncertainty and requires the use of new methods and means of validating and ensuring the properties of reliability and survivability [1-4].

2. Methods
The aim of the present study is to develop a system for validating the properties of reliability and survivability of Unique Mechanical Systems (UMS's) using the strength and resource criteria. The validation system is based on an approach linking a set of logical decision rules, a transdisciplinary model of the UMS technical state and the principles of self-organization [5-9]. The further development of this approach adopts ontological modeling technologies, production expert systems, a model-driven approach, and component development of software systems [10, 11].

3. Results
For the justified assignment of safety factors that provide for the correction of accepted hypotheses, imperfect design schemes, possible deviations in workmanship and operating conditions, we need additional empirical and expert knowledge which is difficult to incorporate into existing models. This knowledge is used in the form of rules that furnish opportunities to justify of the coefficients level, to

determine the likely degradation processes, their speed and damage caused by them, as well as methods and tools that ensure the reliability and survivability of UMS's.

The self-organization principle of the validation system for the properties of reliability and survivability of UMS's, ensuring that the system achieves the goals, is realized not by the control system, but due to the interaction of the system components with each other, initiated by the initial data. The interaction of components is based on the rules describing the relationship between the objectives of the study, the functions of components, and the information processed.

Figure 1 shows the self-organization mechanism of a system for validating the UMS properties. The work of the mechanism begins with the input of initial data reflecting the influencing factors and technical requirements for UMS's, assembly units (AUs) and parts (Ps). Further, the components of the software system interact and perform their functions: self-organization of tasks (Stage 1, Knowledge Base 1 and Scheduler), analytical blocks (Stage 2, Knowledge Base 2, Scheduler, Computing Units and Expert Systems) and finding properties of the object in question (Stage 3, Knowledge Base 3 and all components).

The functioning of these components and ensuring the self-organization of tasks is carried out using the appropriate rules formulated by experts, both on the basis of generally accepted knowledge and the special knowledge that experts have. The rules are stored in the corresponding knowledge bases of expert systems and computing units.

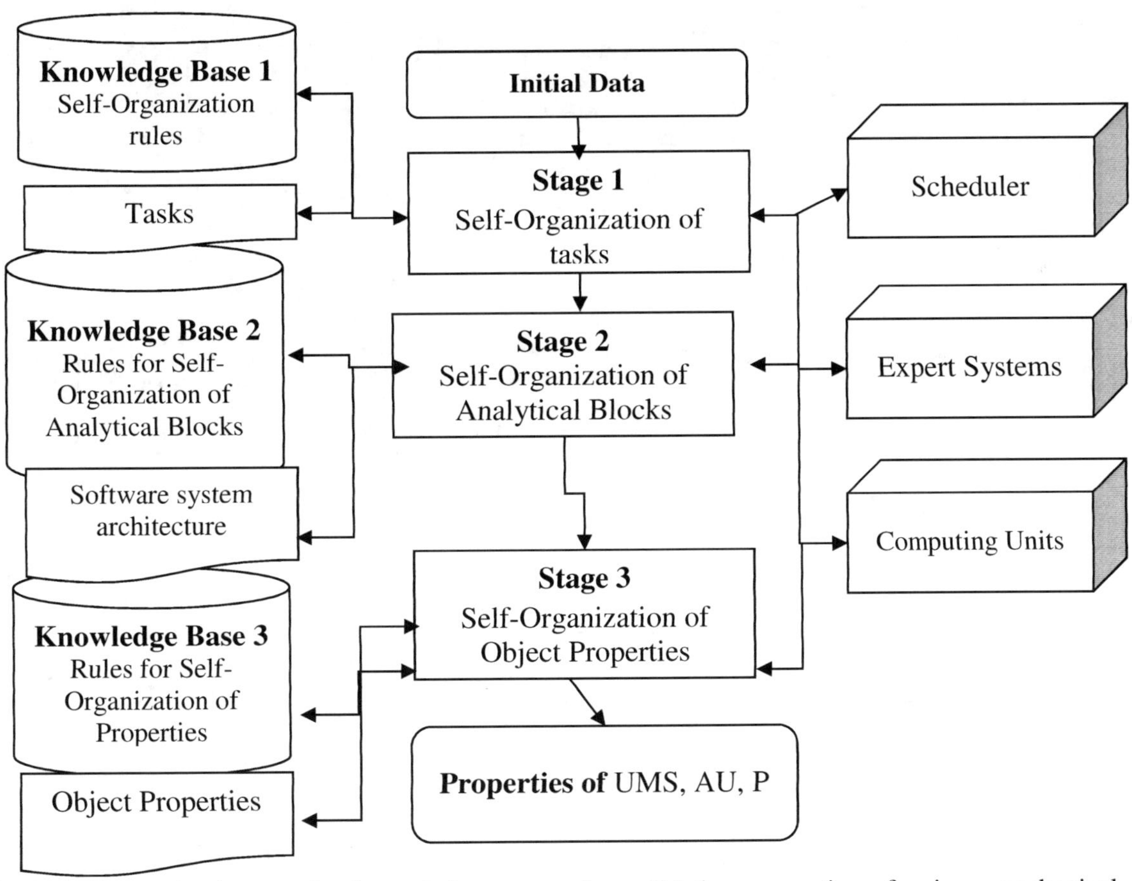

Fig.1. Self-organization mechanism of the system for validating properties of unique mechanical systems (UMS's): AU – assembly unit, P – part

The stage 2 (Figure 1) is followed by the automatic formation of the software system architecture

(Figure 2). The interaction between the components is provided by the Self-Organization Mechanism. The direct validation of properties of UMS's, AUs and Ps is provided by Expert Systems (ES's) and Computing Units (CUs) incorporated into analytical blocks.

Therefore, first, initial data is entered into the system through the user interface. Based on this data, the systems select rules to self-organize (create) tasks, the solution of which provides validation of the properties. Then, for the selected tasks, the appropriate rules (rules for choosing expert systems and computing units) help to choose an ES's and CUs that can solve these tasks and which are combined into analytical blocks. Then, the analytical blocks self-organize (validate) the required properties of the object in question.

For example, we have a set of ES's that help to solve various tasks and CUs that solve formalized problems. Now, depending on the aim and scientific and technical aspects of the problem under study, we choose ES's and CUs to form analytical blocks to address the problem.

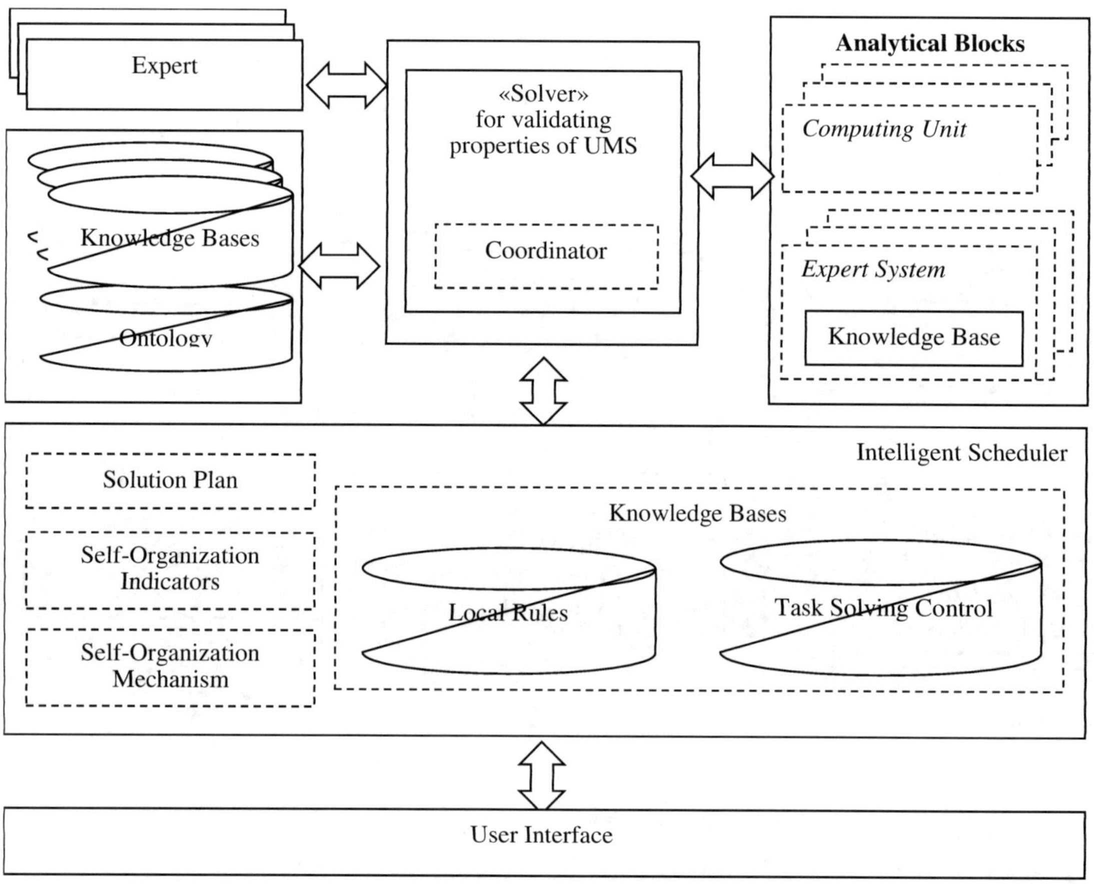

Fig.2. Architecture of the complex software system for validating properties of Unique Mechanic Systems (UMS's)

4. Example

Initial data. Influencing factors. Mechanic (power) load – constant. Contact medium – weakly active liquid at temperature of Q^0C.

Structural, Constructive and Technical Requirements for the Part. *Technical requirements in accordance with the Terms of Reference.* The Part must withstand workloads at a given temperature. The Part's Resource – G hours (G is some variable denoting the number of hours). *Survivabilitry factor* – «leak before destruction». Ensure limited performance and safety of the Part under conditions of unacceptable damage and its possible detection by instrumental and organoleptic methods.

The structure of the rules from Knowledge Base 1 (Figure 1) of self-organization of Tasks incorporated into the expert system and forming a set of tasks, the solution of which will ensure the self-organization of the properties of the Part:

Left-hand part of the rule – *Initial data*; Right-hand part – *Tasks*.

Examples of rules: IF mechanical *Loads* are continuous THEN *Tasks* assess strain-stress state AND justify the choice of the Part's material AND justify the safety factors, bearing capacity and resource AND assess the strength and bearing capacity, etc.; IF *Exposure to a contact working medium* THEN *Objectives* identify possible degradation processes taking into account mechanical impact and determine the rate of degradation processes AND determine the type of damage parameters AND calculate the resource, etc.

The structure of the rules from Knowledge Base 2 (Figure 1) of the self-organization of Analytical Blocks: The left-hand side of the rule is *Task*; The right-hand part is *Analytical Blocks*.

Example of the rule: IF Task complies with Rule 1 THEN corresponding ES's and CUs.

The structure of the rules from Knowledge Base 3 (Figure 1) of the self-organization of the Part's properties based on Expert Systems (ES's) and Computing Units (CUs) incorporated into the analytical blocks: *The left-hand side* – influencing factors; *The right-hand side* – necessary actions and properties.

Examples of the rules for *the ES validating the computational scheme*: IF Part has a shape Φ_i from $\{\Phi_1, \Phi_2, \ldots, \Phi_F\}$ THEN computational scheme for the flat stress-strain state is accepted. *CUs for the VAT calculation.* Computational models correspond to the computational scheme. *ES's for the preliminary choice of the part's material.* IF loads are static AND access to assessment of the technical state is impeded AND operating temperature AND contact medium is weakly active THEN steel is low-alloy AND heat-resistant AND mechanical, physical, chemical properties. *Computing unit.* Calculation of the stress intensity factors, safety factors, etc. *ES's for detecting possible degradation process and its rate, type and damage parameters.* IF long-term design and beyond design constant tensile stresses AND compliance and non-compliance of the contact medium with the technical requirements THEN degradation processes DP_1, DP_2, …, DP_K AND damages Dam_1, Dam_2, …, Dam_M are possible. *Computing unit. Вычислительный модуль.* Calculation of the rate of degradation processes under design and beyond design factors.

The obtained constructive properties of the Part. Material – low-alloy heat-resistant steel. Safety factor for yield strength – 1.6; Resource – 20,000 hours; $K_I < K_{Ic} > 90$ MPa $m^{1/2}$; possible damage: pitting, cracks perpendicular to maximum tensile stresses, cracks that cause depressurization before they become critical, or become visually detectable with a margin of 1.3 to the critical size of the crack.

5. Conclusion

The system for validating the properties of reliability and survivability is self-organizing due to the mechanism for identifying tasks, expert systems, and computing units that ensure their solution.

The mechanism consists of three stages. At the first stage, the initial data, including the requirements for the created object, are compared with the tasks in which they are used to validate the properties satisfying the requirements. At the second stage, to solve the identified problems, the corresponding expert systems and computing units using this data are selected. At the third stage, expert systems and computing units solve tasks based on the corresponding knowledge bases in the automatic mode.

The system delivers validation of the properties of reliability and survivability of UMS's due to the self-organization mechanism, which includes alternative intelligent (expert) and computing systems that provide specified functions based on the rules formulated by experts and to some extent consistent with the adopted development system for UMS's. The system architecture includes a set of components that provide both functional and information support for the process of self-organization of properties.

The logical and computational tasks in the example are presented in a simplified form. However, we can provide their detailed description if necessary. Experts participate in solving tasks during formulation of the rules of knowledge bases (off-line) and online.

The developed mechanism and architecture of an integrated software system are based on logical rules (products) that form the basis of production expert systems. The self-organization of the system and its subsequent use in automatic validation of the properties of UMS is provided without a specific control system that requires high costs for ensuring its reliability and development.

Acknowledgments

This work was supported by the Russian Foundation for Basic Research, Project No №18-08-00560. Project Coordinator – A.F. Berman

References

[1] Makhutov N A 2013 The criterion base of strength, resource, reliability, survivability and safety of machines and human-machine complexes *Problems of Mechanical Engineering and Machine Reliability* (Moscow: Nauka Publishers) **5** pp 25–36

[2] Makhutov N A and Reznikov D O 2016 A multilevel assessment of survivability of complex technical systems, taking into account the scale-structural hierarchy of damage accumulation and destruction processes *Safety in the Technosphere* (Moscow: INFRA-M) **4** pp 3–17

[3] Makhutov N A, Bolshakov A M and Zakharova M I 2019 Risk analysis of accidents of oil and gas industry facilities in arctic climatic conditions *Factory Laboratory. Diagnostics of Materials* (Moscow: Test-ZL) **2** pp 48–54

[4] Botvina L R 2018 On the benefits and difficulties of interdisciplinary research *Physical Mesomechanics* (Tomsk: Pleiades Publishing, Ltd.) vol 21 **6** pp 9–16

[5] Berman A F, Nikolaychuk O A, Yurin A Yu and Pavlov A I 2019 The principles of information technology for solving interdisciplinary tasks of ensuring technological safety on the basis of self-organization *Information and Mathematical Technologies in Science and Management.* (Irkutsk: ESI SB RAS) **2 (14)** pp 5–15

[6] Redko V G 2006 *From behavior models to artificial intelligence* (Moscow: KomKniga) p 315

[7] Kolesnikov A V, Kirikov I A and Listopad S V 2014 *Hybrid intelligent systems with self-organization: coordination and dispute* (Moscow: IPI RAS) p 189

[8] Bernon C 2004 Tools for Self-Organizing Applications Engineering *Lecture Notes in Artificial Intelligence (*vol 2977*)* ed G Di Marzo Serugendo (Springer) pp 283–298

[9] Kolesnikov A A 2012 *Synergetic management methods for complex systems: Theory of system synthesis* (Moscow: Book House "LIBROCOM") p 240

[10] Dorodnykh N O and Yurin A Yu 2019 *The technology of creating production expert systems based on model transformations* ed O A Nikolaychuk (Novosibirsk: SB RAS) p 144

[11] Pavlov A I, Stolbov A B and Dorofeev A S 2019 The Workflow Component of the Knowledge-based Systems Development Platform *CEUR Workshop Proceedings* vol 2463 pp 47–58

IOP Conf. Series: Materials Science and Engineering 1061 (2021) 012008 doi:10.1088/1757-899X/1061/1/012008

Hierarchical structure of synergistic multi-agent system

O E Bezborodova[1], O N Bodin[1], M N Kramm[2], S V Vorobieva[3] and S A Kharitonov[4]

[1] Penza State University, 40, Krasnaya str., Penza, 440026, Russia.
[2]National Research University "Moscow Power Engineering Institute", Moscow, 14, Krasnokazarmennaya str., Moscow, 111250, Russia,
[3] Siberian State University of Telecommunications and Information Sciences, Novosibirsk, 86, Kirov str., Novosibirsk, 630102, Russia
[4]Novosibirsk State Technical University, Novosibirsk, 20, Prospekt K. Marksa, Novosibirsk, 630073, Russia

E-mail: oxana243@yandex.ru

Abstract. The article considers the task of ensuring the efficiency of synergistic information-measuring and control systems. A structural scheme of synergistic information-measuring and control system based on multi-agent technologies has been developed and a decomposition of goals has been presented. For the heterogeneous group of unmanned aerial vehicles involved in search and rescue operations in the emergency zone, control algorithms based on the use of synergies have been proposed. It is shown that the synergistic effect lies in the joint interaction of diverse unmanned aerial vehicles of a heterogeneous group. Calculations are given and graphs are constructed, that illustrate the operation of a heterogeneous group of unmanned aircraft in normal mode and in the event of failure of one of the unmanned aerial vehicles. It has been proven that the group is able to complete the task due to the synergistic effect.

1. Introduction

Improving the efficiency of data collection and processing is necessary to ensure that information-measurement and control systems (IMCS) are guaranteed to work without fail. In this case, we are talking about the automation of data collection and processing processes, the most reliable storage of data, the guaranteed permanent access to it and the possibility of rapid adaptation of the IMCS to the changing demands of the external environment.

In recent years, there has been an increase in interest in the interdisciplinary direction - synergies. Synergetic approaches and models have found wide application not only in philosophy, but also in many branches of science: technology, sociology, pedagogy, philology, etc. The founders in this subject area are Haken G. [1], Prigogin I.R. [2], Kurdyumov S.P. [3], Chernyavskiy D.S. [4], Samarskiy A.A. [5] and others.

A common drawback of synergistic systems in specialized literature on process and object management tasks is the lack of a description of technical implementation: the description of the algorithms of operation is given only at the conceptual level.

The synergistic approach to the creation of management systems is based on systemic principles, the basic of which are the interaction and interdependence of the system and the external environment, the hierarchical structure of the system, the focus of activity, adaptability to the external environment, self-organization, priority of common interests of the system, reliability.

According to the property of emergence, the joint action of several factors almost always differs from the amount of separate effects. It is this difference, commonly referred to as the synergy effect, interaction factor or cooperative effect, that is the quantitative expression of synergy.

The synergistic approach to the design of control systems is to move from the unpredictable behavior of the system by the algorithm of the dissipating structure to the directional movement along the desired invariant diversity - attractors (ψ_s), to which all other parameters of the system (x_n) are adjusted. This is a targeted way of self-organizing synergistic control systems. With this approach, the goal is the attractor $\psi_s(x_1, \dots, x_n) = 0$ - defines the essence of the management process, and its content consists of self-management and directed self-organization in accordance with the specified purpose [6].

All this characterizes the synergistic system as an intelligent, private case of which is a multi-agent system consisting of intelligent agents of various origins (software, hardware, biological). According to the researchers [7 - 9], the practical implementation of synergistic approaches in the creation of specialized systems for managing processes and objects increases the efficiency of their functioning.

2. Statement of the problem

Technogenic objects (TO) are complex systems consisting of a large number of subsystems (agents) that work under changing internal and external conditions, adequately responding to these changes. Therefore, the non-emergency functioning of the TO cannot be ensured without the participation of the IMCS. Failures in the management and control of the TO lead to the emergence of emergencies (emergencies). Options for the effectiveness of the IMCS are their reliability (non-emergency functioning) and the ability to quickly adapt to the changing requirements of the external environment, and in the case of emergency - the time and resources needed to restore the performance of the controlled TO.

To ensure the necessary parameters of effectiveness, the IMCS should be structured so that agents (subsystems) develop independently, ensuring the effectiveness of the IMCS on the target parameters. This behavior, when the simultaneous evolutionary development of several agents leads to the restructuring of the entire decision-making system in the IMCS, can be classified as synergistic effects. Therefore, to ensure technosphere security, synergetic ISIs are required, which consist of a large number of agents in a coordinated interaction that determines the behavior and properties of the system as a whole [1].

In the IMCS distinguishes computational, information, intellectual resources and time resources. In the construction and analysis of the IMCS, it is mainly used by computational and time resources, but this does not mean that other types of resources, such as logistical resources, are ignored for a particular problem. As the functional purpose of the IMCS suggests, its main objective is to organize the information interaction optimally to ensure that the tasks are met.

Optimization refers to the process of finding extremes (global maximum or minimum) that can be rated as the best values (indicators) of a particular target function or choosing the best (optimal) option from a variety of possibilities.

Resource optimization can be done at different levels of the IMCS hierarchy. At the same time, the principle of global optimization must be implemented, i.e. local optimization criteria and decisions based on them should not contradict the global optimum of the IMCS as a whole.

At the same time, any IMCS has limited resources to achieve strategic, tactical or operational goals. Therefore, the task of optimizing resources in the IMCS can be formulated in two ways:

- with limited resources, achieve a global maximum vector target function of the IMCS, i.e. it is necessary to store more data in the allocated space, to make calculations faster, without consuming or releasing too much energy, to transmit more information through existing communication channels, to ensure maximum readiness of the operating subsystems;

- at the planned values of the IMCS indicators to achieve a cumulative minimum of resources used.

This speaks to the importance of developing methods and tools to improve the efficiency of the IMCS in the management of processes and facilities.

The aim of the work is to develop the structure and algorithm of the IMCS, which provide optimal functioning on the basis of synergistic approach and hierarchical principle of building a multi-agent IMCS.

3. Theory

The synergism of the IMCS is manifested in the strengthening of the properties of the system (efficiency E) when combining individual agents, which is expressed by the general formula [10]:

$$E_{1+\cdots+N} > E_1 + \cdots + E_N, \tag{1}$$

where N is the number of agents combined into the system.

The synergistic effectiveness of the unification process is logically defined as systemic efficiency. This efficiency then characterizes the overall effectiveness of non-linear interaction between agents that make up the system. Therefore, synergistic efficiency is a way of taking into account the overall effectiveness of a system that consists of several types of efficiency, when it cannot be determined by simply summing up the effectiveness of individual agents of the system.

The effectiveness of the synergies can be presented as two components of "linear efficiency" and "non-linear efficiency." "Linear efficiency" is created by a simple summing up of the effectiveness of individual agents within the synergistic IMCS, and "non-linear efficiency" is formed by the appearance of the system of properties and parameters not characteristic of individual agents (emergence). Synergistic Es efficiency can be recorded as

$$E_s = E_L + E_N, \tag{2}$$

where E_L is a linear component of efficiency, E_N is a non-linear component of efficiency.

Synergistic efficiency is a characteristic of the health of the system, the reliability of its functioning within the range of basic parameters (time and resources) for a particular system [10].

The basis of synergistic IMCS is the hierarchical multi-agent system (MAS), which is designed to ensure the performance of the IMCS in normal and emergency conditions [11]. This is achieved by a flexible, coherent adjustment of the system in response to changes in the environment, depending on the events (self-organization). As a result, MAS is simultaneously implementing processes for data collection, analysis, planning, optimization, monitoring and control of the state of TO in real time, which ensures that its operation is enhanced.

The main manifestation of self-organization of synergistic IMCS is the reconfiguration of the structure of MAS and the complexity of the system through fluctuations (accidental deviations in the mode of normal functioning of the IMCS or failure of individual agents) states of its agents. Such fluctuations are suppressed by negative feedbacks, ensuring the preservation of the structure, performance and close to the equilibrium state of the system. But in more complex IMCS deviations increase over time, accumulate and lead to either the destruction of the former structure, or to the emergence of a new (the inclusion of duplicating lines and elements, cloning and formation of coalitions of agents).

Self-organization, which results in the reconfiguration of the system, can occur only in high-level systems of complexity, having a certain number of interacting agents, having some critical communication parameters and relatively high probability of their fluctuations. Otherwise, the effects of synergies will not be sufficient to create the collective behavior of agents and thus the emergence of self-organization. Insufficiently complex IMCS are not capable of spontaneous adaptation and development. With excessive influences from the outside lose their structure and irreversibly collapse.

Self-organization of IMCS occurs only when positive feedbacks over negative ones prevail. The functioning of the MAS is based on receiving feedback from sensors carrying information on the state of the TO (environmental IMCS environment) and subsequent adjustment of the parameters of the operation of the TO through the actions of executive mechanisms (decision support system (DSS)). In the self-organizing IMCS, such changes are not eliminated, but are accumulated and amplified by the intellectuality of the agents in the system, which leads to the emergence of a new order and reconfiguration of the IMCS, formed from elements of the previous, modified system.

This process is carried out by cloning agents and forming coalitions within the MAS. If the TO through the sensors receives data indicating the inability of the system to achieve the target (ensuring the normal functioning of the TO), the MAS, after analysis of the situation, without external intervention, gives a command to reconfigure the structure, either by cloning agents, or by redistribution of responsibilities between agents in a coalition of single-functional agents. The choice of self-organization of the system depends on the critical parameters of the functioning of the IMCS and TO. If you need to act quickly (the critical parameter is time), then more resources (agents) can be involved. But, if both time and resources are critical parameters, then due to the synergy of the IMS and the presence of nonlinear efficiency (see formula (2)), it is possible to achieve the goal without attracting additional resources and times, due to the redistribution of duties between single-functional agents [12].

4. Practical implementation

The fundamental principles of self-organization of synergistic IMCS are the ease of information interaction between MAS agents and the ability to quickly reconfigure the structure of the IMCS.

Synergistic IMCS is built according to a standard scheme, which includes the control and control center of objects, sensors and executive mechanisms (figure 1). Moreover, synergistic IMCS can be multifunctional, monitor and manage different objects by nature. But the algorithm for achieving goals is always the same. MAS, working to achieve the goal, breaks the target into tasks, predicts the outcome of each task, draws up a plan to solve each problem, evaluates the capabilities of each agent to solve the problem and distributes them (figure 2). The results of individual tasks are integrated into the goal, on the basis of which a command is given to the executive mechanisms to change the external (environmental) environment.

Let us consider the manifestation of the synergistic efficiency of the IMS on the example of the operation of a heterogeneous group of unmanned aerial vehicles (UAVs) in the emergency zone. In accordance with [13], a heterogeneous group of UAVs, which is part of the emergency (emergency) response subsystem, is sent to the emergency area during search and rescue operations (SRO). The method of carrying out the SRO consists in the implementation of all stages of the SRO: reconnaissance of the emergency zone, search for victims and determining their location, minimizing the levels of exposure to the damaging factors (DF) of emergency situations, rendering medical assistance to the injured, evacuation of transportable victims. All these stages are implemented using a specially equipped heterogeneous group of UAVs, and there can be several UAVs of each specialization.

A heterogeneous group of UAVs is included in the emergency response subsystem in order to minimize the time for detecting victims, providing emergency medical assistance to them, as well as prompt evacuation, detecting and minimizing the sources of damaging emergency factors.

The algorithm of work (figure 3) of a heterogeneous group of UAVs includes the stages of collecting and processing data, as well as the implementation on their basis of actions that change the external environment. These actions are carried out by a synergistic IMCS based on the intellectual interaction of MAS agents (in this case, UAV), which ensures that, in the event of a failure of one UAV, the preservation of the operability of the entire heterogeneous group [12]. Indeed, the structural calculation of reliability shows that if one of the N UAVs fails, the entire heterogeneous group will not lose its performance. The system goes into a more intense mode of operation, in which, having fewer resources, it performs the specified functions. At the same time, the productivity and efficiency of solving the problem are certainly reduced.

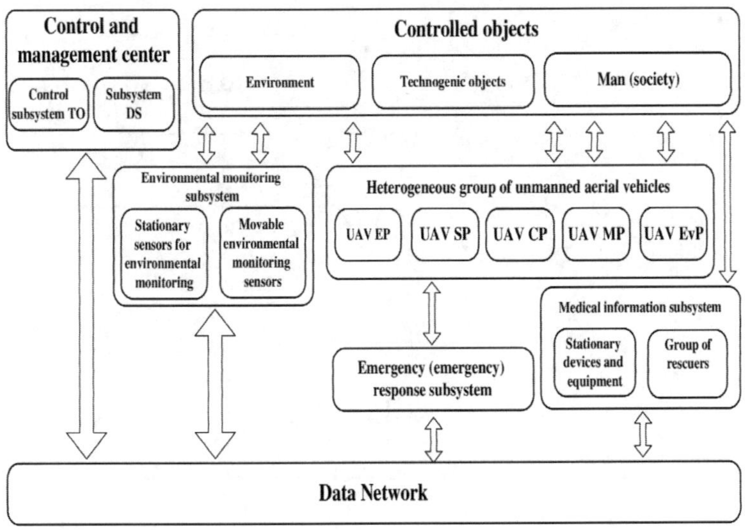

Figure 1. Structural-functional model of synergistic IMCS

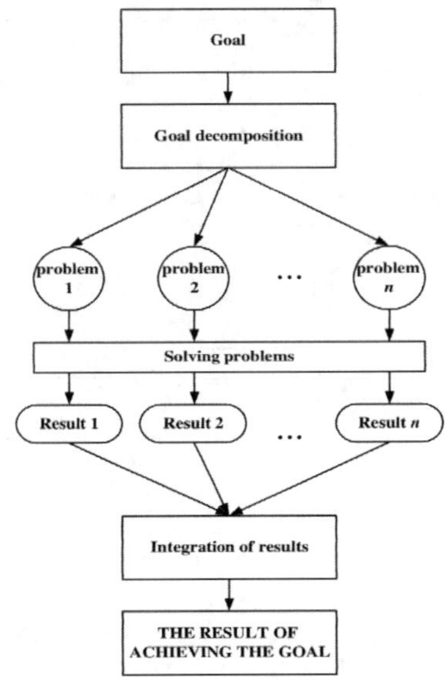

Figure 2. Decomposition of the goal of the MAS for Synergistic IMCS

In the absence of the "Stop" mode, the redistribution of duties of the UAV is not carried out, and the SRO is carried out according to a predetermined work plan. If the "Stop" mode is available, the resource analyzer of the synergistic IMCS realizes the redistribution of the duties of the UAV, according to the algorithm shown in figure 4.

Synergistic IMCS works as follows: first, the agent of the 1st level of the hierarchy analyzes the target, including the parameters of "resources - time" and distributes its solution to a number of parallel tasks to ensure decentralization of management and shorten the time of the SRO. Accordingly, based on the analysis of the original target, the agent of the 2nd level of the hierarchy is switching the resources of the IMCS (distribution of tasks to the agents of the 3rd level). This organization of the work of the IMCS ensures the maximum possible loading of the UAV, based on the features of the algorithm solved by each task and the hardware limitations of each UAV. All UAVs in the heterogeneous group work in parallel and independently, in accordance with the objectives, but exchange the collected data, which reduces the time it takes to solve the tasks before each of them.

The work of the resource analyzer can be considered by the example of the implementation of the formula:

$$(c_{1,}c_2) = \begin{pmatrix} a_{11} & a_{12} \\ a_{21} & a_{22} \end{pmatrix} \cdot \begin{pmatrix} b_1 \\ b_2 \end{pmatrix}, \qquad (3)$$

where c_1 and c_2 are synergistic efficiency of the system taking into account the analyzed parameters "resources - time"; a_{ij} are system agents, for which the i-th parameter is decisive; b_i are system agents, for which the $i+1$-th parameter is decisive.

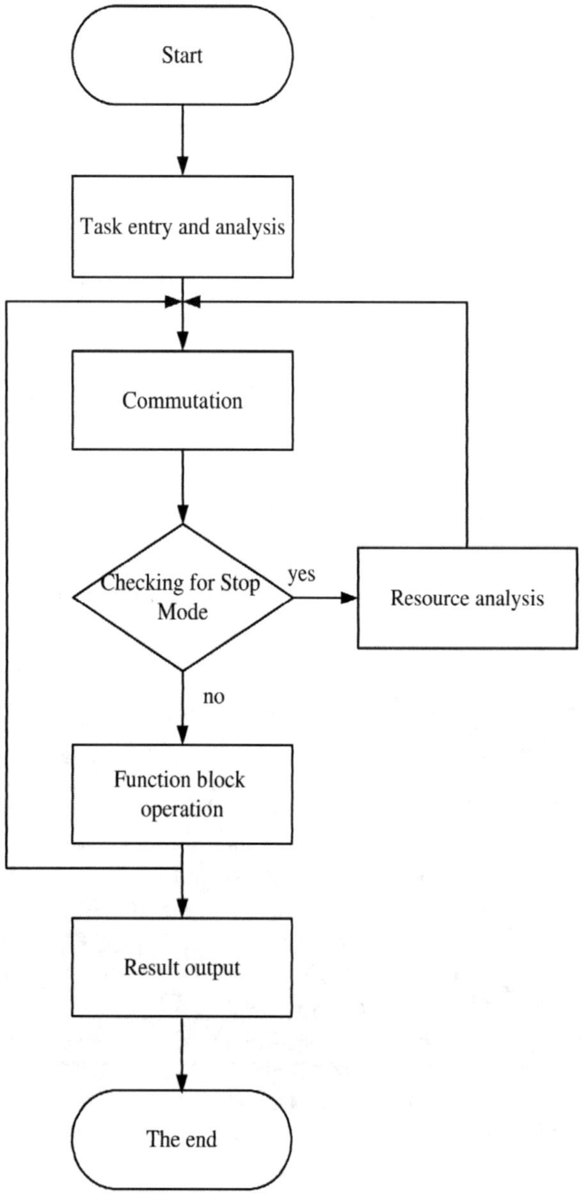

Figure 3. Algorithm of t he Synergistic IMCS

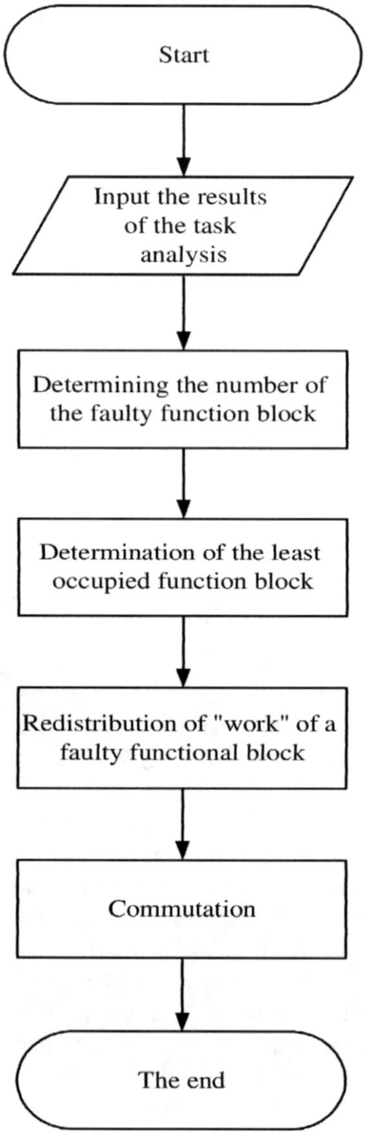

Figure 4. Algorithm of the Resource Analyzer for Synergistic IMCS

Let's say that the heterogenic group of UAVs consists of 6 agents

$$A := \{a_{11}, a_{12}, a_{21}, a_{22}, b_1, b_2\} \quad (4)$$

and all UAV are workable, then the implementation of the formula (3) takes place, as shown in figure 5. Table 1 shows the work of agents (UAVs) on the tacts in normal mode. The resource analyzer in this case monitors the operation of the heterogeneous group of BVS by checking for the presence of the "Stop" mode.

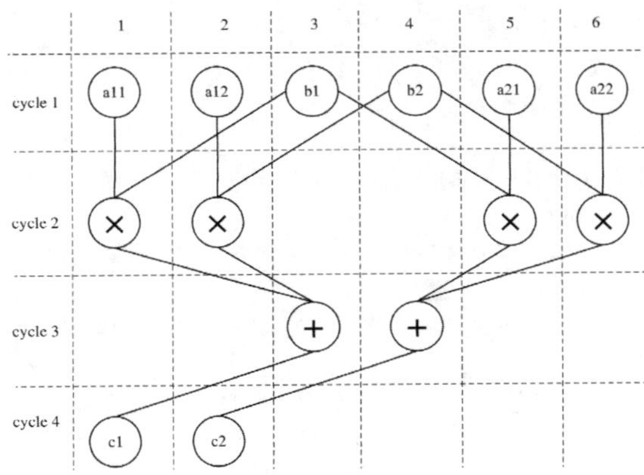

Figure 5. The graph of the SRO when UAV are workable.

Table 1. The work of agents on the tact in normal mode.

Tact	A1	A2	A3	A4	A5	A6
1	rd a11	rd a12	rd b1	rd b2	rd a21	rd a22
2	1×3	2×4			3×5	4×6
3			$1 + 2$	$5 + 6$		
4	dr c1	dr c2				

rd – reading of data; dr – data recoding.

If one of the UAV fails, the resource analyzer carries out recovery procedures in accordance with the above algorithm. For example, the fifth UAVs in the heterogeneous group is out of order, then the implementation of formula (3) will look different, as shown in figure 6. Table 2 shows the work of agents (UAVs) on tacts with an inoperative agent.

As follows from the consideration of figure 6, the fourth UAV takes over the "work" of the fifth UAV. The work of the entire heterogeneous group continues, despite the failure of one of the UAV.

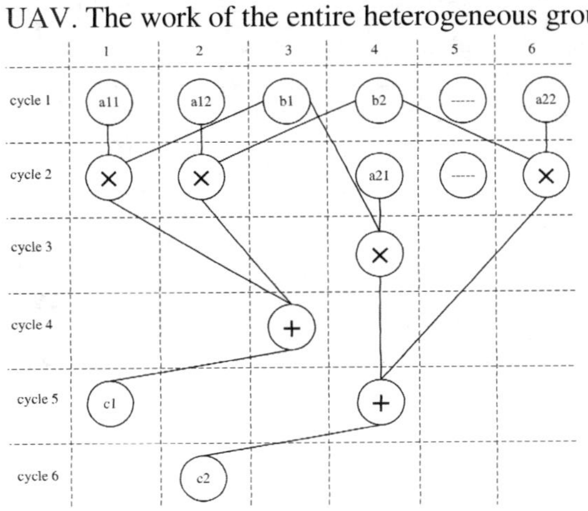

Figure 6. The graph of SRO in case of an inoperable UAV.

Table 2. The work of agents on tacts witn an mode inoperative agent

Tacts	A1	A2	A3	A4	A5	A6
1	rd a11	rd a12	rd b1	rd b2	-	rd a22
2	1×3	2×4		rd a21	-	4×6
3				3×4	-	
4			$1 + 2$		-	
5	dr c1			$4 + 6$	-	
6		dr c2			-	

rd – reading of data; dr – data recording.

Thus, the use of a synergistic approach in the IMCS, based on the intellectual interaction of MAS agents, allows to maintain the effectiveness of the SRO in the failure of the UAV due to the non-linear component of efficiency arising from the synergistic effect in the system.

5. Conclusion

Thus, the structural and functional model of the proposed synergistic IMCS is characterized by a set of structural elements and their functional purpose inherent in most modern IMCS, which makes it invariant about the field of application. Using the proposed synergistic IMCS to manage UAV in an emergency environment allows you to optimize resources through synergies and ensures that the goal is achieved even when the agent fails, as it is regeneratable and maintains health at the failure of the agent (UAV).

References

[1] Khaken G 1985 The hierarchy of instabilities in self-organizing systems and devices (Moscow: Mir Publ.)

[2] Nikolis G and Prigozhin I 1979 *Self-organization in nonequilibrium systems* (Moscow: Mir Publ.)

[3] Kniazeva E N and Kurdiumov S P 2010 *Foundation of synergetics. Synergetic world-view* (Moscow, KomKniga Publ.)

[4] Chernavskii D S 2004 *Synergetics and information: Dynamic information theory* (Moscow: URSS Publ.)

[5] Samarskii A A and Mikhailov A P *Mathematical modeling: Ideas. Methods. Examples* Available at: www.twirpx.com/file/20242/

[6] Kolesnikov A A 2002 Synergistic systems *Software products and systems* **1** pp 3–6

[7] Manova M V and Shirina E V 2015 A synergistic system of intellectual property management *Baltic humanitarian magazine* **2** (11) pp. 150–153

[8] Reshetnikov I S 2011 Synergetic model for building a multi-module information and control system *Bulletin of the Lobachevsky University of Nizhny Novgorod* **3** (2) pp 264–267

[9] Lebedev V I, Lebedeva I V and Shuvaev A V 2018 Synergetic models of socio-economic systems *Fundamental research* **11** pp 256–260

[10] Sukharev O S 2014 *The theory of effectiveness of economy* (Moscow: Kurs: INFRA-M) Available at: https://studref.com/428048/ekonomika/sinergeticheskaya_effektivnost

[11] Bezborodova O, Gromkov N, Bodin O, Bodin A, Baranov V and Trilissky V 2019 Multi-agent Technologies for Comprehensive Monitoring of the State of Territorial Technosphere *2019 8th Mediterranean Conference on Embedded Computing (MECO)* (Budva, Montenegro) pp 1–4 doi: 10.1109/MECO.2019.8760064

[12] Bodin O N, Loginov D S and Tarnopolsky K A 2007 *Patent № 2292075 Russian Federation, Int. Cl. G06F 15/16 (2006.01).* Synergistic computing system : № 2005119236, : Date of filing: 21.06.2005 : Date of publication: 20.01.2007

[13] Sherstnev V V, Bodin O N, Bezborodova O E, Rakhmatullov F K, Gerasimov A I, Ozhikenov K A, Bayanbaj N, Berdibaeva G K 2019 *Patent № 2694528 Russian Federation, Int. Cl. A62B 99/00 (2009.01).* Search and rescue method : № 2018139491, : Date of filing: 07.11.2018 : Date of publication: 16.07.2019

Improving data collection and processing efficiency by using hierarchies of intelligent agents in multi-agent systems

O E Bezborodova[1], O N Bodin[2], D V Paschenko[2]

[1] Penza state University, 40, Krasnaya str., Penza, 440026, Russia
[2] Penza State Technological University, 1a / 11, pr. Baidukova / ul. Gagarina, Penza, 440039, Russia

E-mail: oxana243@yandex.ru

Abstract. The paper considers the challenge of improving data collection and processing efficiency by using various information and measurement systems. A structural scheme of a hierarchical multi-agent control system has been developed, incorporating horizontal (single-level) and vertical (multi-level) connections between intelligent agents. An algorithm is proposed for controlling a heterogeneous group of unmanned aerial vehicles participating in search and rescue operations in an emergency zone based on the hierarchical interaction of intelligent agents, cloning and coalition formation. KPI values have been calculated and the effectiveness of the proposed solution have been proved.

1. Introduction

The challenge of using artificial intelligence to control data collection and processing processes in real time is not new and has been studied for several decades, starting with the classical works of J. Von Neumann, M. Zeitlin, V. Varshavsky, D. Pospelov and others. Nevertheless, its importance increases with each year. Moreover, the application of artificial intelligence is currently associated with the creation and implementation of multi-agent technologies (MAT). A MAT agent implies a software and hardware unit of a certain functional purpose. To enable widespread use of MAT, new development concepts, as well as models for formalization and architecture creation of information, measurement and control systems are required.

Analysis of the capabilities of modern MATs, as well as the end user requirements, shows that the technologies integrating distributed decision making and computation based on formal logical specifications have the greatest prospects in this field. A great contribution to the development of this topic was made by domestic and foreign scientists D. A. Pospelov [1], V. B. Tarasov [2], V. I. Gorodetsky [3], D. Yu. Bugaychenko [4], M. J. Wooldridge [5, 6], M. Pipattanasompornn, H. Feroze, S. Rahman [7], who propose a theoretical justification and practical implementation of multi-agent systems (MAS) for various fields of activity.

Being a part of a certain facility, the multi-agent system is aimed at increasing the efficiency and reliability of its operation. This is achieved by flexible, coordinated fine-tuning of the facility operation in response to changes in the environment, depending on the occurring events. In the end,

MASs simultaneously conduct data collection, analysis, planning, optimization, monitoring and control of the facility state in real time, improving the facility operation efficiency.

2. Statement of the problem

MAT implementation challenges are very popular among researchers and developers. A large amount of research has been conducted on the implementation of MAT in various fields of activity. MAT is proposed for planning, diagnostics, management and monitoring of systems of various purpose. Often MASs used for control consist of a hierarchy of intelligent agents (IA), in which the top-level IA only receives the data it needs for its operation while coordinating the activity of lower-level IAs without directly dictating what actions to perform.

The study [4] proposes a MAS, capable of IA cloning and coalition forming — MASL (MultiAgent System Logic). The disadvantages of MASL are the lack of IA hierarchy, as well as a mechanism for resolving conflicts between IA and the external environment when the IA is unable to fulfill the task.

In study [8] overviews MASs, which incorporate IA hierarchies. The analyzed solutions have the following disadvantages:

- IAs exchange data only by using a bottom-top message board mechanism [9];
- centralization of individual functions in the MAS [10];
- lack of diagnostics and control of IA activity [11];
- lower-level IAs are unable to initiate task changes [12].

The study [8] proposes a MAS, which operation is based on the coordination of key performance indicator (KPI) values taking into account their hierarchy and interconnection.

The purpose of this study is to increase data collection and processing efficiency by using a hierarchical MAS structure, which incorporates IA cloning and coalition forming, on the example of a heterogeneous group of unmanned aerial vehicles (HG UAV) during search and rescue operations (SAR) [13].

To do this, we should solve the following tasks:

- create an improved MAS operation algorithm (MASL+);
- select the parameters describing IA operation, which can be converted to KPI values;
- determine the actual values of these parameters;
- compare the developed MASL+ algorithm with MASL algorithms [4] and multi-level agent interaction algorithm [8] using KPI.

3. Theory

As stated in [14], a multi-level hierarchical control structure has the following basic properties — vertical arrangement of subsystems (vertical decomposition), action priority or upper-level subsystems have the right to change the course of actions, operation of upper-level subsystems depend on the actual execution of their actions by lower-level subsystems.

Multi-level hierarchical control systems include linear and functional elementary organizational structures. They have undeniable advantages, such as:

- unity of management, simplicity and clarity of subordination;
- full responsibility of the upper-level subsystems for the results of the lower-level subsystems;
- fast decision making;
- consistency of actions at all levels;
- lower-level subsystems receive tasks coordinated between each other;
- high competence of lower-level subsystems responsible for solving specific problems;
- lower-level subsystems are specialized in solving certain problem types, eliminating any duplication of their solution.

Based on this, a hierarchy-based MAS shall have a structure shown in figure 1.

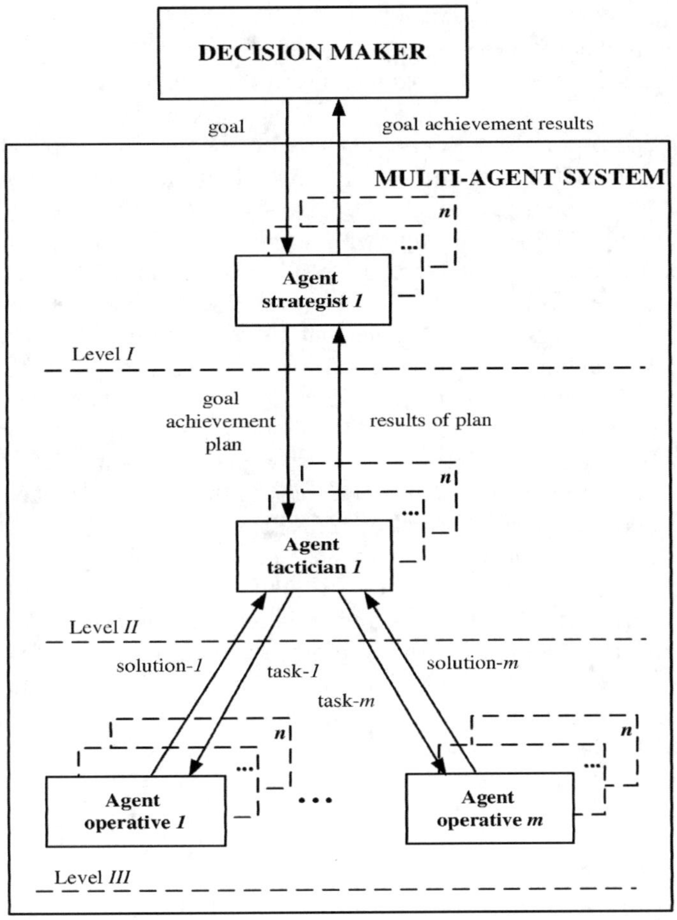

Figure 1. Graphical representation of hierarchy levels in a multi-agent system

According to the degree of subsystem participation in the decision making and implementation process, subsystems are divided into a hierarchy of strategists, tacticians and operators.

Strategists are top-level IAs of the corresponding hierarchy, which direct and coordinate its operation, control and regulate activities to achieve goals and solve problems.

Tactics are specialized IAs that distribute tasks between lower-level IAs, control their activities and act as a link with strategists.

Operators are highly specialized IAs that solve individual tasks.

A balanced KPI system is a tool for assessing system effectiveness, which allows controlling the operation of subsystems, as well as the consequences of their failure [15]. System performance assessment implies finding its effectiveness, that is, the degree to which the goals are achieved and problems are solved, aimed at achieving the goals. Effectiveness is used to measure, control and regulate related activities. It is a clear indicator of the superiority of one system version over another.

A much more detailed is the system of key performance indicators (KPI), which allows determining the most significant system indicators, that can best characterize its effectiveness [16]. KPIs allow determining how compliant is the current system operation with strategic goals.

The structure efficiency criteria for any multilevel hierarchical control system are:

- compliance of the control levels with their functions;
- system levels number (as few as possible);
- all relevant functions concentrated at each level;
- clear identification of how each level participates in the overall process, elimination of function duplication.

The choice of indicators for the KPI list depends on the field of application of a multi-level hierarchical control system and the conditions of its operation, but in most cases, the indicators are resources (financial, material, human, time, etc.) necessary for its operation.

The study [4] describes a MASL algorithm with a linear structure, the main advantages of which are the ability to clone IAs and combine them into coalitions are aimed at solving common problems and achieve the goals. The advantages of a multi-level agent interaction algorithm [8] having a hierarchical structure are the possibility to implement IA interaction both "from top to bottom" and "from bottom to top". The authors have developed a MASL + algorithm with a hierarchical structure, capable of AI cloning and coalition forming (see figure 2), which combines the noted advantages and improves MAS operation KPIs.

The proposed MASL+ algorithm has a Strategist agent controlling HG UAV, and if UAVs encounter an increased load in any area of operation, it has the ability to clone a specialized UAV if one of the UAVs does not cope or fails the task as per KPIs.

Emergency medicine has a "golden hour" concept — the time during which the provided on-site assistance or prompt delivery of the injured to a healthcare unit ensures the maximum chances of survival and the lowest risk of complications after injuries. The human body reacts instantly in an extreme situation by activating compensatory and protective mechanisms at their full force in order to maintain vital activity for about an hour (60 minutes). After that, the blood supply to the main organs — the heart and brain, begins to decrease, while the chances to save the injured are greately reduced. Therefore, the main KPI is the time needed for SAR execution, which depends on how optimal is the HG UAV control algorithm, UAV operation scheme (parallel or serial), as well as the number of UAVs in the heterogeneous group.

4. Practical implementation

The initial data for comparing MAS using KPIs are taken from [13], which describes an HG UAV used during SAR being a part of the rescue team possessing a mobile telemedicine system (MTMS). SAR execution implies conducting all SAR stages — reconnaissance of the emergency situation (ES) zone, search for injured and determining their location, minimization of the levels of impact of the damaging factors (DF) of emergency situations, rendering medical assistance to the injured, recovery of injured, whose state permits them to be transported. All these stages are carried out using a specially prepared group of UAVs controlled by a single MTMS operator. HG UAV includes coordinating UAVs (C-UAV), search UAVs (S-UAV), UAVs for environmental purposes (E-UAV), UAVs for medical purposes (M-UAV), and UAVs for recovery purposes (R-UAV). HG UAV is included in the rescue team in order to minimize the time for finding the injured, providing emergency medical assistance and prompt recovery, as well as to detect and minimize the damaging factor sources of ES.

If MAS control of HG UAV is based on MASL algorithm [4], the MTMS operator separately controls five specialized UAVs with the capability to increase the number of UAVs in each area of operation, if one of UAVs fails to cope or fails the task as per KPI. In such a case, the number of simultaneously observed objects increases when there is a little time for decision making, which creates an additional load on the MTMS operator reducing the likelihood of error-free operation [17].

When the algorithm incorporating interaction of multilevel agents [8] is used to control HG UAV during SAR, the MTMS operator becomes a Strategist agent and controls one C-UAV object (Tactician), which in its turn controls the operation of the entire group (Operator agents). But this algorithm does not have IA cloning, and the Strategist agent cannot add individual Operator agents in problem areas (reconnaissance, assistance to injured, minimization of DF levels) or simultaneously

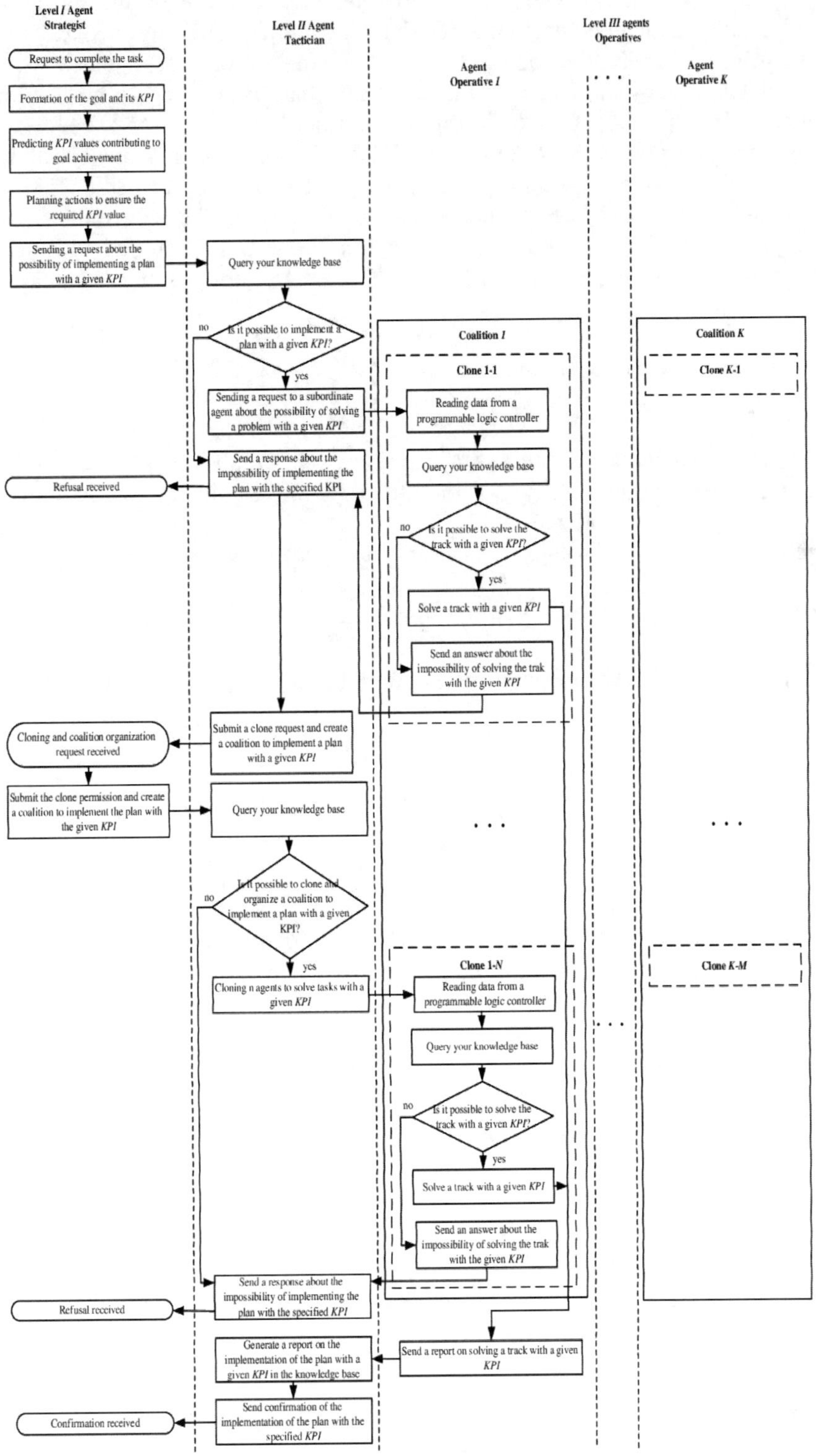

Figure 2. MASL+ algorithm

manage several groups in parallel. This requires a dedicated MTMS operator with his own HG UAV.

Based on this, the list of parameters to be converted to KPI values and used to compare the algorithms shall include parameters that characterize the resources required for SAR. First of all, this should include SAR execution time using UAVs, the number of UAVs used, as well as UAV operation and maintenance costs included. If the operation shall be finished within 1 hour (τ = 1 hour), the average cost of using one UAV c = 10 thousand rubles per hour.

The analysis of the algorithms allowed to determine UAV operation scheme (serial, parallel, serial-parallel). The rest of the parameters were calculated using the method described in [13].

Let us assume that SAR is finished in τ hours with a single UAV. To reduce the SAR duration, the authors propose to use N UAVs with the cost of operation of one UAV c rubles per hour.

The operating and maintenance costs of the UAVs $\boldsymbol{C}$ are determined by the formula

$$\boldsymbol{C} = \boldsymbol{c} \cdot \boldsymbol{N}. \tag{1}$$

UAV total operating time can be defined as

$$\boldsymbol{T} = \boldsymbol{\tau} / \boldsymbol{N} \tag{2}$$

Formulas (1) and (2) demonstrate that as $\boldsymbol{N}$ value grows, $\boldsymbol{C}$ value increases as well and $\boldsymbol{T}$ value decreases. This indicates the inconsistency of the utilized parameters and, consequently, the lack of a single optimal solution to this problem.

It is proposed to form trade-off solutions as a set of Pareto optimal solutions [13] by the formula

$$N = N(\alpha) = \sqrt{\frac{(1-\alpha)\,\tau}{\alpha\, c}}, \tag{3}$$

where $\boldsymbol{\alpha}$ is the parameter of performance criteria convolution, considered in $\boldsymbol{\alpha} \in (\mathbf{0}; \mathbf{1})$ interval and determined by the formula

$$\alpha^* = \frac{\tau}{\tau + c}, \tag{4}$$

where $\boldsymbol{\alpha} \in (\mathbf{0};\ \boldsymbol{\alpha}^*) \in (\mathbf{0}; \mathbf{1})$.

Let us form a grid of convolution parameter values

$$\mathbf{0} < \boldsymbol{\alpha}_1 < \boldsymbol{\alpha}_2 < \cdots < \boldsymbol{\alpha}_i < \boldsymbol{\alpha}_n = \boldsymbol{\alpha}^*. \tag{5}$$

By substituting $\boldsymbol{\alpha}_n$ values sequentially into formula (2), we calculate $\boldsymbol{N}_i$ Pareto optimal values of the optimal number of UAVs for each algorithm, $\boldsymbol{i} = \overline{(\mathbf{1}, \boldsymbol{n})}$.

By eliminating the repeated values from $\boldsymbol{N}_1, \boldsymbol{N}_2, \ldots, \boldsymbol{N}_n$ value set, we get $\boldsymbol{m}$ alternatives for $\boldsymbol{N}_1^0, \boldsymbol{N}_2^0, \boldsymbol{N}_3^0, \ldots, \boldsymbol{N}_j^0, \ldots, \boldsymbol{N}_m^0$ required number of UAVs.

Let us substitute the obtained values into formulas (1) and (2) to calculate cost and operation time alternatives.

Based on these results, a Pareto optimal set of solutions to the problem is formed

$$\boldsymbol{P} = \left\{\boldsymbol{N}_j^0, \boldsymbol{C}_j^0, \boldsymbol{T}_j^0 \mid \boldsymbol{j} = \overline{(\mathbf{1}, \boldsymbol{m})}\right\}, \tag{6}$$

which is issued to the decision-maker (DM) for the analysis and selection of the optimal number of UAVs based on the specific conditions of SAR.

By performing calculations according to formulas (1) – (4), we obtain the dependencies shown in figure 3.

IOP Conf. Series: Materials Science and Engineering 1061 (2021) 012009 doi:10.1088/1757-899X/1061/1/012009

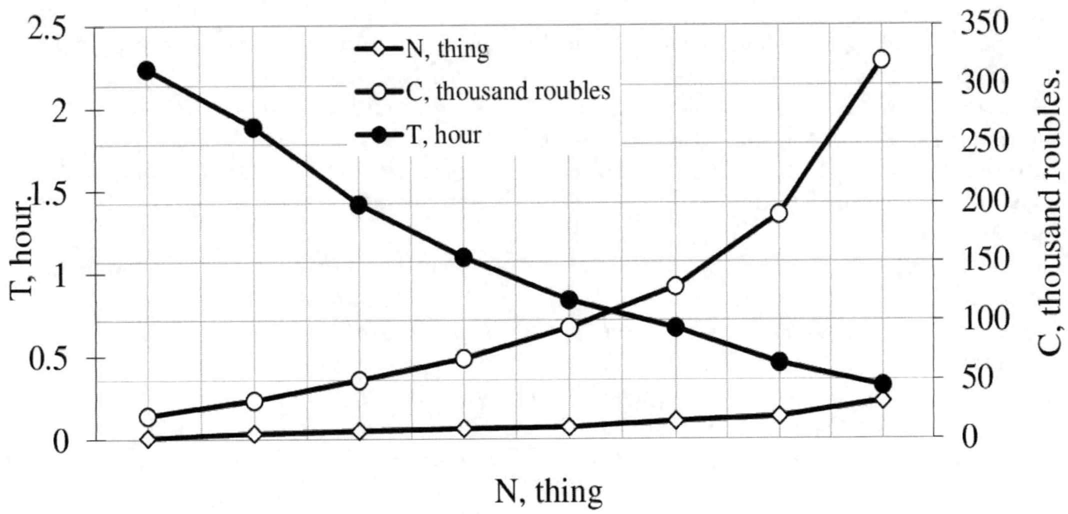

Figure 3. KPI calculation results

Table 1 shows the KPIs of the analyzed MAS algorithms.

Table 1. KPIs of analyzed MAS algorithms

KPI	Algorithm		
	MASL [4]	of multi-level agent interaction [8]	MASL+
BVS operation scheme	serial	parallel	serial-parallel
number of UAVs in SAR	$1 \cdot N$	5	$5 \cdot N$
time to reach the goal, hours	$2.42 \cdot N$	1.89	1 or less
UAV operating and maintenance costs	$10 \cdot N$	50	50 and more

Analysis of the obtained KPI values shows that the proposed MASL+ algorithm is the most efficient among the considered ones since it reduces SAR duration and load on the MTMS operator. This is achieved due to a hierarchical structure, as well as IA cloning and coalition forming capabilities for an operator controlling and monitoring a single UAV (Tactician), which in turn controls and monitors the rest of UAVs (Operators), while the number of UAVs is limited only by the size of the ES zone.

5. Conclusion

To increase the data collection and processing efficiency, the authors proposed an improved MAS (MASL+) operation algorithm based on IA cloning and coalition formation. Parameters have been selected for HG UAV used during SAR that characterize IA operation and can be converted to KPI values. The actual KPI values have been determined. A comparison of KPIs for three algorithms has been carried out. The results show that the algorithm developed by the authors can be recognized as the most efficient, since it allows a single MTMS operator to control multiple HG UAV during SAR, significantly reducing the time it takes to detect the injured and provide emergency medical care, which ensures the maximum chances of surviving and the lowest risk of complications after injuries.

References

[1] Pospelov D A 1998 Multi-agent systems-present and future *Information technologies and computer systems* 1 pp 14–21

[2] Tarasov V B 2002 *From multi-agent systems to intelligent organizations: philosophy, psychology, computer science* (Moscow: Editorial URSS)

[3] Gorodetskii V.I., Karsayev O.V., Samoylov V.V., Serebryakov S.V. 2010 Applied multiagent systems of group control *Scientific and Technical Information Processing* Vol 37 **5** pp 301–317

[4] Bugaychenko D Yu and Soloviev I P 2007 Formal-logical specification of real-time multi-agent systems *Saint Petersburg University Bulletin Ser. 1, Rel. 2, Mathematics* pp 49–57

[5] Wooldridge M J 1992 *The Logical Modeling of Computational Multi-Agent Systems. PhD thesis* (Manchester)

[6] Wooldridge M J 2001 Intelligent Agents *Multiagent Systems* pp 27–79

[7] Pipattanasomporn M, Feroze H and Rahman S 2009 Multi-agent systems in a distributed smart grid: Design and implementation *IEEE/PES Power Systems Conference and Exposition,* (Seattle, WA) pp 1–8 doi: 10.1109/PSCE.2009.4840087

[8] Kovalevsky V E and Onufriev V A 2019 Multi-agent algorithms for the coordination of key performance indicators of the enterprise *SPbSPU scientific and technical statements. Informatics. Telecommunications. Control.* Vol 12 **3** pp 67–80 DOI: 10.18721/JCSTCS.12306

[9] Zhang L and Zhang Y 2010 Research on hierarchical distributed coordination control in process industry based on multi-agent system *2010 International Conference on Measuring Technology and Mechatronics Automation* (Changsha City) pp 96–100 DOI: 10.1109/ICMTMA.2010.683

[10] Zheng G and Li N 2010 Multi-agent based control system for multi-microgrids *2010 International Conference on Computational Intelligence and Software Engineering* (Wuhan) pp 1–4 DOI: 10.1109/CISE.2010.5676818

[11] Shao Y, Wu Y and Chen Y 2014 Design and research of multi-agent control system for central cooling system *2014 IEEE 3rd International Conference on Cloud Computing and Intelligence Systems* (Shenzhen) pp 218–221 DOI: 10.1109/CCIS.2014.7175732

[12] Wei C, Mei J J C and Shaoting G 2016 Design of workshop production management control system based on multi-agent *2016 IEEE 11th Conference on Industrial Electronics and Applications* (Hefei) pp 427–430 DOI: 10.1109/ICIEA.2016.7603622

[13] Sherstnev V V, Bodin O N, Bezborodova O E, Rakhmatullov F K, Gerasimov A I, Ozhikenov K A, Bayanbaj N and Berdibaeva G K 2018 Patent № 2694528 Russian Federation, Int. Cl. A62B 99/00 (2009.01). Search and rescue method : № 2018139491, : Date of filing: 07.11.2018 : Date of publication: 16.07.2019

[14] Mesarovich M, Mako D and Takahara I 1973 *The theory of hierarchical multilevel systems* (Moscow: Mir)

[15] *Implementation of a balanced scorecard* 2005 (Moscow: Alpina Business Books)

[16] Bezborodova O E, Ubiinykh A G, Sherstnev V V and Bodine O N 2020 Evaluation of the effectiveness of improving information-measuring and control systems *Measurement. Monitoring. Control. Control* **3** (33)

[17] Belov P G 1999 *Modeling hazardous processes in the technosphere* (Moscow: Publishing House of the Academy of Civil Protection of the Ministry of Emergencies of the Russian Federation)

IOP Conf. Series: Materials Science and Engineering 1061 (2021) 012010 doi:10.1088/1757-899X/1061/1/012010

Development of optimal modelling of thin-walled elements in the form of a shell made of composite materials

L A Bokhoeva[1], A B Baldanov[1], A S Chermoshentseva[2]

[1] East Siberia State University of Technology and Management, 40v/1, Klyuchevskaya st., Ulan-Ude, 670013, Russia
[2] Bauman Moscow State Technical University, 5/1, 2nd Baumanskaya st., Moscow, 105005, Russia

E-mail: asch@bmstu.ru

Abstract. The paper presents the research of the stress-strain state of a multilayer composite shell. The optimal construction of a multilayer shell made of composite materials (CM) in the ANSYS system is determined. The optimal construction of a multilayer composite shell provides maximum strength and rigidity at specified loads. An automated procedure developed for choose the angle of fiber laying in the layer. According to the calculations and constructed models, an experimental model of a multilayer shell made of laminated fiberglass manufactured. For the experimental manufacture of a multilayer shell by cold forming, the working drawings of the press-form made. We manually put the layers of shell made of fiberglass T-25 (BM) TU 6-11-380-76. We made layer-by-layer laying of layers in strict accordance with the calculation model, while observing the sizes of elements and reinforcement angles along the base in each layer. The surface of each layer is an ED-20 epoxy resin with a hardener. This approach can be widely used in the engineering and manufacture of structural elements and products from composite materials.

1. Introduction

In aerospace, the desire for a fuller use of the carrying capacity of equipment while observing safety requirements is one of the priority factors. A significant increase in the number of spacecraft in orbit leads to a strong increase in the clogging of outer space. This represents a potential threat to the long-term safety of space activities and a direct threat to the safe life of people in near-Earth space [1].

Today, composite materials are high-tech materials used in the aerospace field. A key factor in the use of composites is strength. Composite material is a matrix with a reinforcing material inside it. The use of multilayer composites leads to the need for stress analysis and large deformations. Layered composites themselves represent a structure consisting of layers. The composite layers are specially fixed to each other. Due to the various laying angles, we must combine useful properties and obtain a material that provides high specific stiffness and strength of the power elements. However, to obtain structural elements from multilayer composite materials with high mechanical properties, we must take into account many factors of modern technologies. These factors are the sequence of laying the layers, the arrangement of the layers and their thickness depending on the current loads [2].

This article presents the development of optimal modeling of thin-walled elements in the form of a shell made of composite materials.

2. Modelling of composite shell

In this work, we investigated the stress-strain state of a multilayer shell and solved the following problems:

1. An automated procedure developed for choosing the angle of fiber laying in the layer in the ANSYS system [3];

2. The optimal construction of the multilayer shell in the ANSYS system determined, providing maximum strength and rigidity at given loads.

Shell geometry developed in the SolidWorks system consisting of a multilayer composite shell and a cube, which serves to transfer the model to the Fluid Flow (flient) calculation module. Shell geometry is transferred to the Ansys system. We set the orthotropic properties of the composite material (Figure 1). In the Engineering data module, we set the laying angles of the layers, thickness and properties of the composite material. Data is transmitted to the ACP (Pre) module.

Properties of Outline Row 3: CM1

	A	B	C	D	E
1	Property	Value	Unit		
2	Material Field Variables	Table			
3	Density	1900	kg m^-3		
4	Orthotropic Elasticity				
5	Young's Modulus X direction	5,4E+10	Pa		
6	Young's Modulus Y direction	1,2E+10	Pa		
7	Young's Modulus Z direction	1,2E+10	Pa		
8	Poisson's Ratio XY	0,28			
9	Poisson's Ratio YZ	0,0062			
10	Poisson's Ratio XZ	0,28			
11	Shear Modulus XY	5E+09	Pa		
12	Shear Modulus YZ	5E+09	Pa		
13	Shear Modulus XZ	5E+09	Pa		
14	Tsai-Wu Constants				
18	Puck Constants				
24	Ply Type				
26	Additional Puck Constants				

Figure 1. Material Properties.

In parallel, in the Fluid Flow (flient) subroutine, the imported geometry is divided into finite elements (Figures 2 and 3).

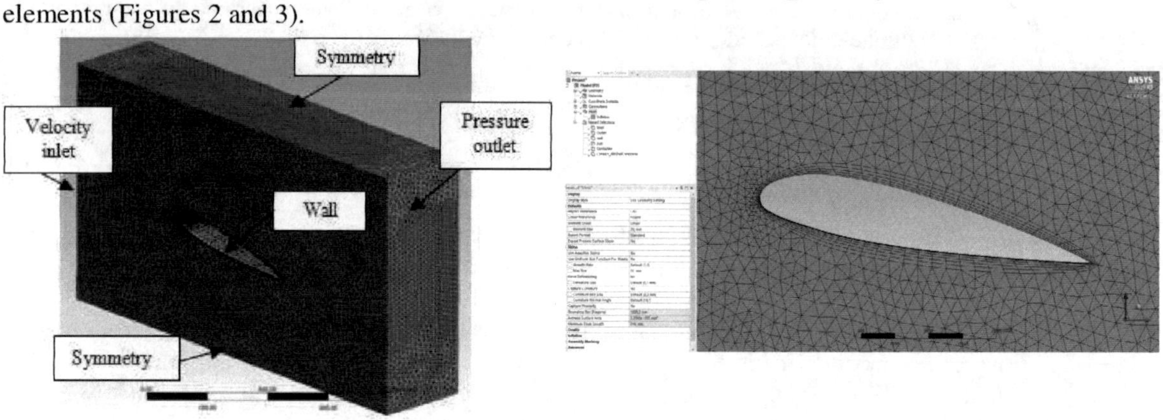

a b

Figure 2. Finite element model constructed by the tetrad mesh method: a – general view; b – front view, fuselage contour.

IOP Conf. Series: Materials Science and Engineering 1061 (2021) 012010 doi:10.1088/1757-899X/1061/1/012010

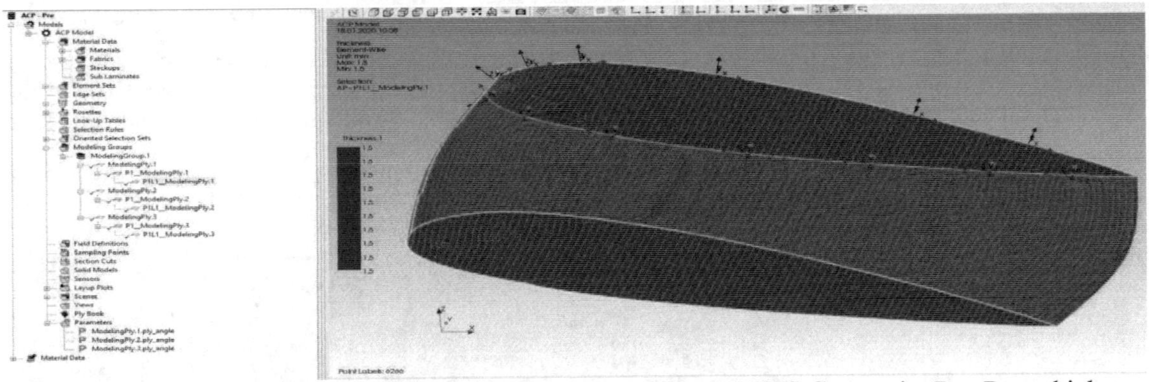

Figure 3. Placement of the first layer of composite material in ANSYS Composite PrepPost, thickness 1.5 mm.

Then we set the load: air velocity (50 m / s) (Figure 4).

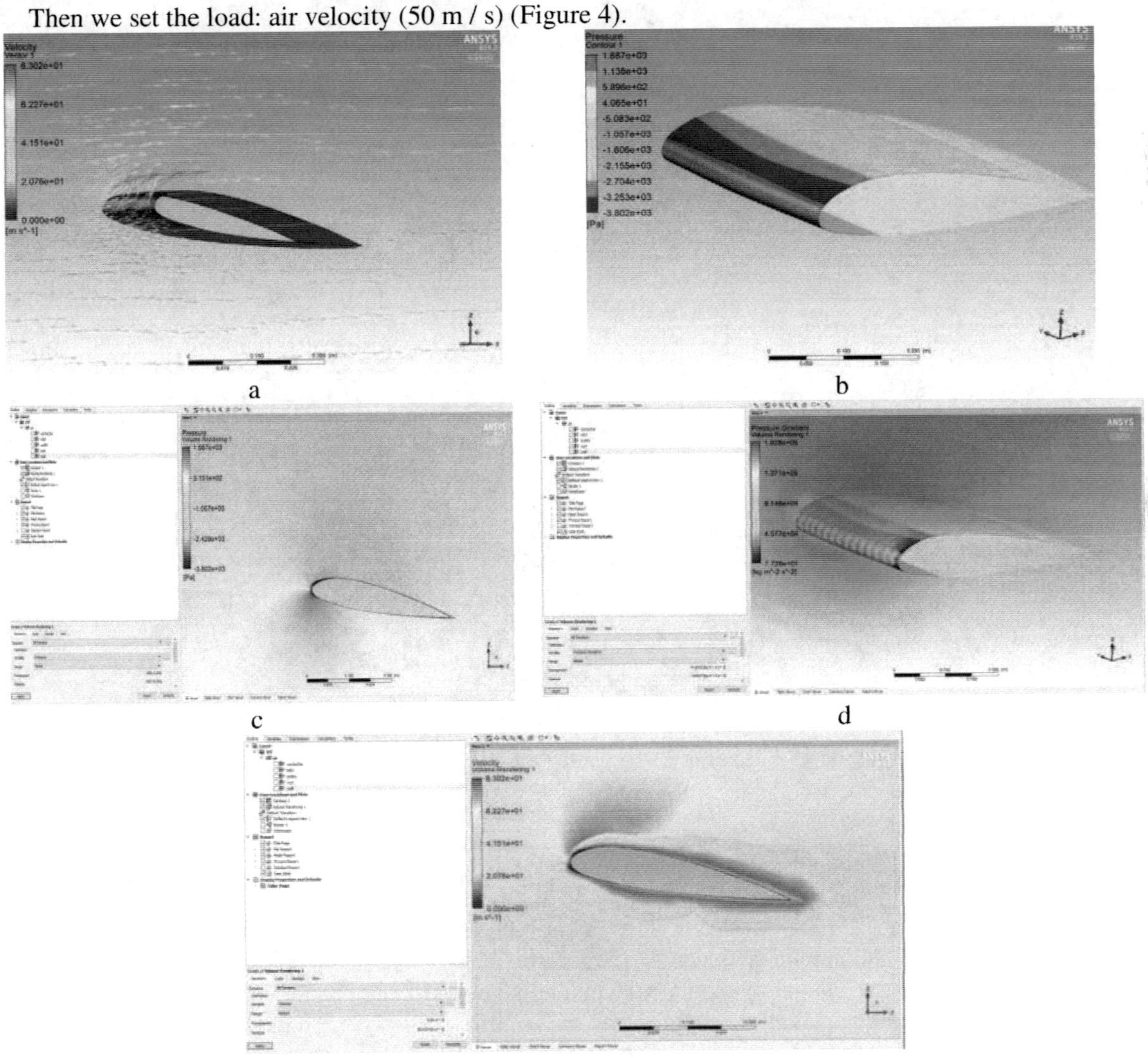

Figure 4. The load assignment: a – air flow velocity vector; b – pressure distribution from the air flow to the fuselage; c – static pressure field; d – pressure gradient; e – components of axial velocities.

The obtained load values from Fluid Flow (flient) and the data obtained by ACP (Pre) are combined in the System Coupling module. We performed the strength calculation using the Static Structural module. Next we set the limits. Figure 5 shows the stiff limit from the hollow sides of the shell. We set the variables for the optimization process - the maximum strain value and the maximum equivalent stresses of each layer.

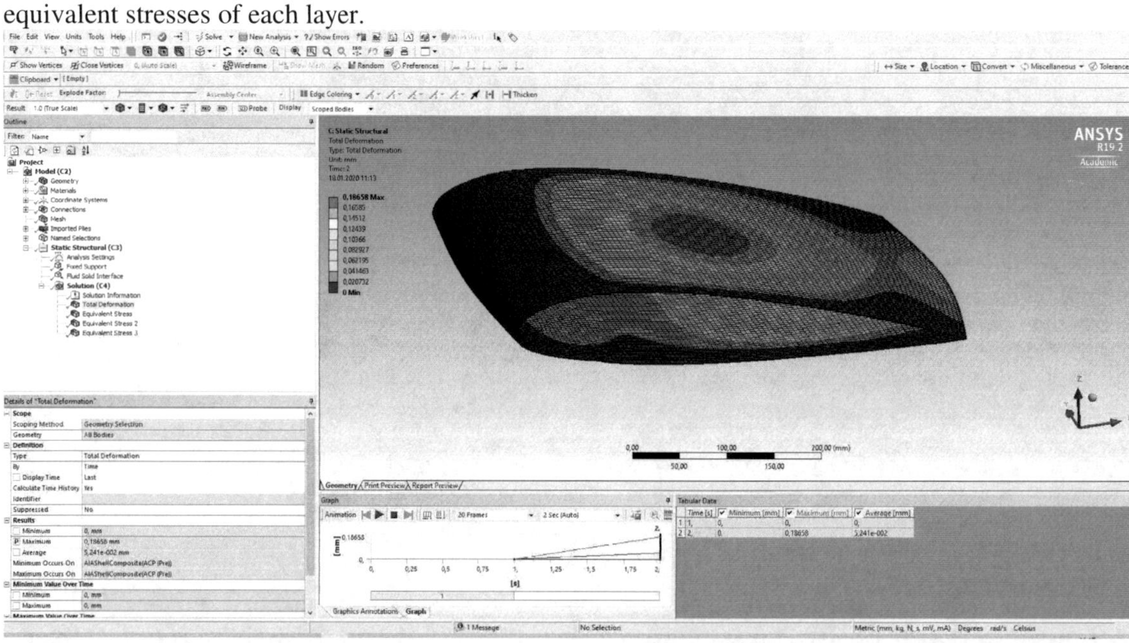

a

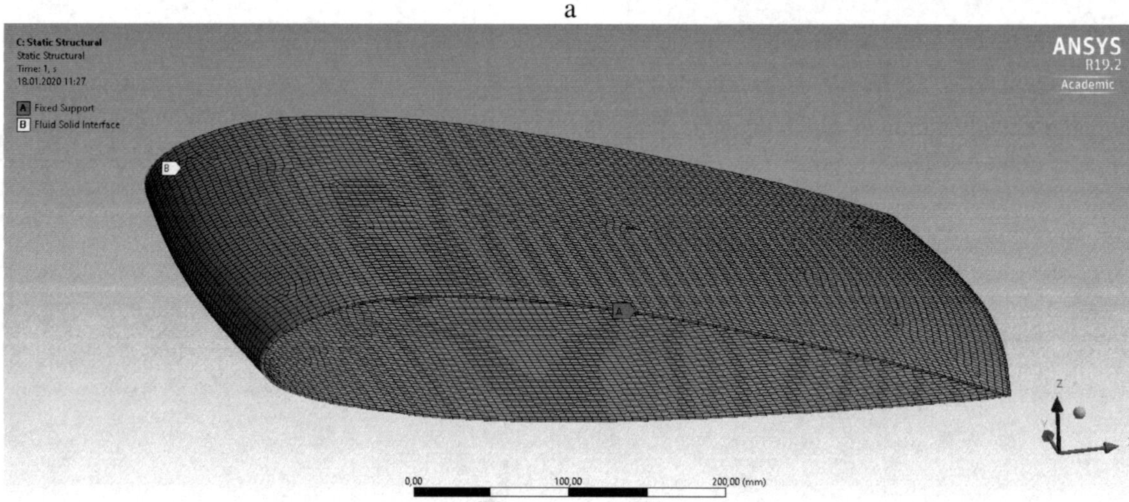

b

Figure 5. The limit of shell: a – computational grid and boundary conditions; b – total deformation.

We determined the optimal construction of a multilayer shell made of composite materials in the ANSYS system, which provides maximum strength and rigidity at specified loads [4]. Figure 6 presents the optimal calculation of the multilayer shell. The shell consists of four complete layers. We developed an automated procedure for choosing the angle of fiber laying in each layer. The calculation begins with the first layer of the shell. Table 1 shows 22 different fiber angles depending on the stiffness of the sheath. From the obtained options, we chose three options for optimal reinforcement.

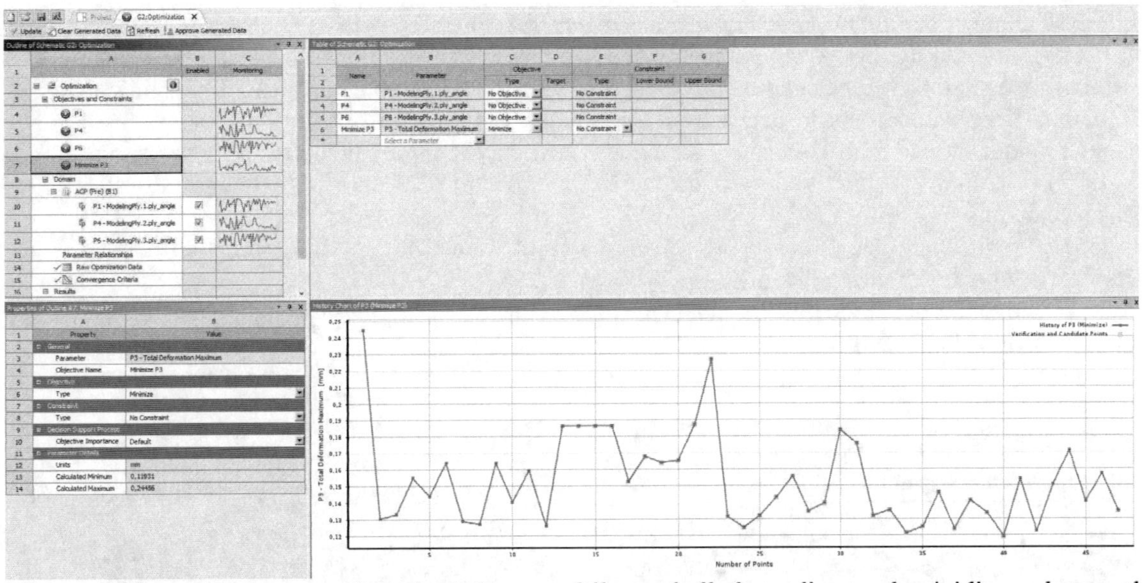

Figure 6. The optimal process for calculating a multilayer shell, depending on the rigidity and strength of the structure.

Table 1. The optimization of first layer

№	The angle of laying of material layers	Maximal load, MPa	Maximal displacement, mm
1	90	21,10183	1,095093801
2	162	9,661041	0,931453943
3	18	9,653471	0,934260152
4	126	14,90961	0,838294026
5	54	14,89833	0,840930263
6	126,4851	14,83455	0,839032358
7	63,55716	16,21259	0,898816012
8	0	9,233469	0,98680513
9	144,4851	12,7972	0,858866131
10	49,75054	14,22289	0,846322747
11	102,5839	21,22485	1,030649697
12	140,4485	13,28761	0,851333735
13	35,30204	12,76257	0,862188449
14	124,2555	15,17309	0,840140706
15	111,404	19,37293	0,957054006
16	134,1003	13,83165	0,846248119
17	125,5222	14,98282	0,837507553
18	113,8144	18,32689	0,915484652
19	119,9556	16,4864	0,880575914
20	129,1943	14,40154	0,842196203
21	137,708	13,54989	0,841213248
22	116,5059	17,74648	0,911944311

The optimal options for the angle of laying of the fibers of the first layer of the shell model:

1. The angle of laying 125^0, maximal displacement 0.84 mm;
2. The angle of laying 140^0, maximal displacement 0.85 mm;

3. The angle of laying 54^0, maximal displacement 0.84 mm.

For increase the rigidity of the multilayer shell, we added an additional three layers over the entire surface of the shell. When calculating the shell, we looked at 11 options for laying fibers in 2 layers, at 47 options for laying fibers in 3 layers, at 41 options for laying fibers in 4 layers. As a result of calculating the optimal multilayer shell, taking into account the minimum deflection, we obtained:

1. The first layer - 150^0, the second layer - 35^0, the third layer - 147^0, the fourth layer - 71^0 (displacement is 0.11 mm);
2. The first layer - 26^0, the second layer - 48^0, the third layer - 131^0, the fourth layer - 62^0 (displacement is 0.12 mm);
3. The first layer - 89^0, the second layer - 53^0, the third layer - 136^0, the fourth layer - 59^0 (displacement is 0.12 mm).

Figure 7 shows the stress state of the shell, consisting of one layer, of two layers, of three layers and four layers.

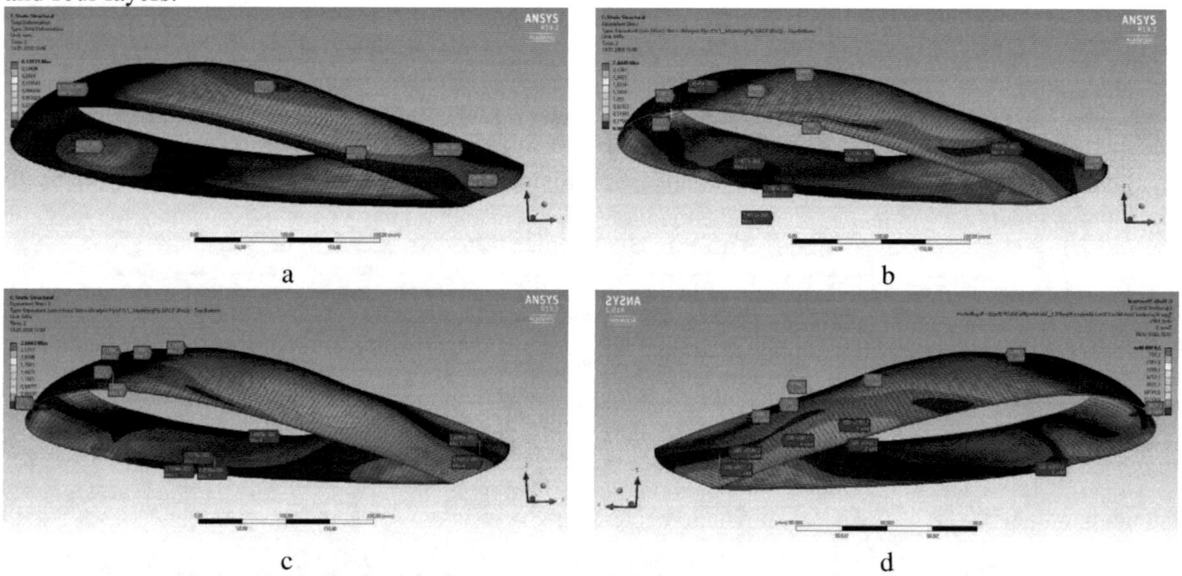

Figure 7. The stress state of the shell: a – one layer; b – two layers; c – three layers; d – four layers.

3. Conclusion

The use of multilayer composites leads to the need for stress analysis and large deformations. As a result, we presented the optimal construction of a composites multilayer shell in the ANSYS system, which provides maximum strength and rigidity at specified loads, and developed an automated procedure for choosing the angle of fiber laying in each layer.

4. Acknowledgments

The reported study was funded by RFBR according to the research project № 18-29-18050\19.

References

[1] Zelentsov V V 2015 *Science and Education: scientific edition of Bauman MSTU* **6** 132–142
[2] Bokhoeva L A 2008 *BMSTU Journal of Mechanical Engineering* **2** 22–27
[3] Basov K A 2006 *The graphical interface of the Ansys complex* (Moscow: DMK Press) p 248
[4] Bokhoeva L A, Baldanov A B and Chermoshentseva A S 2020 *Aerospace MAI Journal* **27(1)** 65–75

Mathematical diagnostic model of brake master cylinder of hydraulic brake system of automobile

A L Borodin, V I Vasiliev and G I Maltseva

Kurgan State University, 63, building 4, Sovetskaya ave., Kurgan, 640020, Russia

E-mail: vviprof@rtural.ru

Abstract. The article considers a dynamic mathematical model of a new method of componentwise diagnosis of the brake master cylinder of a "tandem" type of the hydraulic brake system. We propose a variant of how simulated various faults affect the output parameters of brake process on the basis of mathematical modeling.

1. Introduction

The hydraulic brake actuator is now widely used in passenger cars. Faults in the brake actuator lead to serious traffic accidents. Fault diagnosis before a pre-failure or a failure state of elements of the braking system is an important task for engineers, involved in the service and design of vehicles. Researches, aimed at increasing the range of detected faults and improving the diagnosis of the hydraulic brake system, are relevant.

Currently, the diagnosis of the hydraulic brake system is carried out mainly on the basis of the general signs of "in good-in bad order". Carrying out componentwise diagnosis of the entire system and, in particular, its individual units and elements will help to determine specific faults, identify a pre-failure state and reduce the labor intensity, diagnosing the brake system.

In connection with the labor intensity and costs of conducting experiments, it is better to use a theoretical experiment [1, 2]. Simulation of faults on a computer will subsequently reduce the number of natural experiments; help to choose a set of informative, single-value and stable diagnostic parameters.

The greatest labor intensity of fault diagnosis in a hydraulic brake system is found in the brake master cylinder (BMC), which controls braking systems, equipped with electronic auxiliary braking systems such as ABS / ESC. The componentwise diagnosis of BMC is an important task for vehicle owners and diagnosticians.

2. Material and methods

Recommendations and methods for determining faults and mathematical modeling of dynamic processes in the brake master cylinder, proposed in the literature [3, 4, 5], in some cases make it difficult to use them for componentwise diagnosis of BMC. Therefore, in the mathematical model, some formulas have been put in a form that allows revealing the dependence of specific structural parameters on the technical state of individual elements.

The processes, occurring during the retardation of the hydraulic brake actuator, are studied insufficiently. In early works brake diagrams and diagrams of the dependence of the force of pressing the brake pedal or deceleration because of time are presented with a great degree of simplification [7, 8, 10, etc.]. At the same time, the increasing force on the brake pedal during the retardation of the brake actuator is not taken into account. But later works [5] note an actual nonlinear change in the force of pressing the brake pedal and pressures in the brake circuits when the brake system is actuated. Of course, the force of influence is set by "Technical regulations" [9], but the drag force to pressing the brake pedal should reflect the change in all power characteristics of the BMC operation in dynamics and should show the behavior of the change in these parameters, depending on the change in the technical state of the system elements in service.

The new developed method of componentwise diagnosis of the BMC of the hydraulic brake system allows determining the state of BMC elements and making a conclusion about their serviceability, according to the drag force of pressing the brake pedal in the insensitivity section of the brake actuator, in various modes of influence on the drive brake pedal [14].

The method is based on the hypothesis: carrying out diagnostics in the insensitivity section of the brake actuator in various modes of influence on the brake pedal will allow determining the faults of the elements of the brake master cylinder by the difference between the drag force pressure and the standard values.

While developing the method, the emergence of the hydraulic brake system has been analyzed, the evaluation parameters have been derived and the characteristics that affect the change in the performance of the elements have been identified. For this, an experimental complex has been developed, which makes it possible to assess the operation of the system in all its elements according to the main operating parameters [6]; mathematical model and algorithm has been designed for determining faults; software has been developed to evaluate the operational characteristics of the brake actuator, to identify faults and a pre-failure state of elements to diagnose the resource, timely maintenance and repair.

3. Calculating theory

The dynamic processes of the hydraulic braking system, consisting of subsystems, are most accurately described by a system of linear and nonlinear, algebraic and differential equations, functionally related to each other. Each subsystem is described by a corresponding differential or algebraic equation. A simplified dynamic system of the hydraulic circuit is shown in Figure 1. This figure shows the brake master cylinder - double-circuit, of a "tandem" type with the fixation of the floating piston by a pre-compressed spring with brake fluid compliance, concentrated in two volumes, the first and second circuits of the BMC. Calipers and rear brake cylinders are shown in a simplified manner and are connected to one piston and a connecting pipe for each circuit, respectively.

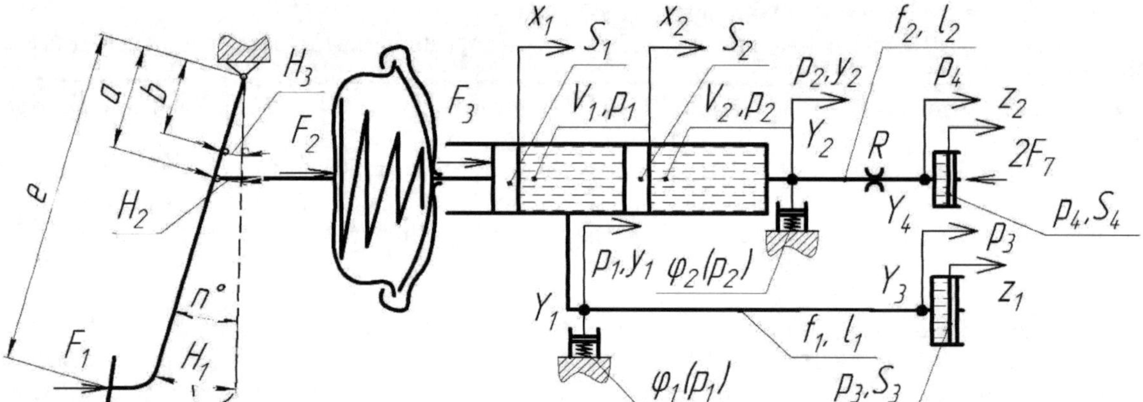

Figure 1. Simplified dynamic system of hydraulic circuit of brake system

The control of the brake system is due to the change in fluid pressure which is going on because of the actuation of the pedal; this change in pressure is created by the movement of the pistons in the brake master cylinder. In this case, an input force is applied to the first piston rod, which is a function of time.

A mathematical model of the hydraulic circuit with two separately concentrated volumes of fluid describes the dynamic processes, occurring in such a circuit with a known mode of pressing the brake pedal. This model by a system, consisting of the following nonlinear differential equations [5, 3]:

$$\begin{cases} F_{\text{ШТ}} + R_1 = F_{\text{пр}1} + F_{\text{И}1} + F_{p1} + \left(F_{\text{Д}1} + F_{\text{тр}1} + F_{\text{тр}2}\right)\cdot sgn\left(\frac{dx_1}{dt}\right); \\ F_{\text{пр}1} + F_{p1} + R_2 = F_{\text{пр}2} + F_{\text{И}2} + F_{p2} + \left(F_{\text{Д}2} + 2\cdot F_{\text{тр}2}\right)\cdot sgn\left(\frac{dx_2}{dt}\right); \\ m_{\text{п}1}\frac{d^2z_1}{dt^2} + F_{sup} + F_{\text{тр}4} sign\frac{dz_1}{dt} - p_3S_3 = 0; \\ m_{\text{п}2}\frac{d^2z_2}{dt^2} + F_{ztc} + P_{\text{тр}3} sign\frac{dz}{dt} - p_4S_4 = 0; \\ a_1\frac{d^2z_1}{dt^2} + a_2\frac{dz_1}{dt} + a_3\left(\frac{dz_1}{dt}\right)^2 sgn\frac{dz_1}{dt} + \frac{F_{\text{тр}4}}{F_3} sgn\frac{dz_1}{dt} + p_3 = p_1; \\ a_1\frac{d^2z_1}{dt^2} + a_2\frac{dz_1}{dt} + a_3\left(\frac{dz_1}{dt}\right)^2 sgn\frac{dz_1}{dt} + p_4 + \frac{F_{\text{тр}3}}{F_4} sgn\frac{dz_1}{dt} = p_1; \\ \frac{dx_1}{dt} = \frac{S_3}{S_1}\frac{dz_1}{dt} - \frac{dx_2}{dt} + \left[(x_{max1} - x_1) + \frac{S_3}{S_1}(z_{min1} + z_1) + \frac{f_1l_1}{S_1}\right]\psi_1(p_1)\frac{dp_1}{dt}; \\ \frac{dx_2}{dt} = \frac{S_4}{S_2}\frac{dz_2}{dt} + \left[(x_{max2} - x_2) + \frac{S_4}{S_2}(z_{min2} + z_2) + \frac{f_2l_2}{S_2}\right]\psi_2(p_2)\frac{dp_2}{dt}. \end{cases} \quad (1)$$

where $F_{ШТ}$ – control force on the master cylinder rod,
F_{np1} – first piston spring force,
$F_{И1}$ – inertial force of the first piston,
F_{P1} – force from pressure in the primary area,
$F_{Д1}$ – damping force (force of resistance to the movement of the first piston created by the fluid), H;
F_{mp1} – friction force from the first type compression ring,
F_{mp2} – friction force from the second type compression ring,
R_1 – first cylinder support pressure,
x_1 – movement of the primary piston,
t – current time of system operation,
F_{np2} – second piston spring force,
$F_{И2}$ – inertial force of the second piston,
F_{P2} – force from pressure in the secondary area,
$F_{Д2}$ – damping force of the second piston,
F_{mp2} – friction force from the second type compression ring,
R_2 – second cylinder support pressure,
x_2 – movement of the secondary piston,
m_{n1} – mass of moving parts, connected to the support piston,
F_{sup} – force of resistance to the movement of the support piston,
F_{mp4} – inertial force of support piston,
z_1 – movement of the support piston,
p_3 and S_3 – pressure and surface area of the support piston.
m_{n2} – mass of moving parts, connected to the piston of the rear brake cylinder,
F_{mp3} – inertial force of the piston of the rear brake cylinder,
F_{ztc} – force of resistance to the movement of the piston of the rear brake cylinder,
z_2 – movement of the piston of the rear brake cylinder,
p_4 and S_4 – pressure and surface area of the rear brake cylinder.
m_{n1} – mass of moving parts, connected to the support piston,
p_3 – pressure in the support piston,

p_1 – pressure in the first area of the brake master cylinder,

sgn – piecewise constant function of the actual argument (the amount of movement of the first and second pistons),

l_1 and f_1 – length and cross-sectional area of the pipe,

S_3 – surface area of the support piston,

k_ε – approximation coefficient, the value of which depends on the relative roughness ε of hydraulic lines,

ρ – brake fluid density,

υ – kinematic viscosity coefficient,

ζ_1 – local resistance coefficient.

m_{n2} – mass of moving parts, connected to the piston of the rear brake cylinder,

p_4 – pressure in the piston of the rear brake cylinder,

p_2 – pressure in the second area of the brake master cylinder,

l_2 and f_2 – length and cross-sectional area of the pipe,

S_4 – surface area of the piston of rear brake cylinder,

$\psi_1(p_1)$ – non-linear coefficient of compliance of the primary circuit.

$\psi_2(p_2)$ – non-linear coefficient of compliance of the secondary circuit.

Let us consider the main faults and their mathematical implementation in a mathematical model.

The amount of air in the brake fluid affects the braking performance of a vehicle. In mathematical modeling, the amount of air in a liquid characterizes the compliance coefficient [5]. The determination of the coefficient of compliance depending on the percentage of air in the brake fluid has been made on the basis of experimental data. According to the researches, the curves of the compliance coefficient for the brake fluid have been constructed with air content of 1%, 5%, 10%, 15%, 20%. On the curves trend lines have been plotted with 99% accuracy. Formulas, reflecting the dependence of the compliance coefficient on the percentage of air in the brake fluid and pressure, are used in the mathematical model.

Changes in the stiffness of the springs of the primary and secondary circuits directly affect the compression forces of the springs in the circuits.

The force of the first spring:

$$F_{\text{пр}1} = C_{\text{пр}1} \cdot (b_1 + x_1 - x_2), \qquad (2)$$

where C_{np1} – stiffness of the return spring of the primary piston, Н/м;

b_1 – pre-compression of the primary piston spring, м;

x_1 – movement of the first piston, м;

x_2 – movement of the second piston, м.

The force of the second spring:

$$F_{\text{пр}2} = C_{\text{пр}2} \cdot (b_2 + x_2), \qquad (3)$$

where C_{np2} – stiffness of the return spring of the secondary piston, Н/м;

b_2 – pre-compression of the secondary piston spring, м;

x_2 – movement of the second piston, м.

By introducing different stiffness into the mathematical model, it is possible to trace the change in the operation of the entire system in dynamics.

The wear of the compression rings of the primary and secondary circuits will lead to a change in two parameters in the operation of the BMC - the damping force and the friction force. The damping force given in formulas 4 and 5 is influenced by the fluid flow through the brake pipe and the fluid flow through the hole between the compression ring and the cylinder. The calculation of the damping force is detailed in the following literature sources [12, 13, 14, 15, 16].

Guided by the Poiseuille formula, the second rate of fluid flow through the pipe (provided a laminar fluid flow) will be equal to:

$$Q = \frac{\pi \cdot R^4}{8 \cdot \eta \cdot l} \cdot (p_1 - p_2) = \frac{\pi \cdot d^4}{128 \cdot \eta \cdot l} \Delta p, \quad (4)$$

where p_1 - p_2=Δp – pressure fall at the ends of the capillary, Pa;
Q – volumetric fluid flow rate, m³ / s;
R – capillary radius, m;
d – capillary diameter, m;
η – dynamic viscosity coefficient, Pa • s;
l – brake line length, m.

The damper is the pistons of the primary and secondary areas of the BMC. When the piston moves at a speed V, it presses on the brake fluid located in the lower area of the cylinder and pushes it through the gap δ between the piston and the cylinder into the overflow tank, and also pushes the fluid into the front and rear circuits of the brake system through tubes with a radius of 1.5 mm.

Dynamic viscosity for cooled brakes at a brake fluid temperature of $50^\circ C$ is given in reference sources [19, 20].

The fluid flow through the gap comprises:

$$Q_2 = \frac{\pi \cdot R_{CP} \cdot \delta^3 \cdot \Delta p}{6 \cdot \eta \cdot l_2}, \quad (5)$$

where Δp – pressure difference at different ends of the compression ring, Pa.
η – dynamic viscosity coefficient, Pa • s;
$R_{CP}=0,5(R_1+R_2)$ – the average radius of the annular slot between the cylinder radius and the piston radius, m;
l_2 – compression ring width, cm;
δ – gap between the piston and the cylinder (slot width), cm.

If the pressure difference in the sealed system is zero, the flow rate $Q1$ will also be zero. It can be assumed that if there are leaks of brake fluid through the working brake cylinders, the pressure difference will begin to differ from zero and $Q1$ will become more than zero.

$$Q = Q_1 + Q_2 = \Delta p \cdot \left[\frac{\pi \cdot r^4}{8 \cdot \eta \cdot l_1} + \frac{\pi \cdot R_{CP} \cdot \delta^3}{6 \cdot \eta \cdot l_2}\right] = \Delta p \cdot (k_1 + k_2). \quad (6)$$

The amount of fluid, flowing from one part of the cylinder to another is equal to:

$$Q = S_1 \cdot V_П, \quad (7)$$

where S_1 – cylinder surface area, m.
$V_П$ – piston speed, m/s.

$$\Delta p = \frac{S_1 \cdot V_П}{(k_1 + k_2)}. \quad (8)$$

The fluid pressure creates force $F_Д$, which opposes the movement of the piston and is equal to:

$$F_Д = \Delta p \cdot S_2, \; S_2 = \pi \cdot R^2, \quad (9)$$

where Δp – fluid pressure, influencing the piston;
R – piston radius.

$$F_Д = \frac{S_2 \cdot S_1 \cdot V_П}{(k_1 + k_2)} = \frac{(\pi \cdot R^2)^2 \cdot V_П}{(k_1 + k_2)}. \quad (10)$$

The opposing force with the consideration of formula (4.3) and the expression for the volume of the overflowing fluid, is equal to:

$$F_Д = C_{Д1} \cdot V_П, \tag{11}$$

where $C_{Д1}$ – damping coefficient (specific damping force), defined as

$$C_{Д1} = \frac{(\pi \cdot R^2)^2}{(k_1 + k_2)}; \; k_1 = \frac{\pi \cdot r^4}{8 \cdot \eta \cdot l_1}; \; k_2 = \frac{\pi \cdot R_{CP} \cdot \delta^3}{6 \cdot \eta \cdot l_2}. \tag{12}$$

Coefficients *k1* and *k2* depend only on the construction sizes of the damper. In case of wear of the compression rings, coefficient *k2* is additionally calculated for the gap between the ring and the BMC body.

The second component that is directly affected by the wear of the rings is the friction force of the compression rings against the cylinder walls, and the increase of the compression rings in diameter, due to swelling, will also affect the friction force. In the first case, with a decrease in the structural parameter of the diameter of the compression ring, the friction force will decrease almost to zero, in the second case it will increase until the seizure of the BMC pistons.

An increase in the outer diameter of the compression rings occurs due to the use of a brake fluid of a non-recommended type, the ingress of gasoline and other operating fluids into the brake fluid, increased rubber aging will be accompanied by a change in the size dimensions of the compression rings with a subsequent increase in the friction force.

The friction force in the compression fittings when starting from a place (friction at rest) is 1.5 ... 2.5 times greater than the friction force during steady motion; with an increase in the pressure of the compressed medium from 2 to 15 MPa, the friction force additionally increases from 0.2 to 0.25% [12, 21].

The frictional force of a circular rubber ring is defined as the component of two forces, the friction force arising from the pressure of the working fluid, and the friction force arising from the preliminary compression of the ring.

For round rubber rings, located in the piston groove, the friction force is:

$$F_{тр} = T_e \cdot \pi \cdot D + T_n \cdot \pi \cdot (D^2 - D_k^2)/4. \tag{13}$$

where D – inner diameter of the cylinder in cm,

$D_к$ – ring groove diameter in cm,

$D_ш$ – outer diameter of the rod,

T_n – the friction force, arising from the pressure of the working fluid per 1 cm2 of the contact surface of the ring, is determined graphically.

T_e – the friction force, arising from the preliminary compression of the ring by 1 cm of its length, is determined graphically, depending on the preliminary assembly compression.

$$w = \frac{\gamma}{d} \cdot 100\%, \tag{14}$$

where $\gamma = d - b_1$, here d – ring section diameter;

b_1 – groove depth, taking into account the gap between the sealing surfaces.

To determine forces T_n , T_e plots were made in Excel and trend lines and formulas for theoretical calculation were selected, the error was less than 1%.

$$T_n = 2{,}93 \cdot 10^{-7} \cdot p. \tag{15}$$

$$T_e = 0{,}0185 \cdot w + 0{,}032. \tag{16}$$

To reconcile the data received with the SI system, we carry out the necessary transformations and obtain the following formula:

$$F_{тр} = 9.81 \cdot (T_e \cdot \pi \cdot D + T_n \cdot \pi \cdot (D^2 - D_k^2))/4. \tag{17}$$

These equations for calculating friction forces are applicable in a mathematical model corrected to friction of rest and friction of motion. When changing the diameter of the ring section within the permissible limits indicated above, we obtain the effect of changes in structural parameters in real service conditions on the output diagnostic parameters.

The introduction of the subsystem of differential equations for correcting the numerical values of the structural parameters into the algebraic expressions gives a change in the characteristics of the dynamic system of differential equations at the output. By changing the diagnostic parameters, it is possible to determine the faults of the BMC.

4. Results

As a result of the conducted research, the following results have been received: we have updated the mathematical model of the operation of the main double-circuit brake cylinder with the fixation of the floating piston by a pre-compressed spring. The possibility to introduce the most common faults of BMC into the mathematical model has been obtained. The program for modeling BMC states has been developed and worked out. Optimal set of diagnostic parameters has been chosen and the algorithm for determining faults has been designed. A comparison of the mathematical and experimental models has been carried out, the general discrepancy was 1.42%.

5. Discussions

The application of the method of diagnosing the BMC of the hydraulic brake system by the drag force to the force of pressing the drive pedal at the actuation of the brake system, with various modes of influence on the brake pedal, shows the operation of the brake actuator most objectively. After choosing a methodology for evaluating the characteristics of a brake actuator with precise definitions of faults and developing an algorithm for determining faults in the BMC, we were able to determine faults directly in the nodes of the brake system and determine their service life to failure.

6. Conclusions

At present, research in this area has been minimized, much attention is paid to the development of new braking assist systems, although the whole service of the systems often depends on the correct and stable operation of the BMC itself. Studies of the patterns, occurring in the hydraulic brake actuator, help to identify faults in the brake system at all periods of service. The use of research results when diagnosing the brake system will reduce the labor intensity of diagnostic operations and will increase the reliability of the diagnosis.

References

[1] Miroshnikov L V, Boldin A P and Pal V I 1977 *Diagnostics of the technical condition of vehicles at automobile transport enterprises* (Moscow: Transport). p 263

[2] Migal V G 2016 *Fundamentals of technical diagnostics of cars* (Kharkov: Maidan) p 372

[3] Schuklinov S G 2013 *Theoretical foundations of adaptive braking control of wheeled machines. Dissertation of doctor of technical sciences* (KHNADU: Kharkov) p 428

[4] Kosarev V N, Ametov S A, Volgin S N and Kozlov P L 2005 *Operation maintenance and repair manual VAZ 21213 NIVA* (Moscow: Astrel) p 241

[5] Metlyuk N F and Avtushenko V P 1980 *Dynamics of pneumatic and hydraulic drives of cars* (Moscow: Engineering) p 231

[6] Govorushchenko N Y 1970 *Diagnosing the technical condition of cars* (Moscow: Transport) p 256

[7] Aleksandrov M P, Lysyakov A G, Fedoseev V N and Novozhilov M N 1985 *Brake devices. Handbook* (Moscow: Engineering) p 312

[8] Petin Y P 1974 *Study of the possibility of diagnosing brake systems of cars in road conditions by changing the statistical information on the movement of the brake pedal and driving force. Abstract of a thesis of candidate of technical sciences* (Moscow: MADI) p 22

[9] Customs Union commission 2011 *Technical regulations of the customs union (On the safety of wheeled vehicles)* (Moscow) p 318

[10] Bashty T M, Alekseeva T V, Babanskaya V D and Zagrebalniy V I 1989 *Technical diagnostics of hydraulic drives* (Moscow Engineering) p 264
[11] Borodin A L, Vasiliev V I, Sharypov A V and Cherepanov A P 2015 Development of a method for diagnosing hydraulic brake systems of vehicles at service brake applications *Bulletin of the Kurgan State Agricultural Academy (Technical science* vol 4 (**16**) (Kurgan: Publishing house of the Kurgan State Agricultural Academy) pp 29–32
[12] Yavlenskiy K N, Timofeev B P, Chaadaeva E E, Verkovich G A, Golovenkin E N and Golubkov V A 1989 *Handbook of a designer of precision instruments industry* (Leningrad : Engineering) p 729
[13] Rainkina L N 2005 *Hydromechanics. Tutorial for solving problems* (Moscow: Russian state university of oil and gas named after I. M. Gubkin) p 119
[14] Babaeva N F 1967*Calculation and design of elements of gyroscopic devices* (Moscow: Engineering) p 480
[15] Tikhmenev S S 1956 *Elements of precision instruments (Guidelines for calculation and design*) (Moscow: State publishing house of the defense industry) p 361
[16] Anisimov A V, Kondrashev V L, Likhoded K A, Shoshiashvili M E 2012 *Dynamics of hydraulic systems (a tutorial)* (Novocherkassk: South-Russian state polytechnic university) p 131
[17] Shkolnikov V M, Anisimov I G, Badyshtova K M and Bnatov S A 1999 *Fuels, lubricants, technical fluids. Assortment and application. Reference book* (Moscow: Publishing center Techinform) p 596
[18] Nosova E V and Rudnykh N 2008 Brake fluids (expert evaluation) *Automarket plus Sport* (**12**) (Irkutsk: Automarket plus Sport) p 26
[19] Melik-Gaikazov V I, Podgorny Yu P and Samusenko M F 1968 *Hydraulic drive of heavy lifting machines and self-propelled aggregates* (Moscow: Engineering) p 264
[20] Livotov V S, Prosvirov A C and Napalkov A V 2002 *Technological calculations of elastic elements. Part 1. Verification calculations of coil springs and spring rings* (Volgograd) p 16
[21] Petrenko A M and Zvekov A T 2009 *Cargo manipulators of special vehicles: training manual* (Moscow: MADI (State Technical University)) p 90

The automated special software for component testing of the satellite communication station

O V Drozd and D V Kapulin

Siberian Federal University, 79, Svobodny Ave., Krasnoyarsk, 660041, Russia

E-mail: odrozd@sfu-kras.ru

Abstract. Research tests of satellite communication station components are distinguished by a special variety of tasks to be solved, associated both with a wide range of components of the station, and a variety of estimated technical and functional characteristics. The whole set of characteristics can be subdivided into three categories: frequency and quality characteristics of components, operation parameters of the station and parameters of electromagnetic compatibility. These measurement operations of these parameters should be carried out using an automated measuring complex, which includes a measuring signal generator, vector spectrum, signal and network analyzers and operated under control of a specialized software. In the paper the architecture, logic algorithms and some application results of the software are shortly presented. The special software consists of the individual modules for control technical characteristics of low-noise and power amplifiers, cable assemblies, transceiver signal path element, and documenting the results. The program also includes modules that implement measurement instrumentation control used to assess the operational parameters and electromagnetic compatibility of the communication equipment. The developed specialized software allows automatically performing the entire cycle of operations related to the analysis of frequency and spectral characteristics, as well as quality characteristics of the devices under test in the automatic mode.

1. Introduction

Research tests of Satellite Communication Earth Station (SCES) components are distinguished by a special variety of tasks to be solved, associated both with a wide range of components, and a variety of estimated technical and functional characteristics. Assessment of the main functional parameters of such components as low-noise amplifiers, power amplifiers, cable assemblies and transceiver signal path elements, as well as their functional assemblies, is associated with significant amount of measurements, their considerable duration with the need to record the results associated with the specific product [1–3].

Thus, the problem of ensuring the measurement automation of the SCES component main parameters in an automated or automatic mode with the formation of reporting documentation arises. In general, the research test automation in microwave technology is caused by an increase the complexity level of modern radio engineering systems, in particular, components of radio frequency paths for satellite communications, navigation, and radar systems [3–5]. The whole set of recorded characteristics and researches can be subdivided into three categories [6–9]:

- frequency and quality characteristics of the SCES components: amplitude-frequency and signal attenuation characteristics, voltage standing wave ratio, etc.;

- parameters of the SCES functioning: the ability to adjust the output signal power, the signal transmission at the fixed frequency, the irregular wave transmission coefficient of the station transceiver path, etc.;
- electromagnetic compatibility in the presence of external radio-frequency interference: ensuring the station receives a carrier signal, noise emission of the station transceiver path, blocking on image channels, etc.

These operations should be carried out using specialized control and measuring equipment, combined into a single Automated Measuring Complex (AMC). The structure of the AMC includes a vector network analyzer, a measuring signal generator, a vector spectrum and signal analyzer, etc. [10]. In this case, the specified equipment should be considered not only as a collection of separate units, but also as a component of a test environment, operating under control of a specialized software. This approach makes it possible to develop the AMC as a laboratory information and testing system, which can be used in the research of various parameters in a variety of projects performed by different teams in the framework of current production tasks [11].

2. Architecture and logic algorithms of the automated measuring complex

In the work the AMC special software for monitoring, configuring and conducting research tests of the satellite communication station components is considered. The main task of the software is the remote control of the AMC measurement instruments, automation of tests and recording their results.

The special software was developed using the integrated software development environment Microsoft Visual Studio and the C# programming language. It is intended for installation on the AMC workstation and control of the following measurement equipment:

- Rohde & Schwarz SMA100A signal generator;
- Rohde & Schwarz FSW vector spectrum and signal analyzer;
- Rohde & Schwarz ZVA24 vector network analyzer;

The specific method of connecting devices under test and instrumentation depends on the type of the research test being carried out. Possible connections with the vector network analyzer are shown in Fig. 1.

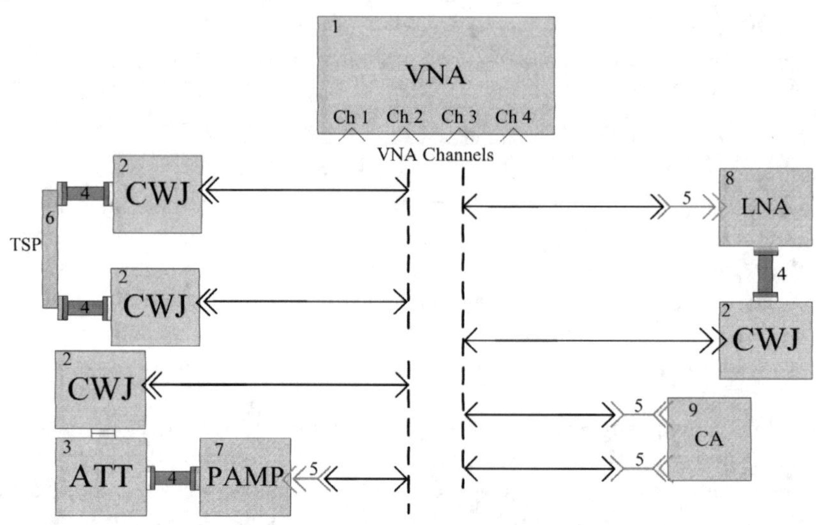

Figure 1. Variants of connection the devices under test to the vector network analyzer: 1 – Vector Network Analyzer (VNA), 2 – Coaxial-Waveguide Junction (CWJ), 3 – Waveguide Fixed Attenuator (ATT), 4 – Waveguide Sections, 5 – Radio-Frequency Transitions; devices under test: 6 – Transceiver Signal Path (TSP), 7 – Power Amplifier (PAMP), 8 – Low Noise Amplifier (LNA), 9 – Cable Assembly (CA).

The component structure of the special software includes a number of separate modules for measuring the SCES main parameters: low-noise amplifiers, power amplifiers, modems, cable assemblies and the transceiver signal path as a whole and fixation of measurement results.

The program also includes modules that control certain measurement equipment (signal generator, vector spectrum and signal analyzer, vector network analyzer) used in the estimation of the operational parameters and communication station electromagnetic compatibility. The scheme relationship between the developed software modules, third-party software, drivers and instrumentation is shown in Fig. 2.

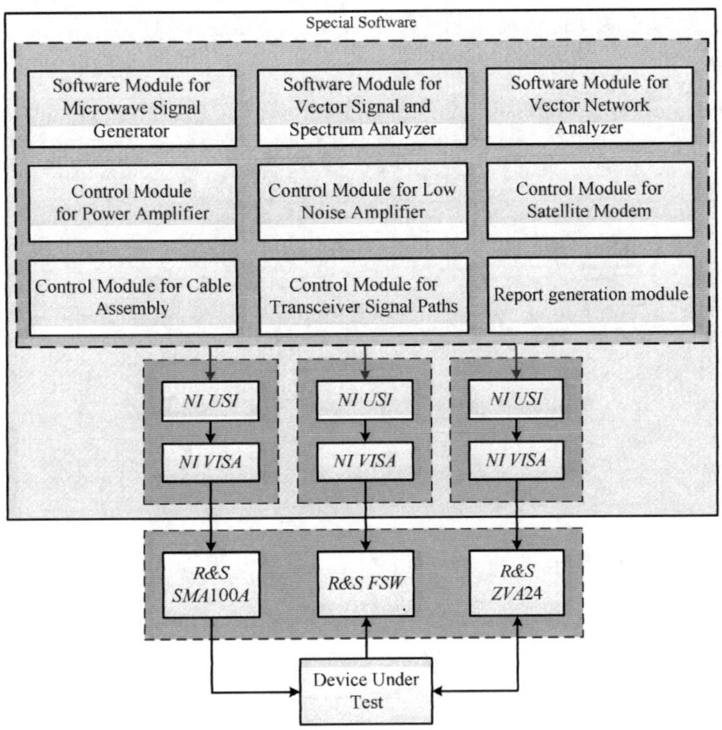

Figure 2. The structural relationship of the special software modules with third-party software, drivers and instrumentation.

For correct operation of the special software data import drivers in universal format (NI USI), tools for implementing the software architecture of virtual instruments (NI VISA) and the Microsoft .NET Framework software platform version 4.0 or higher are installed on the control workstation. Fig. 3 shows the algorithm of the software, explaining the procedure for analyzing the frequency and quality characteristics of the satellite communication station components.

The generalized process of the program functioning can be represented as follows. When the software is started, the measurement equipment is initialized. Then the parameters of the input test signal and spectral analysis are adjusted. Next, a spectrum diagram of the signal passing through the radio frequency transmitting and receiving path of the SCES is formed. After obtaining a characteristic form of the signal spectrum, it is appropriate to the analysis of the signal spectrum diagram depending on the type of the test being carried out, assessment with the technical requirements and generate reporting documentation.

Further, the analysis of the frequency and quality characteristics of the device under test is carried out. For this, first of all, the procedure of forming the test input signal according to set parameters is performed. Next, two groups of characteristics are formed separately: frequency characteristics (frequency response and signal attenuation) and quality characteristics (voltage standing wave ratio,

VSWR). For each of the specified parameters, a diagram is plotted both for the full frequency range supported by the vector network analyzer and for the bandwidth set in the Terms of reference and Technical requirements for specific components of the transmitting, receiving and data processing equipment of the communication Earth station. The generated diagrams of frequency and quality characteristics of the tested devices are displayed on the corresponding tabs of the software interface, depending on the type of the device.

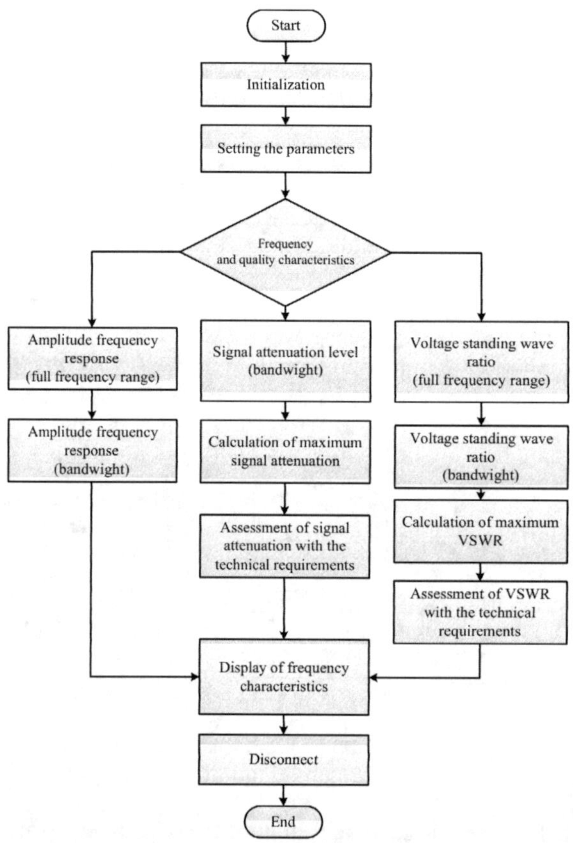

Figure 3. Block diagram of the frequency response and quality characteristics analysis algorithm of SCES components.

The method for analyzing the signal spectrum diagram passing through the radio frequency transmitting and receiving path of the communication station can be different depending on the task of the operation parameters verification or the assessment of electromagnetic compatibility.

Assessment for compliance with the technical requirements for the main operation parameters involves the following checks:

1) ability to adjust the output power of the carrier information signal within the specified limits;

2) providing the station to transmit signals at the fixed frequency in a given frequency range;

3) calculation the irregular wave transmission coefficient of the station transceiver path;

4) permissible deviation of the station output power at the constant level of the intermediate frequency signal;

5) receiving the carrier frequency signal, allocated for the operation of the input frequencies of the receiving path.

Verification of the requirements for the electromagnetic compatibility assumes an assessment the out-of-band emission level of the communication station, noise emission of the station transceiver

IOP Conf. Series: Materials Science and Engineering 1061 (2021) 012012 doi:10.1088/1757-899X/1061/1/012012

path, frequency deviations for transmission and blocking on image channels of the SCES signal. The measurement result protocols are saved in text (PDF) and tabular (XLS) formats on the disk space of the automated measuring complex workstation.

The developed special software uses remote control methods of the measuring equipment by the National Instruments VISA software architecture for unified testing, debugging and remote control of the equipment via the LAN interface. The control workstation sends a request command and waits for a response from the equipment. An example of the request command is measuring the values of frequency response, frequency response, switching to spectral or quality analysis. The report comes about the status of the instrument or the results of the measurements. The results are updated in determined intervals.

3. Assurance of the satellite communication station components testing

The devices under test are the components of the radio frequency transmitting and receiving path of the satellite communication station with the following characteristics:

- frequency range, for receive – 17.7…21.2 and 19.2…19.8 GHz, for transmit – 27.5…31.0 and 43.9…44.5 GHz;
- equivalent isotropic radiated power, 58.3 dBW for the Ka Band and 63.0 dBW for the Q Band;
- antenna gain in the Ka/Q Band range, for receive – 44.6 dB, for transmit – 51.6 dB;
- antenna gain in the Ka/Ka Band range, for receive – 43.3 dB, for transmit – 47.2 dB;
- attenuation of the signal received through the image channel and the intermediate frequency – 60 dB;
- modulation methods, for receive – TDM and SCPC, for transmit – TDMA and SCPC.

Fig. 4 shows an example of the signal generator and the vector spectrum and signal analyzer remote control, and the resulting frequency characteristics of the communication station transceiver signal paths.

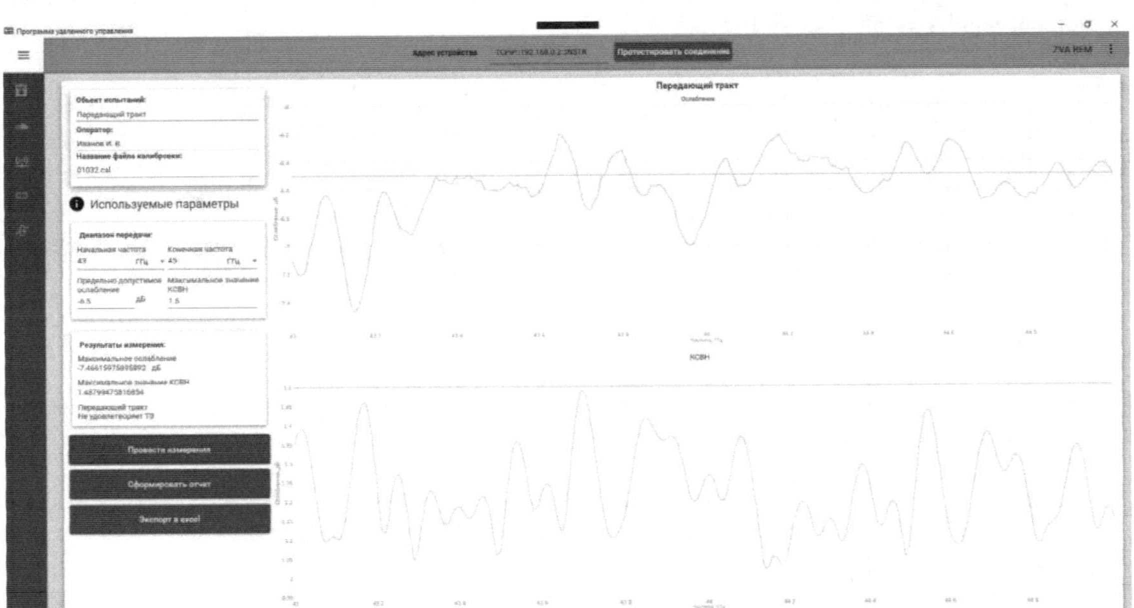

Figure 4. Displaying the frequency characteristics of the SCES transceiver signal paths (in Russian)

According to the results of the measurements it is possible to draw conclusions about the compliance of the technical characteristics of communication station components with the requirements. It is also possible to draw a conclusion about the correct operation of the software as part of the automated measuring complex. The developed specialized software allows automatically performing the entire cycle of the operations related to the analysis of frequency and spectral characteristics, as well as the quality characteristics of the devices under test in the automatic mode.

4. Conclusion

The developed special software for the automated measuring complex can be used to provide experimental studies of the Earth satellite communication station components. In addition, it is possible to assess the operation parameters of the station as a whole and the characteristics of electromagnetic compatibility in the presence of external electromagnetic interference.

The special software consists of the individual modules for control parameters of such components the communication station as low-noise amplifiers, power amplifiers, cable assemblies, transceiver signal paths elements and the transmission path as a whole, and documenting the measured results. The program also includes the modules that implement measurement instrumentation control used to assess the operational parameters and electromagnetic compatibility of the tested communication equipment.

Automated measuring complex includes the Rohde & Schwarz control and measuring equipment, which is used for testing and verifying the main characteristics of the satellite communication station components. The developed special software allows obtaining numerical values and spectrum diagrams, found extrema of measured signals, frequency characteristics of the devices under test and other signal parameters determined by the used measurement technique.

Acknowledgments

This work was financially supported by the Ministry of Science and Higher Education of the Russian Federation in the implementation of the integrated project «Creation of a production of earth stations of advanced satellite communications systems to ensure the coherence of hard, northern and Arctic territory of Russian Federation», implemented with the participation of the Siberian Federal University (agreement number 075 -11-2019-078 dated 13.12.2019).

References

[1] Awang Z 2014 *Microwave Systems Design* (Singapore: Springer Singapore) 3–9
[2] Su D, Xie S, Dai F, Liu Y and Jia Y 2019 *Theory and Methods of Quantification Design on System-Level Electromagnetic Compatibility* (Singapore: Springer Singapore) 78–94
[3] Rouphael T J 2015 *Wireless Receiver Architectures and Design* (Elsevier) 61–80
[4] Drozd O V., Russkikh P A and Kapulin D V. 2019 Special software for Automated Measuring Complex based on Rohde & Schwarz equipment *2019 International Siberian Conference on Control and Communications (SIBCON)* (IEEE) pp 1–4
[5] Lienig J and Bruemmer H 2017 *Fundamentals of Electronic Systems Design* (Cham: Springer International Publishing) 181–190
[6] Seokjin Kim, Elkis R and Peckerar M 2009 Device Verification Testing of High-Speed Analog-to-Digital Converters in Satellite Communication Systems *IEEE Trans. Instrum. Meas.* **58** 270–80
[7] Gladyshev A B, Dmitriev D D and Tyapkin V N 2019 Software and Hardware Development and Testing of Repeaters of Communication Satellites *2019 International Siberian Conference on Control and Communications (SIBCON)* (IEEE) pp 1–5
[8] Li J and Yang W 2014 Research on the Protection Distances for the Interference from the Fixed-Satellite Earth Stations to the Space Research Earth Stations *2014 IEEE International Conference on Computer and Information Technology* (IEEE) pp 842–5
[9] Dassano G, Orefice M, Fresia F and Schurig F 2018 New techniques for the evaluation of antenna performances in ground stations for satellite communications *2018 IEEE International Symposium on Antennas and Propagation & USNC/URSI National Radio Science Meeting* (IEEE) pp 1647–8

[10] Dunsmore J P 2012 *Handbook of Microwave Component Measurements* (Chichester, UK: John Wiley & Sons, Ltd) 52–62
[11] Arpaia P, De Matteis E and Inglese V 2015 Software for measurement automation: A review of the state of the art *Measurement* **66** 10–25

Influence of inertial links on the distribution of vibration amplitudes of points of the working body of a technological vibrating

A V Eliseev[1], N K Kuznetsov[1], S V Eliseev[2], Q T Vuong[2]

[1] Irkutsk National Research Technical University, 83, Lermontov St., Irkutsk, 664074, Russia
[2]Irkutsk State Transport University, 15, Chernyshevskiy St., Irkutsk, 664074, Russia

E-mail: eavsh@ya.ru

Abstract. Features of problems of dynamics and formation of dynamic states of working bodies of vibrating technological machines are considered. The purpose of the work is to develop a method for constructing mathematical models of vibration machines that provide the ability to evaluate, control and adjust the forms of movement of working bodies by introducing additional links into the structure of the system, forming a closed internal contour. Methods of structural mathematical modeling are used in the construction of original structural formations of their typical elements. The direction of constructive and technical solutions in the implementation of approaches to configure and adjust the structure of the vibration field of the working body is proposed. The method is implemented using the analytical apparatus of the automatic control theory. The features of dynamic states formation based on the application of transfer functions of structural mathematical models of mechanical oscillatory systems are studied. The use of transfer functions of interpartial connections for the formation of vibration fields of working bodies is proposed. Analytical relations are obtained that determine the possibility of implementing specific dynamic modes, including the decoupling of movements by individual coordinates, taking into account the features of the connectivity of movements by characteristic points of the working body, the manifestation of oscillation nodes or modes of dynamic damping. Features of using structural mathematical modeling methods in problems of dynamics of mechanical oscillatory systems with additional connections are discussed. The work is aimed at specialists in the field of theoretical and applied mechanics, machine science and mechanical engineering.

1. Introduction

The use of additional connections in the structures of mechanical oscillatory systems, considered as calculation schemes of technical objects for various purposes, is a fairly developed direction in solving problems of machine dynamics. In many cases, structural and technical forms of additional connections are implemented as an introduction to the structure of mechanical oscillating systems of various mechanisms: gear, lever, including screw non-self-braking devices. The presence of mechanisms as specific mechanical chains in the structure of oscillatory systems allows us to solve a fairly wide range of dynamics problems, in which changes in the dynamic properties of the system, as a whole, are achieved due to dynamic effects that occur in the interactions of the system elements[1-3]. The article develops a method of structural mathematical modeling in applications to solving problems of evaluating the features of dynamic interactions when connecting devices for conversion of movement (DCM) with mass-inertia and elastic typical elements of a mechanical oscillatory system, which is a design scheme of a technological machine[4].

2. The features of technical object. Problem statement.

The technical object whose design scheme is shown in figure 1 is a mechanical oscillatory system with two degrees of freedom. A special feature of the approach is to take into account the connection features of typical elements of a mechanical system [5,6].

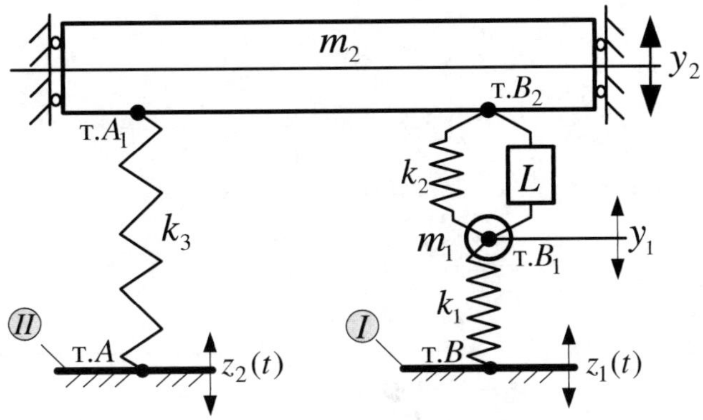

Figure 1. Schematic diagram of a suspension with a device for converting movement (DCM).

An inertial element of mass m_2 has a support spring of stiffness k_3 and rests in p.(A) on the support surface II (figure 1). in addition, the element m_2 rests on an elastic complex element (or a structural formation – a quasi-spring) consisting of a spring of stiffness k_1 with contact points (B) and (B_1), respectively, with a pair of elements k_2 and L in pp.(B_1), (B_2). At the point (B_1), the mass m_1 is concentrated. The second part of the structural block is a parallel connection of elastic elements with stiffness k_2 and DCM with an intermediate mass m_1. Contact with the inertia element m_2 occurs in p.(B_2). The position of the inertial element m_2 is determined by the y_2 coordinate, and the intermediate mass m_1 is determined by the y_1 coordinate, respectively. It is assumed that the system has linear properties and makes small fluctuations relative to the static equilibrium position. The movement of the reference surfaces I and II (figure 1) is determined by the known harmonic functions $z_1(t)$ and $z_2(t)$ of the same frequency; there is a functional relationship between the external perturbations

$$\bar{z}_2 = \alpha \bar{z}_1, \qquad (1)$$

where α is the constant coupling coefficient of external disturbances.

The reduced mass of the control unit, designated as L in Figure B.1, is determined by the expression

$$L = \frac{J}{r_{av}^2 \operatorname{tg}^2 \gamma}. \qquad (2)$$

where J is the moment of inertia of the flywheel nut, r_{av} is the average radius of the screw, and γ is the angle of elevation of the helix.

The position of the system is considered in coordinates y_1 and y_2 in a fixed basis. To build a mathematical model, we will make expressions for the kinetic and potential energies:

$$\mathrm{T}=\frac{1}{2}m_1\dot{y}_1^2+\frac{1}{2}L(\dot{y}_2-\dot{y}_1)^2+\frac{1}{2}m_2\dot{y}_2^2, \tag{3}$$

$$\Pi=\frac{1}{2}k_1(y_1-z_1)^2+\frac{1}{2}k_2(y_2-y_1)^2+\frac{1}{2}k_3(y_2-z_2)^2. \tag{4}$$

Using the Lagrange equation of the second kind, and then the Laplace transform under zero initial conditions, we write a system of differential equations in operator form [6]:

$$\bar{y}_1(m_1+L)p^2+\bar{y}_1(k_1+k_2)-\bar{y}_2(Lp^2+k_2)=k_1\bar{z}_1, \tag{5}$$

$$\bar{y}_2(m_2+L)p^2+\bar{y}_2(k_2+k_3)-\bar{y}_1(Lp^2+k_2)=k_3\bar{z}_2, \tag{6}$$

The structural mathematical model or block diagram of the automatic control system is shown in figure 2. the structure (figure 2) consists of two partial systems that are connected by inertia-elastic, which makes it possible to "zeroing" the connection at the frequency

$$\omega_0^2=\frac{k_2}{L}. \tag{7}$$

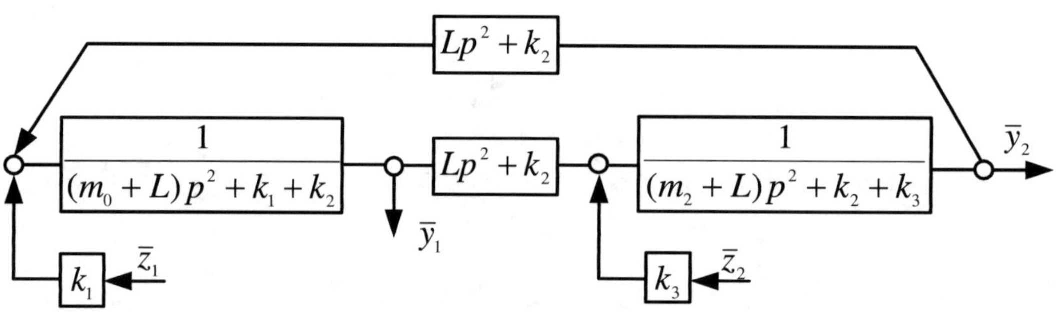

Figure 2. Block diagram of a system with a device for converting motion (DCM).

The transfer functions of the system under kinematic perturbation from the base are defined by the expressions

$$W_1(p)\underset{\bar{z}_1\neq 0,\,\bar{z}_2=0}{=}\frac{\bar{y}_1}{\bar{z}_1}=\frac{k_1[(m_2+L)p^2+k_2+k_3)]}{A(p)}, \tag{8}$$

$$W_2(p)\underset{\bar{z}_1\neq 0,\,\bar{z}_2=0}{=}\frac{\bar{y}_2}{\bar{z}_1}=\frac{k_1(k_2+Lp^2)}{A(p)}, \tag{9}$$

$$W_3(p)\underset{\bar{z}_1=0,\,\bar{z}_2\neq 0}{=}\frac{\bar{y}_1}{\bar{z}_2}=\frac{k_3(k_2+Lp^2)}{A(p)}, \tag{10}$$

$$W_4(p)\underset{\bar{z}_1=0,\,\bar{z}_2\neq 0}{=}\frac{\bar{y}_2}{\bar{z}_2}=\frac{k_3[(m_1+L)p^2+k_1+k_2)]}{A(p)}, \tag{11}$$

where $A(p)=[(m_1+L)p^2+k_1+k_2)][(m_2+L)p^2+k_2+k_3)]-(Lp^2+k_2)^2$ (12)

– is the characteristic frequency equation of the system.

The diagram (figure 1) shows that m1 is a low-inertia element through which the cascade $(Lp^2 + k_2)$ is connected to the elastic element k_1 (in this case, the point (B_1) is introduced into consideration).

The task of the study is to evaluate the dynamic features introduced into the mechanical oscillatory system by a device for converting motion, as well as by introducing an intermediate mass-inertia element m_1, which creates two cascades in the suspension system, sequentially connected to each other.

3. Assessment of the dynamic properties of the system upon addition DCM

Usually, the point (B_1), in which three typical elements are simultaneously in contact with the transfer

functions k_1, k_2 and Lp^2, is not considered as a characteristic point. The authors interpret such a nodal point; its position is determined by the coordinate $\overline{y}_1$. The limiting case is the situation when m_1 tends to *0*, which predetermines the approach to assessing the dynamic stiffness of a two-stage series connection of system elements, the value of m_1 affects the distribution of dynamic bond reactions in pp.*(B)*, *(B₁)* and *(B₂)*.
Using the structural diagram in Figure 2, we write down the transfer functions, taking into account the connectedness of external influences, defined by the expression (B.1)

$$\underset{\overline{z}_2=\alpha\overline{z}_1}{W_1''(p)} = \frac{\overline{y}_1}{\overline{z}_1} = \frac{k_1[(m_2+L)p^2+k_2+k_3]+\alpha k_3(Lp^2+k_2)}{A(p)}, \tag{13}$$

$$\underset{\overline{z}_2=\alpha\overline{z}_1}{W_2''(p)} = \frac{\overline{y}_2}{\overline{z}_1} = \frac{\alpha k_3[(m_1+L)p^2+k_1+k_2]+k_1(Lp^2+k_2)}{A(p)}. \tag{14}$$

Figures B. 3, a, and b show converted block diagrams with the allocation of elements m_1 and m_2 covered by the corresponding negative feedbacks. In the physical sense, the transfer functions of such connections correspond to (or determine) the dynamic rigidity of structural formations.

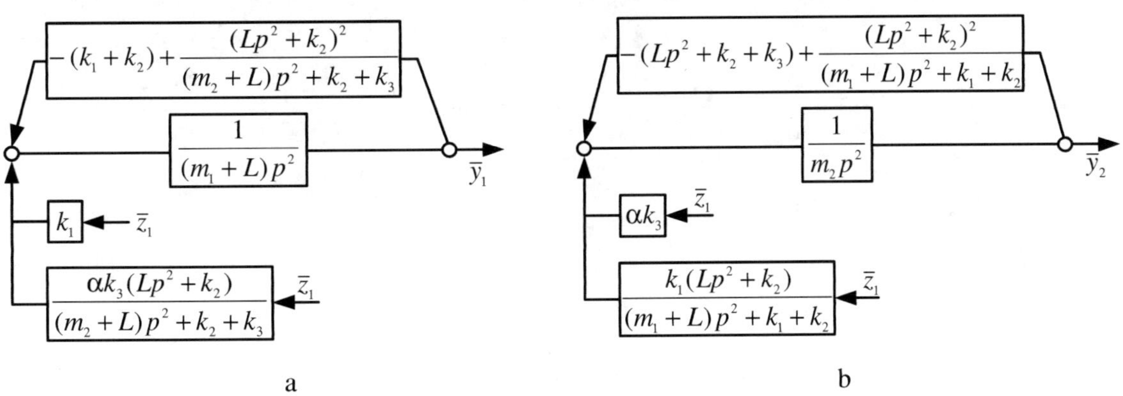

Figure 3. Structural diagrams of the system, expanded relative to the mass-inertia elements m_1 and m_2: a-m_1-coordinates y_1; b-m_2-coordinates y_2

Dynamic reactions at points of contact with the reference surface are defined by the expressions

$$\left|\overline{R}_{\text{sup}}\right| = \left|\overline{R}_A\right| + \left|\overline{R}_B\right| = k_3\cdot\overline{z}_1\cdot W_2''(p) + k_1\cdot\overline{z}_1\cdot W_1''(p) = \frac{k_3\cdot\overline{z}_1\{\alpha k_3[(m_1+L)p^2+k_1+k_2]+k_1(Lp^2+k_2)\} + k_1\cdot\overline{z}_1\{k_1[(m_2+L)p^2+k_2+k_3]+\alpha k_3(Lp^2+k_2)\}}{A(p)}, \tag{15}$$

or

$$\left|\overline{R}_{\text{sup}}\right| = \frac{\overline{z}_1\{[\alpha k_3^2(m_1+L)+k_1^2(m_2+L)+k_1k_3L(1+\alpha)]p^2 + \alpha k_3^2(k_1+k_2)+k_1^2(k_2+k_3)+k_1k_2k_3(1+\alpha)\}}{A(p)}. \tag{16}$$

The dynamic response $\left|\overline{R}_{\text{sup}}\right|$ is “zeroing” at the frequency of external influence

$$\omega_{00}^2 = \frac{\alpha k_3^2(k_1+k_2)+k_1^2(k_2+k_3)+k_1k_2k_3(1+\alpha)}{\alpha k_3^2(m_1+L)+k_1^2(m_2+L)+k_1k_3L(1+\alpha)}. \tag{17}$$

The dynamic response on the m_2 «object» is defined through dynamic stiffness, which can be found using structural transformation rules

$$\bar{k}'_{\text{пр}} = \frac{(Lp^2+k_2)[(m_1+L)p^2+k_1+k_2]-(Lp^2+k_2)^2}{(m_1+L)p^2+k_1+k_2} = \frac{(Lp^2+k_2)(m_1p^2+k_1)}{(m_1+L)p^2+k_1+k_2}. \tag{18}$$

Thus

$$\begin{gathered}\left|\bar{R}_{m_2}\right| = \left|\bar{R}_{A_1}\right| + \left|\bar{R}_{B_2}\right| = \left|\bar{R}_{A_1}\right| + \bar{k}'_{\text{пр}}\cdot W_2''(p)\cdot\bar{z}_1 = \\ \bar{z}_1\{\alpha k_3[(m_1+L)p^2+k_1+k_2]+k_1(Lp^2+k_2)\}\times \\ = \frac{\times\{k_3[(m_1+L)p^2+k_1+k_2]+(Lp^2+k_2)(m_1p^2+k_1)\}}{[(m_1+L)p^2+k_1+k_2]A(p)}.\end{gathered} \tag{19}$$

We find the relation of dynamic reactions $\left|\bar{R}_{m_2}\right|$ and $\left|\bar{R}_{\text{sup}}\right|$, using the techniques described in [6]

$$N(\omega) = \frac{\left|\bar{R}_{m_2}\right|}{\left|\bar{R}_{\text{sup}}\right|} = \frac{\{\alpha k_3[(m_1+L)p^2+k_1+k_2]+k_1(Lp^2+k_2)\}\times \times\{k_3[(m_1+L)p^2+k_1+k_2]+(Lp^2+k_2)(m_1p^2+k_1)\}}{\{[\alpha k_3^2(m_1+L)+k_1^2(m_2+L)+k_1k_3L(1+\alpha)]p^2 + +\alpha k_3^2(k_1+k_2)+k_1^2(k_2+k_3)+k_1k_2k_3(1+\alpha)\}\times \times[(m_1+L)p^2+k_1+k_2]}. \tag{20}$$

4. Discussion

The ratio of reactions of dynamic connections at characteristic points of the source system can be considered as a parameter that reflects the dynamic state of the system, taking into account the features defined by a set of standard elements, external disturbance conditions, and the choice of parameter values.

In a conventional mechanical oscillatory system with two degrees of freedom ($m_1 \neq 0$, $L = 0$, $\alpha = 0$), the amplitude-frequency response in the coordinate has a mode of dynamic damping of vibrations at the frequency

$$\omega_{1dyn}^2 = \frac{k_2+k_3}{m_2}, \tag{21}$$

what follows from (8) for $L = 0$. If $m1 = 0$, $L = 0$, then the original system is reduced to a system with one degree of freedom; in this case, the coordinate is canceled; the frequency of natural vibrations of the system with one degree of freedom in this case is determined by the expression

$$\omega_{\text{nat}}^2 = \frac{k_1k_2+k_2k_3+k_1k_3}{m_2(k_1+k_2)}. \tag{22}$$

To assess the influence of parameters of typical elements, a model problem is used, which considers changes in α (0, 0.5, 2); m_1 = 100 kg; m_2 = 1000 kg; k_1 = 100 kN/m; k_2 = 200 kN/m; k_3 = 300 kN/m; L (0, 100 kg).

In general, the system can display resonant modes at the natural oscillation frequencies determined from the frequency characteristic equation (12). In the transition to the amplitude-frequency characteristics (AFC) of coupling reactions, which is defined by the expression (20), it should be noted that the following features will be observed in the form of AFC: at ω →∞, the ratio $\frac{|\overline{R}_{m_2}|}{|\overline{R}_{sup}|}$ will tend to a certain limit.

Frequency characteristics of the bond reaction are shown in figure 4.

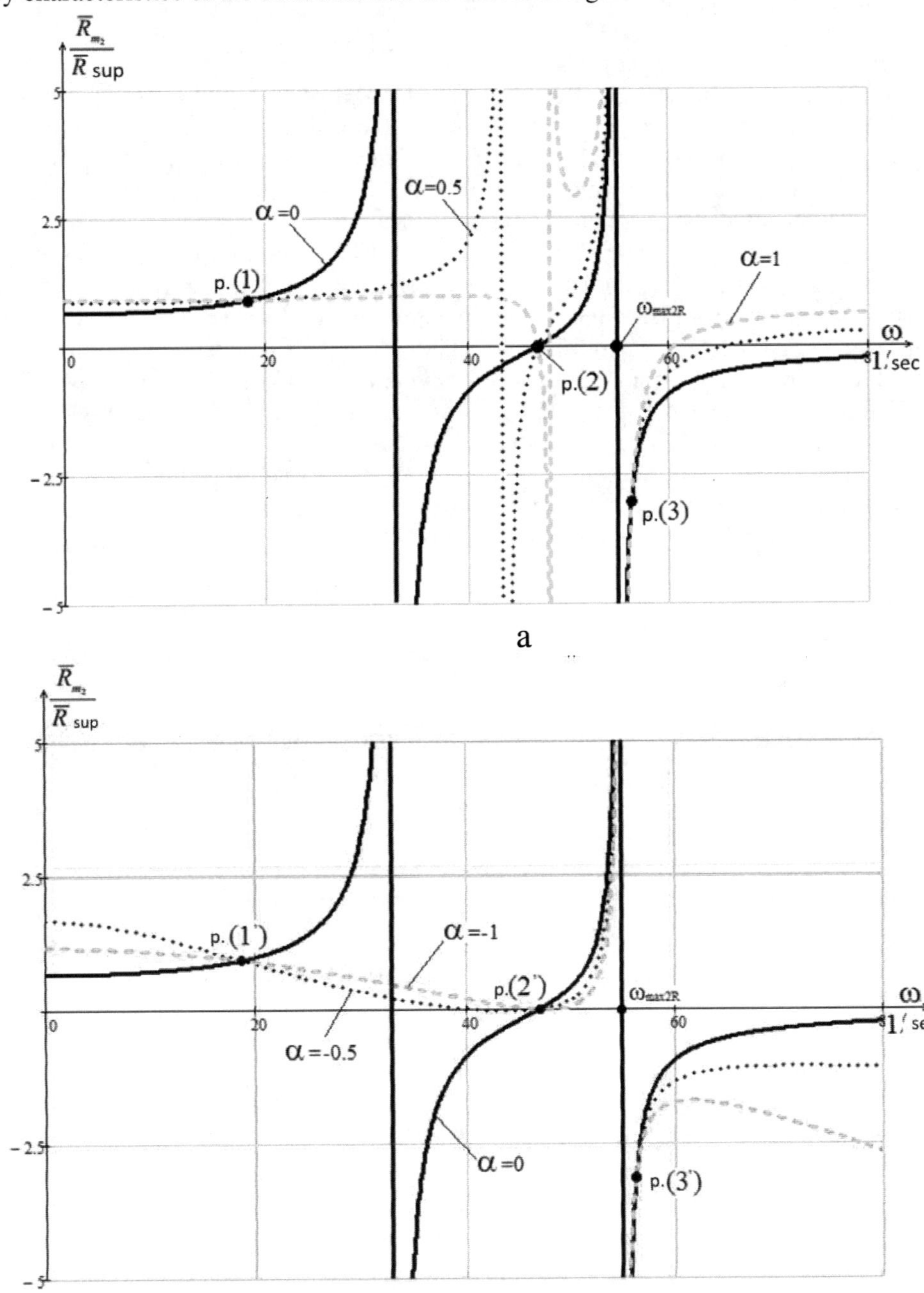

Figure 4. Frequency characteristics of the bond reaction ratio at m_1 = 100 kg, L = 0: a-α = 0, 0.5, 1; b - α = 0, -0.5, -1

At the frequency ω_{dyn2R} the graph, the relations of bond reactions as previously mentioned intersect with the abscissus axis (p. (2) in figure 4,a and p. (2') in figure 4, b) regardless of the coefficient α, as determined from (b. 24'). In figure 4, and all graphs of the relation of bond reactions with positive values of α still pass through common points (p. (1) and p. (3)), also respectively p. (1') and p. (3') in figure 4, b for negative values of α.

5. Conclusion

A method for constructing mathematical models of a system of transport suspensions is proposed, which allows evaluating their dynamic capabilities under the action of various external perturbations, including the simultaneous action of several factors. The idea of the method of analysis and dynamic systems is to use a structural mathematical model in the form of a block diagram of an automatic control system. It is shown that a structural diagram of a suspension with a selected object whose dynamic state is evaluated can be considered as a basis for determining the dynamic stiffness of the system at characteristic points. This allows, in turn, to determine the dynamic reactions of connections as the product of dynamic stiffness on the dynamic displacement of the point in question, determined by the transfer function of the system.The paper introduces the concept of amplitude-frequency characteristics, in which the «output signal» is the dynamic response of connections. The analysis shows that changes in the dynamic reactions of connections can differ significantly from the ideas about the dynamic properties of systems obtained in the usual way. Setting up dynamic suspensions and evaluating their dynamic properties can be performed using such dynamic characteristics as the ratio of bond reactions occurring on the support surface and on the protected object. In this case, the dynamic state can be controlled by changing the system's stiffness ratio.

6. Acknowledgments

The reporting study was carried out with the financial support of the Ministry of science and higher education of the Russian Federation in the framework of project No. 0667-2020-0037 «Development of scientific foundations of effective processes obtaining valuable elements, substances and materials from waste and man-made and natural multicomponent mixtures, including methods of design and technological support».

References

[1] Karnovsky I A, Lebed E 2016 *Theory of vibration protection* (Switzerland: Springer) p 708
[2] de Silva C W 2000 *Vibration. Fundamentals and Practice* (Boca Raton, London, New York, Washington, DC: CRC Press) p 957
[3] Harris S M, Piersol A G 2002 *Shock and Vibration Handbook* (New York: McGraw - Hill Book Co) p 1457
[4] Tarasik V P 2004 *Mathematical modeling of technical systems* (Minsk: Dizayn PRO Publ.) p 640
[5] Eliseev A V, Kuznetsov N K, Moskovskikh A O 2019 *The Dynamics of Machines. System conception, structural schemes and connections of elements* (Moscow: Innovative mechanical engineering) p 381
[6] Eliseev S V, Eliseev A V 2020 *Theory of Oscillations. Structural Mathematical Modeling in Problems of Dynamics of Technical Objects* (Springer International Publishing) p 521

IOP Conf. Series: Materials Science and Engineering 1061 (2021) 012014 doi:10.1088/1757-899X/1061/1/012014

The task of increasing the number of training situations as one of the areas for improving the ergatic software and hardware systems "Aviation simulator"

V R Roganov[1], E A Asmolova[1], N S Esimova[2], R U Omirbekov[2], O Kuvshinova[3] and G K Aidarbek[1]

[1] Penza State Technological University, 1a, Gagarina ave, Penza, 440039, Russia
[2] Humanitarian and Technical Institute "Akmeshit", 43, G.Muratbayeva ave, Kyzylorda, 120000, Kazakhstan
[3] Penza State University of Architecture and Construction, 28, Titova, Penza, 440039, Russia

E-mail: vladimir_roganov@mail.ru

Abstract. Currently, the ergatic software and hardware complex, called the "flight simulator", is used worldwide for professional training of crew members in piloting an aircraft in special flight conditions. Flight and integrated flight simulators include a visual environment simulator that synthesizes for a pilot a three-dimensional image of the external environment behind the simulator's cabin, which allows pilots to be trained in visual landing on a runway model, while solving aeronautical problems. Simulators of the visual environment developed earlier could not synthesize recognizable areas of the area more than 400 × 400 km in size necessary for training in solving problems of self-driving. Therefore, simulators for the training of navigators were separately developed, teaching the determination of the location of the model of an aircraft in the model of the environment according to the models of radio engineering means of aircraft. Modern optical-technical-software systems for the synthesis of three-dimensional models of the external environment allow us to develop new and modernize existing flight and complex flight simulators by adding training to the crew of the aircraft in solving problems of aircraft navigation. The analysis carried out by the authors of this article showed that for this it is necessary to solve the problem of selecting objects that should become benchmarks and their placement according to the model of a large area. Solutions are proposed that allow us to solve the tasks and develop a model of a recognizable piece of the earth's surface that is sufficient to solve the problems of piloting and self-driving training on an aircraft simulator.

1. Introduction

The modern development of hardware-software complexes of machine synthesis allows the real-time synthesis of terrain images on a monitor screen similar in quality to images on a television screen obtained with the use of video cameras. [1]. Certain euphoria from the results obtained often allows us to erroneously interpret the results as a solution to the problems of synthesis and modelling of a three-dimensional image [2]. Such an interpretation is usually based on the results obtained when viewing the "non-planar" image obtained by processing the lines on the monitor screen (figure 1) or when considering moving models of three-dimensional objects whose faces are illuminated taking into account the rules of computer graphics (for example, when a cube model is moved on the screen, or a word written in "volume" font). 3D modelling systems are used in various sectors of the national economy: for training drivers of vehicles, while improving the conditions for operations using endoscopic racks and in other cases [3]. In the future, we will consider flight simulators (figure 2), allowing training pilots professional skills in piloting an aircraft and partially solving navigation problems on the ground [4].

Figure 1. Example of a “non-planar image”

2. Main part

To obtain a model of a three-dimensional external environment around the simulator cabin, at first, on a flat-screen, a video sequence is synthesized from 2D projections of the part of the 3D external environment model that fell into the surveillance camera (figure 3)

Figure 2 - Cabin of the flight simulator with a display device with a "wide pupil" [5]

A 2D projection of a part of the 3D external environment is synthesized on a flat-screen by specialized software and hardware systems for machine synthesis [6]. For a long time, it was believed that it was enough to connect a display device for synthesizing 3D images to such a screen and such a system can be used as a “visual environment simulator” or as a complex that allows you to train a pilot of visual landing on a runway model [7].

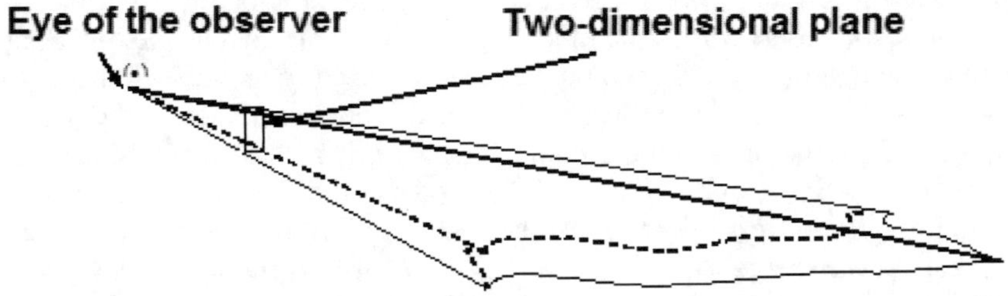

Figure 3. Possible position of the surveillance camera in the model of the environment

The technology for obtaining 3D images is as follows. On a flat screen, a 2D projection of a part of a three-dimensional model of the external environment is synthesized with an interval of no more than 120 ms. Next, use one of two possible schemes with optical devices.

Conventionally, the first system is a system with stereo glasses. Usually, these are two screens, each of which is connected to its specialized software and hardware system for machine synthesis. One synthesizes an image for the left eye. One synthesizes an image for the right eye. The advantage of such a system is that a trained observer sees a 3D image in any case if the solid angles of the surveillance cameras coincide with the angles between the observer's eye and the corresponding screen on which the corresponding image is synthesized. The disadvantage is that there is a time interval during which the visual apparatus of a particular person learns to see a 3D image of object models with quality that allows a person to visually determine the distance to the selected model. Usually, any person wearing stereo glasses immediately sees a sharp rather than a blurry image on the screens (among people there are exceptions when a person starting to view images through stereo glasses can lose orientation in space or get a severe headache after the session of such work). Moreover, not everyone can immediately visually determine the distance to the models in question, just as in real life they visually determine the distance to the selected objects. The results of experiments in which the authors of the article participated in conducting the research work "Research and optimization of manufacturing technologies for beam splitting plates and spherical mirrors, for single-channel and multi-channel glasses-free indicators of pseudo-volume images with a narrow pupil" (state contract No. 8009p / 8265 of 04/30/2010) (1 and 2 stages) between Video3 LLC and the Federal State Budgetary Institution "Fund for the Promotion of the Development of Small Forms of Enterprises in the Scientific and Technical Field" showed that the training interval for a specific person's visit to see a 3D model and visually determine the distance to it can last up to 5 months, with daily work with such a system for at least 20 minutes a day. In this case, all researchers immediately saw a clear image, but could not accurately determine the distance to visible models.

So, when teaching the human visual apparatus to see the 3D model when using systems with steroid glasses, negative effects may appear:

- headache;
- short-term loss of orientation in space during the first sessions of working with such systems, which was observed in two researchers (with a total number of researchers of 78 people).

Some researchers working with such systems had negative effects. They disappeared after long sessions of working with such systems, which is explained by the training of the human visual apparatus to see a 3D model. Most researchers have not observed negative effects when working with systems with stereo-glasses.

The second system used in the flight simulator is a frameless 3D image modelling system when a collimator is placed between the flat screen and the human eye (such a system affects the accommodation and convergence of the human visual apparatus). For observing 3D images, when using such systems, in addition to matching the corners of the visibility pyramid with the viewing angles of the 3D image modelling system used. To activate accommodation and convergence of the human visual apparatus, the observer must be mobile, or moving objects must be in the frame.

Both considered systems for modelling 3D images are referred to as three-dimensional systems for modelling three-dimensional images. The three-dimensional model is created in the human mind due to the targeted influence of special optics on the human visual apparatus of special images, which are 2D projections of three-dimensional external environment models on the plane of the screen (screens). We are all different, and this is confirmed by the results of our studies, but these features of human vision allow all people to see a 3D image model with sufficient quality to train an eye, but there may be an interval during which a person learns to see a 3D image. When used without spectacle systems, of all researchers, there was one of those who did not see the 3D image at the first moment of time.

In the 20th century, pseudo-volume systems were the main ones in the simulators of the visual environment of flight simulators. Their effectiveness is confirmed by the ability to train pilots in the complex process of visual landing on the runway model, including on the shortened model of the runway deck of the aircraft carrier. Previously, it was believed that the disadvantages without spectacle systems for modelling 3D images are:

- 3D viewing conditions only with a moving observer;
- the distance between the observer's eye and the nearest visible three-dimensional model is more than 80 meters;
- the effect of observing a three-dimensional model is possible only in complete darkness, since the optical system absorbs a lot of light distance between the observer's eye and the nearest visible three-dimensional model is more than 80 meters.

Glasses-free 3D model synthesis systems are categorized into two classes: "narrow-pupil" systems and "wide-pupil" systems. In a "narrow-pupil" system there is a limited space (figure 4), where the eyes of the observer should be, and only one pilot can comfortably observe the 3D model generated by it. In a "wide-pupil" system, the 3D model observation zone covers the area of a wide-body aircraft cabin (figure 2). In a "narrow-pupil" system, the observer's head is placed within the observation area of the 3D image (figure 5). A single glasses-free indicator simulates for an observer a 3D model of the external space in the viewing angles of 40° horizontally and 30° vertically. To expand the visual angles, several such indicators are installed (figure 2, figure 5). To give an example, in figure 5 there are 3 indicators that add up to 120° horizontal and 30° vertical visual angles.

In all mentioned 3D image synthesis systems, a trained person can visually assess the distance to the observed objects. All considered systems are pseudo-volume. Unlike the true-volume systems (based on a rotating screen), these systems use a video sequence of 2D projections, and the human visual apparatus builds it up to a 3D model.

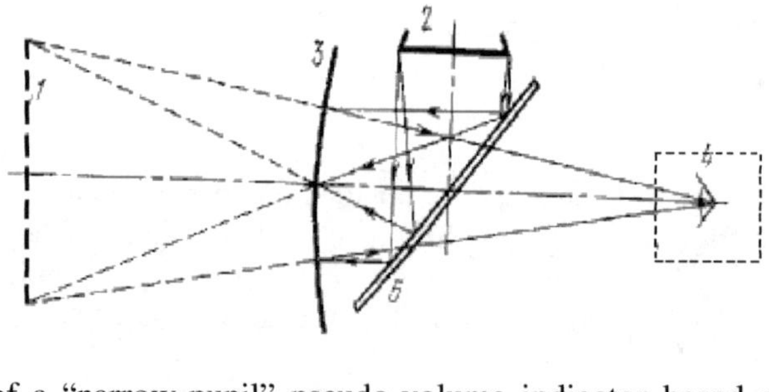

Figure 4. Scheme of a "narrow-pupil" pseudo-volume indicator based on a mirror collimator: 1 virtual image screen; 2 intermediate image screen, a TV screen where 2D projections of 3D objects are synthesized; 3 concave spherical mirror; 4 eye of the observer; 5 beam splitter plate.

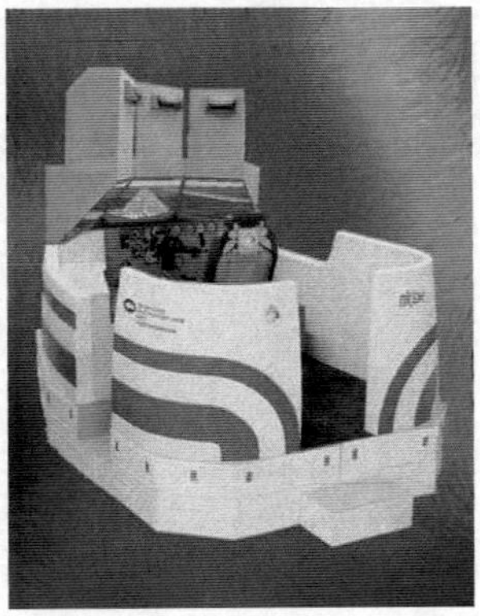

Figure 5. OKU (an optical communication device), a three-channel indication device with a "narrow pupil" manufactured by Penza design modelling Bureau installed on a flight simulator.

The following effects were noticed during the analysis of pseudo-3D indicators of the OKU-type produced by NPP or RELLI, ELVIRA, SVETLANA A/15/M produced by Video3 LLC:

- There is a volume of space in which the observer's eye or eyes see the three-dimensional image of the virtual space models, looking at two-dimensional changeable models which appear on the screen with the intermediate image;
- the image is bright enough to be seen by the observer in the natural light;
- the pseudo-3D image is a little "blurred", but distinct enough, which does not affect the possibility of using such systems in aviation simulators, in endoscopy towers and in other cases when it is necessary to simulate 3D- images of objects;
- there is a "window wall zone" between the observer and the closest virtual space model, which, based on our experiments, is approximately 8 m, while, according to [1,3] it is at least 80 m.

The analysis of technical solutions proves that there exist good enough solutions to a problem of modelling the 3D-image of the environment model, which can be classified (figure 6). This is the result of the development of aviation simulators as training systems for creating professional piloting skills, when such systems had to synthesize the 3D-image of the runway model at the stage of learning the visual landing of the aircraft.

IOP Conf. Series: Materials Science and Engineering 1061 (2021) 012014 doi:10.1088/1757-899X/1061/1/012014

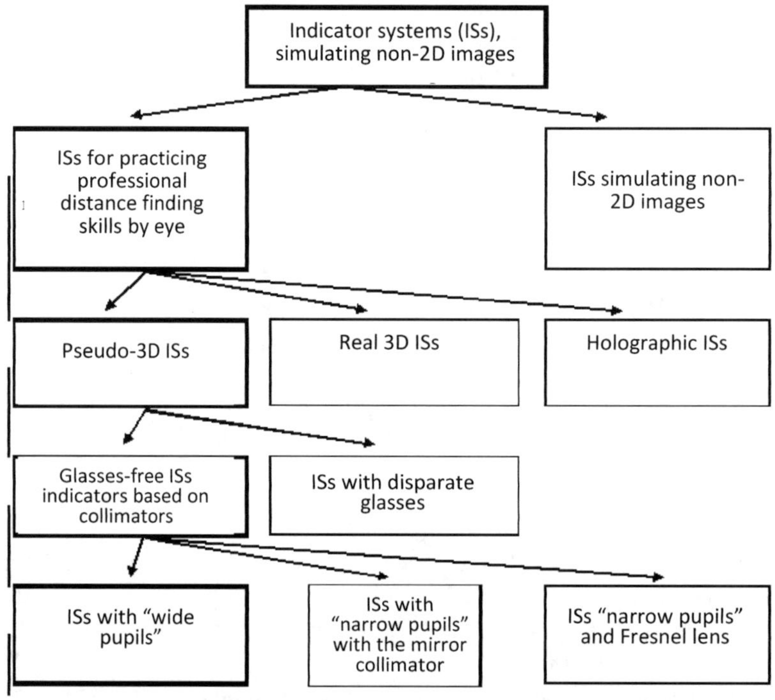

Figure 6. The classification of indicator systems simulating non-2D images (the systems used in building aviation simulators are in bold type).

The pilot flying a real aircraft has to solve flight direction and navigation tasks. At present navigation tasks are solved to the maximum degree only during the flight above the airfield model. For this purpose, nearly all pilot-centred and complex aviation simulators create a recognizable 15×15 km land surface model with the model of the main runway of a particular airfield in the middle. The Canadian company CAE, which has been a world leader in creating aviation simulators, is now developing visual environment simulators which allow theoretical modelling of the 1500×1500 km flight area. In order to solve navigation tasks, we have to solve the problem of creating a model of a real area with the models of levelling objects. This model is called a visualization scene. In the past, the solution of this problem was limited by the insufficient memory capacity and efficiency of the software and hardware which were used for developing computer image generators. These limitations have been overcome. The following problems remain to be solved.

The first one is which requirements the visualization scene must meet. It must have the models of levelling objects M_j^V which can be seen via the visual environment imitator of any type and, if necessary, the pilots must solve all the navigation tasks using them.

$$P_{t_i}^V = \sum_{j=1}^{N_{t_i}^V} I^V(M_j^V), \qquad (1)$$

where $P_{t_i}^V$ is the place of the aircraft model in the environment model, determined by the pilot during the flight on the aviation simulator using the visual situation imitator.

$I^V(M_i^V)$ – information about the position of the aircraft model in the environment model obtained by the pilot when aiming the aircraft model by the visually observable j-th model M_i^V through the cockpit window of the aviation simulator;

$N_{t_i}^V$ – the number of levelling objects, which at the time moment t_i are seen in the visual situation imitator.

The second one – the navigation tasks when flying a real aircraft can be solved not only while watching the levelling objects through the cockpit window, but also while watching the levelling objects through thermal scope M_j^Q and radio locator M_j^R.

$$P_{t_i}^Q = \sum_{j=1}^{N_{t_i}^Q} I^Q(M_j^Q), \qquad (2)$$

where $P_{t_i}^V$ is the position of the aircraft model in the environment model, determined by the pilot when using the aviation simulator based on the information from the thermal scope imitator,

$I^Q(M_i^Q)$ is the information about the position of the aircraft model in the environment model, obtained by the pilot when aiming the aircraft model by the visually observable j-th model M_i^Q on the screen of the thermal scope imitator;

$N_{t_i}^Q$ is the number of levelling objects, which at time moment t_i are seen through the thermal scope imitator.

$$P_{t_i}^R = \sum_{j=1}^{N_{t_i}^R} I^R(M_j^R), \qquad (3)$$

where $P_{t_i}^R$ is the position of the aircraft model in the environment model, determined by the pilot when using the aviation simulator based on the information from the radio locator imitator,

$I^R(M_i^R)$ is the information about the position of the aircraft model in the environment model, obtained by the pilot when aiming the aircraft model by the visually observable j-th model M_i^R on the screen of the radio locator imitator;

$N_{t_i}^R$ is the number of levelling objects, which at time moment t_i are seen through the radio locator imitator.

The third navigation task can be solved using the radio-technical facilities of air navigation (radio compass, radio station of short radar navigation, etc.)

$$P_{t_i} = F_k(I_k), \qquad (4)$$

where P_{t_i} – is the position of the aircraft model in the environment model, determined by the pilot when using the aviation simulator based on the information $F_k(I_k)$ from the k-th imitator of the radio locator.

When developing different aviation simulators of the environment model with the possibility of a pilot's solving navigation tasks: the visual environment imitator, the thermal scope imitator and the radio locator imitator, three visualization scenes are developed. The visualization scene for the visual environment imitator. The visualization scene for the thermal scope imitator. The visualization scene for the radio locator imitator. Bearing in mind the facts that during a flight a pilot obtains up 90% of information through the visual analyser and that it is impossible to create a complete model – when developing visualization scenes we must remember that the number of models of levelling objects, placed in the corresponding scenes are a subset of real objects in the geographical area selected as the visualization scene prototype. In this case allowance must be made for the fact that the number of models of levelling objects in the whole visualization scene must be enough for solving navigation tasks:

$$N_S^V = \frac{N^V}{S^V}, \qquad (5)$$

$$N_S^V \geq K_S^V, \qquad (6)$$

where N_S^V – the number of models of levelling objects of the visualization scene of the visual environment imitator per unit area;

N^V – the number of models of levelling objects in the visualization scene of the visual environment imitator;

K_S^V – the minimal number of models of levelling objects per unit area of the scene of the visual environment imitator allowing navigation tasks to be solved.

$$N_S^Q = \frac{N^Q}{S^Q} \quad , \tag{7}$$

$$N_S^Q \geq K_S^Q \, , \tag{8}$$

where N_S^Q – the number of models of levelling objects of the visualization scene of the thermal scope imitator per unit area;

N^Q – the number of models of levelling objects in the visualization scene of the thermal scope imitator.;

K_S^Q – the minimal number of models of levelling objects per unit area of the scene of the thermal scope imitator allowing navigation tasks to be solved.

$$N_S^R = \frac{N^R}{S^R} \quad , \tag{9}$$

$$N_S^R \geq K_S^R \, , \tag{10}$$

where N_S^R – the number of models of levelling objects of the visualization scene of the thermal scope imitator per unit area;

N^R – the number of models of levelling objects in the visualization scene of the thermal scope imitator;

K_S^R – the minimal number of models of levelling objects per unit area of the scene of the radio locator imitator, allowing navigation tasks to be solved.

As it was shown above, the visual environment imitator gives the maximum amount of information to a pilot training on the aviation simulator: besides determining the location of the levelling point in the environment model and solving navigation tasks simultaneously, this imitator allows improving the skill of judging by eye. Therefore, when developing visualization scenes for the thermal scope imitator and for the radio locator imitator the visualization scene for the visual environment imitator is used as a basis, or

$$N^V \supset N^Q \, , \tag{11}$$

$$N^V \supset N^R \, . \tag{12}$$

3. Conclusion

1. Optical-software-hardware systems of modern imitators of environment models synthesized for the aviation simulator allow solving the task of the synthesis of a recognizable area of land up to 1500×1500 km. Possible number of a number of models of levelling objects allowing are solved navigation tasks and determine visually the distance to the selected models of levelling objects.
2. In order to upgrade pilot-centred and aviation simulators by including training situations connected with solving navigation tasks it is necessary to develop visualization scenes for visual situation imitators, the thermal scope imitator and the radio locator imitator. In this case, all the models of levelling objects in the visualization scene for the thermal scope imitator are a subset of models of models objects of the visualization scene of the visual environment imitator. All the models of levelling objects in the visualization scene for the radio locator imitator are a subset of the models of levelling objects of the visualization scene for the thermal scope of the visual environment imitator.

References

[1] Roganov V R, Miheev M Y, Roganova E V, Bolat A, Nurgozhin I and Fillipenko V 2017 Main provisions for formation of cognitive model of visually observable environment synthesized for aircraft simulator *Advances in Eng. Research, Act. Issues of Mech. Eng. (AIME 2017)* (Published by Atlantis Press) pp 671–676

[2] *Design of optical systems* 1983 ed R Shannon (Moscow: Mir) p 430

[3] Roganov V R, Miheev M Y, Roganova E V, Grintsova O and Lavendels J 2018 Modernisation of Endoscopic Equipment Using 3D Indicator in Applied Computer Systems pp 75–80

[4] *Training systems* 1981 ed Shukshunov V E (Moscow: Engineering) p 256

[5] https://yandex.ru/images/search?text=авиационный%20тренажёр%20картинки&stype=image&lr=49&source=wiz&pos=4&img_url=https%3A%2F%2Fpbs.twimg.com%2Fmedia%2FECF29lMWkAA9brR.jpg%3Alarge&rpt=simage (date of the application 02.04.2020).

[6] Roganov V R, Miheev M Y, Asmolova E A and Zhashkova T V 2016 Visual simulators for training simulators for vehicle drivers *(Proceeding of the soft Int. Symp. Reliability and Quality)* pp 326-328

[7] Modeling in the training of flight personnel of civil aviation 1992 *(Results of science and technology. Ser. air transp. VINITI)* **25** p 136

IOP Conf. Series: Materials Science and Engineering 1061 (2021) 012015 doi:10.1088/1757-899X/1061/1/012015

Regulation of the stop points throughput capacity in urban public transport

A I Fadeev, S Alhusseini

Siberian Federal University, 79, Svobodny Avenue, Krasnoyarsk, 660041, Russia

E-mail: fai@ak1967.ru

Abstract. The article considers the problem of determining the throughput capacity of bus lines, which is usually determined by the limited throughput capacity of stopping points. Currently, the throughput capacity of a stop point is usually defined as the inverse mathematical expectation of the average service time for one transportation vehicle. To account such factors as random disturbances in the transport process, appropriate correction coefficients are applied. However, it has not been proved that the considered coefficients provide acceptable results for practice in all cases. In this article, in order to study the throughput capacity, the work of a stopping point is described as a multi-channel single-phase queuing system (QS). An algorithm is proposed for simulating a public transport stopping point model based on the principle of consecutive submitting requests. The recommendations formulated for the throughput capacity regulation of stop points are illustrated by using an example of a specific stop point.

1. Introduction

One of the most important parameters of the transport system is the throughput capacity of lines, which is usually determined by the stop points throughput capacity in urban public transport system [1, 2, 3, 4, 5]. The throughput capacity of a bus lane depends on the critical stop points throughput capacity. As a rule, it is the stop point with the largest passenger traffic [2].

Currently, the throughput capacity of a stop point is usually defined as the inverse mathematical expectation of the average service time for one transportation vehicle. To account such factors as random disturbances in the transport process, appropriate correction coefficients are applied. For example, in the work [6] the coefficient of the incoming request flow unevenness is used, which is defined as the ratio of the maximum recorded intensity of the transportation vehicles flow to the average intensity for the calculation period. In the work [2, see 27-13] the coefficient of service request refusal probability and the coefficient of passengers service time variation at the stop point are used. However, it has not been proved that the considered coefficients pro-vide acceptable results for practice in all cases.

In the works [5,6,8,9,10] it was established that the stopping places at the stop points have unequal productivity: the first place has the highest productivity, after that the productivity of each subsequent place decreases. Therefore, the concept of "effective number of stopping places" is applied, which corresponds to the conditional stopping places number with the throughput capacity of the first stopping place.

2. Problem description

In order to study the throughput capacity of a stop point, we will describe it as a multi-channel single-phase queuing system (QS) (figure 1). To the system, a flow of service requests with intensity λ arrives (passengers boarding and alighting the vehicles). The service channel is the stopping place of the fleet at the stop point. μ_i - performance (service intensity) of the i channel (stop place). We will not impose restrictions on the queue length and waiting time in this QS.

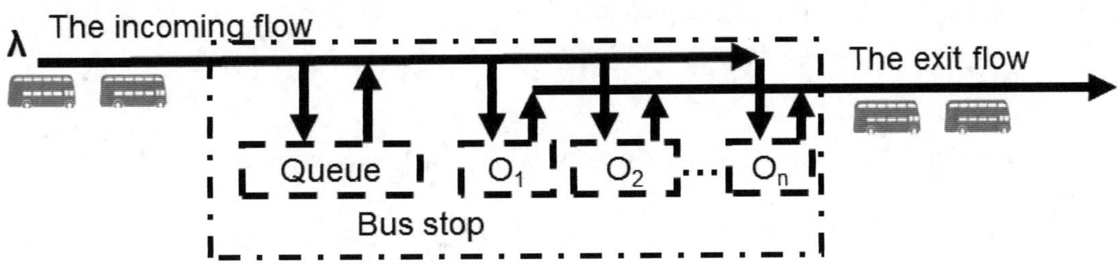

Figure 1. Diagram of a stop point as a QS
O_1,..., O_n – service channels (stop places)

The performance (intensity of service) of the service channel has a random character, which is described by one of the statistical laws, for example, gamma distribution, log-normal distribution [5], exponential distribution [7, 8], Erlang distribution [3], etc.

Given the unequal performance of stopping places, the flow of service requests has an intensity:

$$\mu = \sum_{i=1}^{m} \mu_i = k_m \mu', \tag{1}$$

where: m - number of occupied service channels;

μ_i - performance (service intensity) of the i channel;

k_m - reduction factor to the first channel intensity of service (μ') with n occupied service channels ($k_1 = 1; k_2 \leq 2; ...$).

Previous studies have shown that the random process of service requests in the QS under consideration is not the simplest, since it is described not by an exponential distribution, but by a gamma distribution. In this regard, to calculate the parameters of the QS we use simulation modeling [10,11,12].

3. Problem solution

We formulate the methodology of the simulating model of public transport stop point, based on the principle of consecutive requests submission [10].

There are two random processes in the system: requests received and requests served.

The time to submit *i* request is calculated as follows:

$$t_i^{\lambda} = t_{i-1}^{\lambda} + \tau_i^{\lambda}, \tag{2}$$

τ_i^{λ} - random time interval between submitted requests in the system, determined in accordance with the given distribution law.

The time to serve the request completion:

$$t_j^{\mu} = t_i^{b} + \tau_j^{\mu}, \tag{3}$$

where: t_j^b - service start time (submit a service request);

τ_j^μ - random required time to serve request, determined in accordance with the given distribution law.

Figure 2 shows a diagram of the change in the state of the QS. Events that change the state of the system are the submission and completion of service requests.. A queue in the system occurs if the number of unserved requests exceeds the number of service channels. The requests set A located in the system is described by the following relation:

$$A(I,T^i,T^s,T^o), \tag{4}$$

Where: I - Id service request;

T^i,T^s,T^o - time of the request submission, inception of the request service and request service completion.

In the system n service channels, t_j^μ - The time required for j service channel to become available ($j=\overline{1,n}$). For each channel, there are specific statistical laws of submitting and serving requests.

The simulation algorithm for the stop point functioning:

1. System initialization:

number of service channels;

the distribution law of the requests' incoming flow;

the distribution law of required time to serve request for each channel;

time for channel to become available is set to 0 (all channels are free).

2. To determine the random point time (appropriately distributed) when the request is submitted, the calculation of requests cortege attributes is carried out a_k as follows:

2.1. time of request submission $t_k^i = t_i$.

2.2. Determining the time when serve the request is begun. To do this, select the service channel with the least time to become available $t_j^\mu = Min(t_i^\mu), i=\overline{1,n}$.

Thus, the service start time:

$$t_k^s = \begin{cases} t_k^i, if\ t_k^i \geq t_j^\mu \\ t_k^\mu, if\ t_k^i < t_j^\mu \end{cases}, \tag{5}$$

2.3. To determine the time of service request completion.

$$t_k^o = t_k^s + \tau_j^\mu, \tag{6}$$

where: τ_j^μ - the required time to serve request at j channel, determined in accordance with a specific random law.

time for j channel to become available $t_j^\mu = t_k^o$.

3. Verification of the simulation completion condition $t_i > T$. If the condition is met, the simulation is completed; start calculate simulation results.

Otherwise, go to step 2.

Calculation of the functioning parameters of QS according to the results of simulation modeling is carried out as follows:

1. The probability of requests absence in the system:

$$P_0 = {}^{T_0}\!/_{T}\,, \tag{7}$$

where: T_0 - the total time where the system was in the state S_0 during the simulation;
T – total duration of the simulation.
2. Queue probability in QS:

$$P_{wt} = \sum_{i>n} T_i / T\,, \tag{8}$$

where: T_i - the total time where the system was in the state S_i
3. Average queue length:

$$l_{wt} = \frac{\sum_{i>n} iT_i}{\sum_{i>n} T_i}\,, \tag{9}$$

4. The total time when the system was in the state S_i

$$T_i = \sum_{s_{j-1,j}=i} (t_j - t_{j-1})\,, \tag{10}$$

4. Results analysis

Using the example of a specific stop point, we will formulate recommendations for its throughput capacity regulation. Based on the request service time sample, the parameters of the vehicles random service process at the stop point are given in table (1). It was found that, as in other studies, service channels have unequal performance: The average service time for a request increases with increasing channel number. The random process submits to gamma distribution.

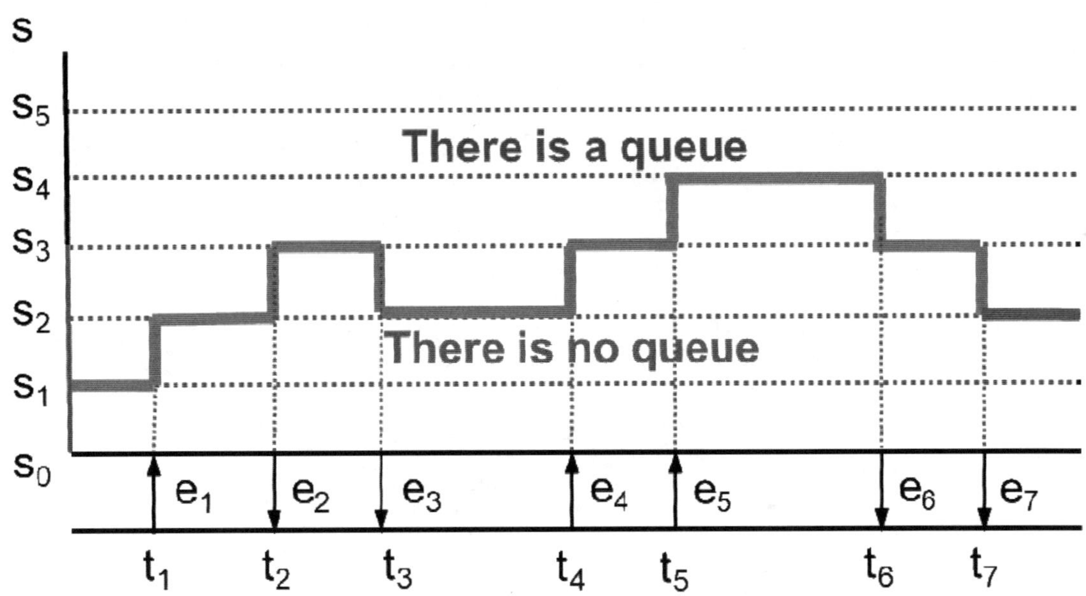

Figure 2. Diagram of changes in the state of the QS (with three channels);
where: s_j, $j = \overline{1,\infty}$ - QS status determined by the number of requests in the system;

For certainty, we consider the case of Poisson flow of requests arriving at the QS under consideration, although in general terms the proposed simulation model is applicable for any distribution law of random vehicles flow arriving to a stop point.

We will use the following machine algorithm to simulate random processes of submitting requests and their serving [8]:

$$x=\sum_{i=1}^{b}(-\frac{1}{\lambda}\ln R_i)=-\frac{1}{\lambda}\sum_{i=1}^{b}\ln R_i=-\frac{1}{\lambda}\ln(\prod_{i=1}^{b}R_i), \tag{11}$$

where: R_i - random numbers with uniform distribution.

We will use this machine algorithm (11) for both exponential (the simplest random process) and gamma distribution. In the case of the exponential distribution, b = 1.

Table 1. The distribution parameters of serving vehicles time at the stop point

Distribution parameters	Rate	Calculated value		
		Place I	Place II	Place III
Alpha (a)	9,098	8,9	9,2	9,6
Beta (b)	4,893	5,0	5,0	5,0
Expected value, c.		44,51	46,22	48,10
Standard deviation	14,6			
Dispersion	218,0			

Considering the simulation results of a stop point with varying the number of stopping places from 1 to 3. Figure 3 shows the dependences of the probability of the requests absence in the system P_o and the queue occurrence probability P_s for the considered options of stop point when the requests intensity varies from 5 requests per hour to the throughput capacity limit. The following limit values are obtained for the throughput capacity of the considered configuration options for a stop point: 80 units per hour for one stopping place, 155 units per hour for two places and 230 units per hour for three places.

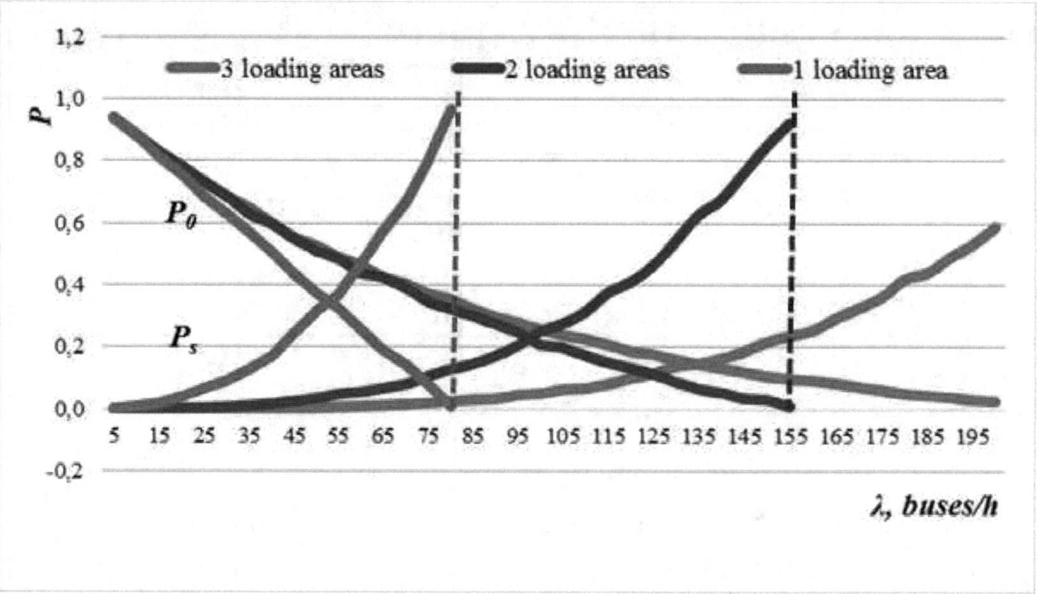

Figure 3- Parameters of the stop point functioning for a different number of stopping places (from 1 to

3 stopping places), where:

P_o - probability of requests absence in the system; P_s - the probability of queue occurrence.

Therefore, in practice, the application of the throughput maximum capacity of a stop point is unreasonable. At the stop point, a line of vehicles will constantly be present, which will block the road network. Wherefore, the stopping point must have a reserve of throughput capacity for the smooth functioning of the routes network.

Obviously, the ideal (most effective) mode of functioning is the absence of a queue at the stop point. The lines of vehicles at the stop point lead to negative processes that significantly re-duce the throughput capacity of the road network and the level of road safety. It is proposed to use the criterion of the queue absence to determine the stop point throughput capacity by using standard statistical probability levels (values) 10%, 5% and 1% of queue occurrence. A similar approach was applied in HCM 2010 [2], which regulate the probability of vehicle service denial (all stopping places are occupied) under the condition of normal distribution of passenger service time.

Table 3 shows the parameters of the throughput capacity of a stop point, obtained by the results of simulation, for the configuration of a stop point with one, two or three stopping places and probability levels 0.01, 0.05 and 0.1 of queue occurrence. the probability of the requests absence in the system ($\boldsymbol{P_o}$) and the probabilities of the queue of two ($\boldsymbol{P_{s2}}$)three ($\boldsymbol{P_{s3}}$) and four ($\boldsymbol{P_{s4}}$) requests are given. From the table we can see that the throughput capacity of a stop point at a probability level of queue occurrence $\boldsymbol{P_{s1}} = 0.01$ ensures the functioning of the stop point virtually without queues. Only 1% of the time at the stop point will be a queue, the length of which does not exceed one vehicle. At the probability level of queue occurrence $\boldsymbol{P_{s1}} = 0.05$ in the 1% of the time system, a queue of two requests will be observed. At the $\boldsymbol{P_{s1}} = 0.1$ assumes the system will operate with a queue in 10% of the time, at 2-3% of the time the queue will consist of two requests, and at 1% of the time - from three requests.

Thus, the throughput capacity of the system, based on their probability level of queue occurrence at 1%, should be established for cases where the queue of vehicles at the stop point has a significant effect on road traffic. For example, if the stop point is located immediately after the intersection, there is no possibility of forming a queue of vehicles on the road network. The probability level of queue occurrence of 10% is recommended to be used in cases where the queue of vehicles at a stop point does not lead to the occurrence of congestion on the road network.

To formulate the procedure for regulation the throughput capacity of stop points during the formation (optimization) of the transit program (Public transit supply):

- determine a list of critical stop points (stop points with the highest passenger traffic) for each of the directions of the route network;
- run a survey of these stop points to calculate the parameters of random processes of submitting and serving requests;
- use the analytical or simulation model, to calculate the throughput capacity parameters of the considered stop points;

Evaluate the calculation results for routes with insufficient stop points throughput capacity. The decision to be made is to adjust the transit supply, increase stopping places or reduce traffic intensity by increasing the capacity of fleet.

Table 2. Parameters of the stop point throughput capacity

Intensity λ, unit/h	Probability of requests absence P_o	Probability of queue occurrence			
		One or more requests P_{s1}	Two or more requests P_{s2}	Three or more requests P_{s3}	Four or more requests P_{s4}
One place					
11	0,872	0,01	0,001	0,000	0,000
22	0,726	0,05	0,008	0,001	0,000
31	0,617	0,10	0,024	0,006	0,001
Two places					
34	0,641	0,01	0,001	0,000	0,000
57	0,469	0,05	0,012	0,003	0,001
74	0,358	0,10	0,031	0,010	0,003
Three places					
62	0,447	0,01	0,002	0,000	0,000
100	0,256	0,05	0,015	0,004	0,001
120	0,189	0,10	0,037	0,014	0,005

5. Conclusion

1. The developed mathematical model, which consider the stop point of urban public transport as a multi-channel single-phase queuing system, allows us to determine the dependence of the stop point throughput capacity on the number of stopping places and the parameters of the random process of fleet serving, furthermore, regulate the intensity of fleet traffic over network sections.

2. The methodology for regulating the throughput capacity of urban public transport stop points, based on modeling the processes of boarding and alighting passengers by using the apparatus of queuing theory by means of the proposed method of simulation, allows us to solve this problem with various variants of the considered random processes.

3. According to the results of the calculations, it is recommended that the stop point throughput capacity is set according to the criterion of the queue absence, using standard statistical levels (values) of 10%, 5% and 1%.

References

[1] Efremov I S, Kobozev V M and Judin V A 1980 *Theory of urban passenger traffic: Textbook. Manual for universities* (Moscow: Higher School)

[2] Highway Capacity Manual 2000 *Transportation Research Board, National Research Council* (Washington, D.C., USA)

[3] Gudkov V A, Mirotin L B, Vel'mozhin A V and Shirjaev S A 2006 *Passenger road transport: the textbook for high schools* (Moscow: Hotline Telecom)

[4] Zedgenizov A V 2007 Estimation of the time of release of a stop point of urban passenger transport *Irstu Bulletin*, **4 (32)** pp 145–151

[5] Lipenkov A V, Kuzmin N A and Erofeeva L N 2015 Mathematical model of the capacity of a stop point in the absence of maneuvers to overtake each other's buses *Bulletin of the Orenburg University* **4 (179)** pp 87–94

[6] Dimova I P and Grachev V V 2008 Determining the capacity of stopping points at the current stage of passenger transport development *Bulletin of Irkutsk state technical University* **4 (36)** pp 66–70

[7] Fadeev A I and Fomin E V 2012 The issue of stopping capacity *Integrated Logistics, Publisher: All-Russian Institute of Scientific and Technical Information of the Russian Academy of Sciences* Moscow **2** pp 7–11

[8] Fomin E V and Fadeev A I 2012 The methodology for determining the throughput capacity of stopping points GPT *Bulletin of Irkutsk State Technical University* **4(63)** pp 117–124
[9] Eliseev M E, Lipenkov A V and Maslova O A 2012 On approaches to modeling bus dead time at bus stops in urban passenger transport *World of Transport and Technological Machines* **3** pp 84–93
[10] Sochnev A N 2012 Simulation of the movement of shuttle buses *Modern scientific research and innovation* **5** available online: http://web.snauka.ru/issues/2012/05/12603
[11] Buslenko N P 1978 Modeling complex systems *The main edition of physical and mathematical sciences* (Moscow: Science)
[12] Aliev T I 2009 Fundamentals of modeling discrete systems (SPb: SPbSU ITMO)

IOP Conf. Series: Materials Science and Engineering 1061 (2021) 012016 doi:10.1088/1757-899X/1061/1/012016

Mathematical model for studying the braking distance of a car equipped with ABS in winter conditions

A I Fedotov [1], V O Gromalova[1]

[1] Irkutsk National Research Technical University, 83 Lermontov st., Irkutsk 664074, Russia

E-mail: fai@istu.edu

Abstract. The article aims to develop a mathematical model for analytical studies of the braking process of a car equipped with an anti-lock braking system in order to identify the relationship between the stopping and braking distances of a car and pedestrian visibility in dark winter conditions, when headlights are contaminated with anti-icing materials. The mathematical model of the braking process of a car equipped with ABS was analyzed. Experimental studies of the braking process of a car with running and disabled ABSs were carried out on winter roads. The experimental studies of headlight contamination with chemical anti-icing materials (CAIM) and changes in luminous intensity were carried out in real road conditions in winter. The results are presented in the article. The results of the experimental studies of pedestrian visibility in dark conditions with contaminated headlights are presented. The adequacy of the mathematical model of the braking distance of a car with running and disabled ABSs in winter conditions was assessed. The trends characterizing the effect of headlight contamination on the pedestrian visibility in dark winter conditions are described. The results of the study are presented in tables, diagrams and equations. The values of the stopping distance were determined at different initial braking speed rates. The visibility distance for a car with contaminated headlights was determined in dark winter conditions.

1. Introduction

The task of improving vehicle safety on winter roads is crucial. According to the official statis-tics of the State Traffic Safety Inspectorate of the Russian Federation, every third traffic accident occurs at night [1].

The climate of many Russian regions is a factor of slippery roads. In winter, most of the roads are covered with snow and ice. One of the methods for eliminating snow-ice deposits (SID) is the use of CAIM [2]. At low temperatures, a "sandwich" from ice, snow, anti-icing materials and salt brine is formed on the roads [3].

In conditions of heavy traffic, anti-icing materials mixed with snow, salt brine and dirt are risen into the air by wheels and form an aerosol suspension contaminating car headlights. In dark conditions, the road illumination and pedestrian visibility are reduced.

The car has no systems informing drivers about the level of headlight contamination. Reduced visibility can be assessed intuitively and subjectively. This problem is aggravated by a low coeffi-cient of tire grip to the road covered with CAIM which increases the braking distance [4,5]. These factors reduce vehicle safety on winter roads covered with CAIM in dark conditions. This problem is especially acute on winter roads of Siberia with a sharply continental climate and a daily tem-perature difference of 20-25 ° C.

Using the developed mathematical model and experimental methods, the article aims to study the braking distance of cars on winter roads. The trends of the influence of headlight contamination on pedestrian and road infrastructure visibility are presented. The stopping distance of a car equipped with an anti-lock braking system is analyzed at varying initial braking speed rates.

2. Materials and methods

An analysis of papers studying the influence of CAIM on vehicle safety and operational properties shows that the effect of CAIM on the headlight contamination is not taken into account. Their influence on the road illumination, road infrastructure visibility and vehicle safety on dark winter roads is not taken into account [3, 6].

The results of 2018 and 2019 experimental studies conducted by the authors on winter Irkutsk roads covered with CAIM were published earlier. TOYOTA FUN CARGO and NISSAN QASHQAI were used in the experiments [7].

The results formed the basis of the method which can be used to study the effect of headlight contamination with CAIM on the possibility of stopping the car in front of an obstacle.

It should be noted that the braking distance is a component of the stopping distance along with the distance of the driver's response and the response time of the brake system (Fig. 1)

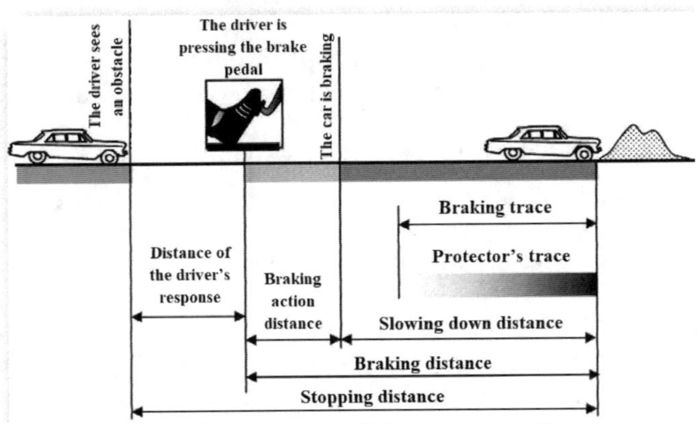

Figure 1. Components of the stopping distance

An experimental study of the braking performance was carried out on winter roads with a low grip coefficient, both with running and disabled ABS using TOYOTA FUN CARGO with a set of winter non-studded TOYO STUDLESS GARIT G4, 185/65 R14 tires. The studies were carried out on packed and not packed snow at a temperature of -20-25 °C, using the Effect device at initial speed rates of 10, 20, 40, 60 km/h [8]. The results are presented in Figure 2.

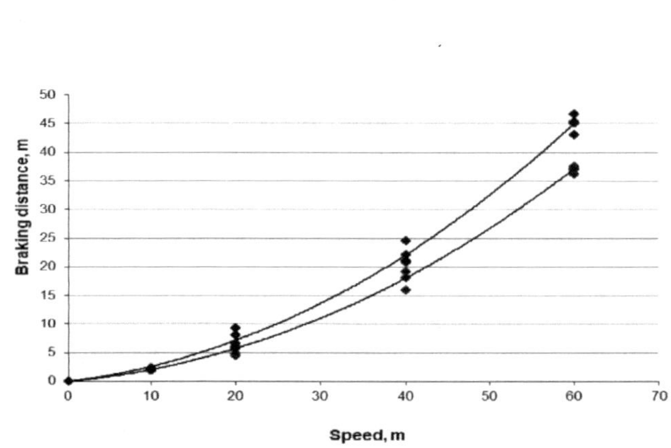

Figure 2. The braking distance graphs for Toyota Fun Cargo car with 185/65 R14 TOYO STUDLESS GARIT G4 tires on packed snow roads at -25 ° C: The upper graph - with the working ABS; the bottom graph - with the ABS disabled; ♦ experiment; ——— - calculation

An approximating dependence of the stopping distance of a car with a running ABS on the initial braking speed V_o was determined for the packed snow surface.

$$S_{T\,ABS} = 0{,}0099 \cdot V^2_0 + 0{,}1587 \cdot V_0 \qquad (1)$$

where $S_{b\,ABS}$ – the braking distance of a car equipped with ABS.

The approximating dependence (1) was obtained at the confidence approximation coefficient $R^2 =$ 0,99.

When a car with a disabled ABS is braking, the length of its stopping distance S_T can be described by approximating expression

$$S_{b\,ABS} = 0{,}0083 \cdot V^2_0 + 0{,}1183 \cdot V_0 \qquad (2)$$

The approximating dependence (2) was obtained at the confidence approximation coefficient $R^2 =$ 0,99.

The studies were conducted to analyze the braking performance of vehicles on winter roads covered with CAIM. For this purpose, the Automobile Transport department of Irkutsk National Research Technical University developed a mathematical model which presents the moving mass of the car as an oscillating system with a sprung mass in the form of a solid body with six degrees of freedom (Fig. 3, 4, 5).

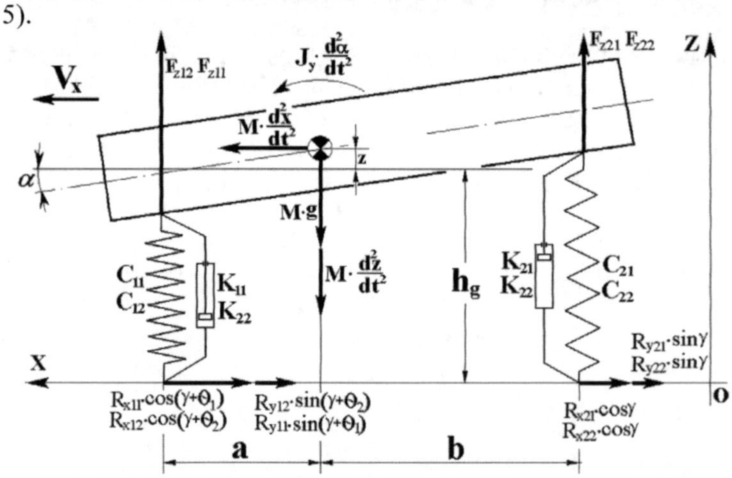

Figure 3. Design diagram of the car braking process (XOZ)

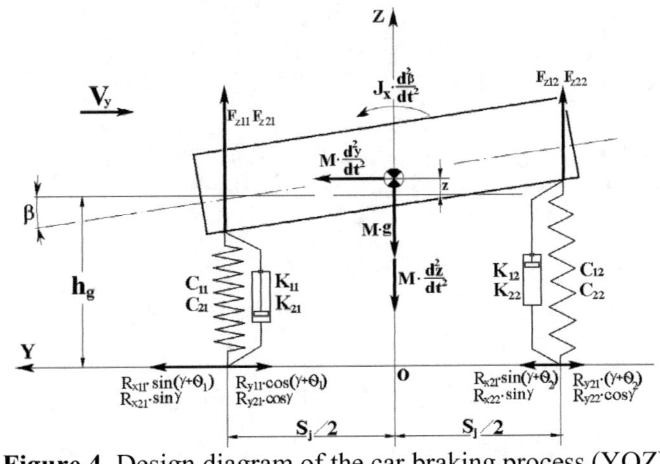

Figure 4. Design diagram of the car braking process (YOZ)

The equations describing the movement in space are built in relation to the moving (X, Y, Z) and fixed (X′, Y′, Z′) coordinate systems. The fixed coordinate system X′, Y′, Z′ is rigidly fixed to the road. The plane X, O, Y of the moving coordinate system lies in plane X′, O, Y′ and moves along it together with the car. The position of the moving system X, Y, Z with respect to the fixed system X′, Y′, Z′ is determined by coordinates x′, y′ and angle γ of its rotation relative to the OZ axis.

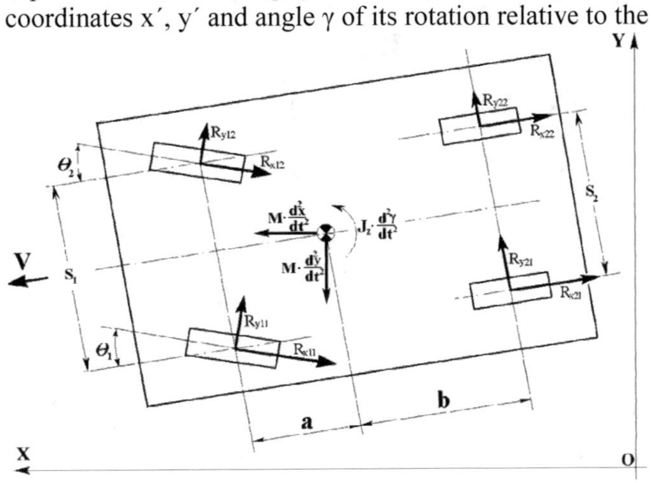

Figure 5. Design diagram of the car braking process (XOY)

The car body is connected with the moving coordinate system X, Y, Z so that its center of mass can perform only linear movements along the OZ axis, and the sprung mass M can rotate relative to the axes OX - (by the roll angle) and OY - (by the differential angle). Moreover, the roll center coincides with the mass center.

The position of the car body relative to the moving coordinate system (X, Y, Z) is determined by the coordinates:

α – the angle of rotation of the car body relative to the OY axis - (trim);

β – the angle of rotation of the car body relative to the OX axis - (roll);

z – the coordinate of the center of mass relative to the OZ axis.

The car suspension is shown in the form of stiffness C and shock absorbers having asymmetric compression and rebound characteristics. Damping of shock absorbers was determined by coefficients K_c – compression and K_r - rebound.

The differential equations of movement of the body mass written in relation to the moving X, Y and Z axes are as follows:

$$\frac{d^2x}{dt^2}=\frac{R_{x11}\cdot\cos\theta_1+R_{x12}\cdot\cos\theta_2+R_{x21}+R_{x22}+R_{y11}\cdot\sin\theta_1+R_{y12}\cdot\sin\theta_2}{M}-\frac{dy}{dt}\cdot\omega_z$$

$$\frac{d^2y}{dt^2}=\frac{R_{y11}\cdot\cos\theta_1+R_{y12}\cdot\cos\theta_2+R_{y21}+R_{y22}-R_{x11}\cdot\sin\theta_1-R_{x12}\cdot\sin\theta_2}{M}-\frac{dx}{dt}\cdot\omega_z \qquad (1)$$

$$\frac{d^2z}{dt^2}=\frac{F_{Z11}+F_{Z12}+F_{Z21}+F_{Z22}-M\cdot g}{M}$$

where: - R_{x11}, R_{x12} and R_{y11}, R_{y12} – projections of the longitudinal reactions of the right and left front wheels on the X and Y axes;

R_{x21}, R_{x22} and R_{y21}, R_{y22} – projections of the longitudinal reactions of the right and left rear wheels on the X and Y axes;

F_{Z11}, F_{Z12}, F_{Z21}, F_{Z22} – forces in shock absorbers and elastic suspension elements of each wheel;

M – mass;

θ_1 and θ_2 – angles of rotation of the right and left wheels.

ω_z – angular speed of rotation of mass M relative to the Z axis.

The differential equations of the angular movement of the mass relative to X, Y, and Z axes after solving them with respect to the highest derivatives:

$$\frac{d^2\alpha}{dt^2}=\{(F_{Z22}+F_{Z21})\cdot b-(F_{Z12}+F_{Z11})\cdot a+(R_{x11}\cdot\cos\theta_1+R_{x12}\cdot\cos\theta_2+R_{x21}+R_{x22}+R_{y11}\cdot\sin\theta_1+ R_{y12}\cdot\sin\theta_2)\cdot(h_g+Z)\}\ /\ J_y \qquad (2)$$

$$\frac{d^2\beta}{dt^2}=\{(F_{Z12}-F_{Z11})\cdot S_1/2+(F_{Z22}-F_{Z21})\cdot S_2/2+(R_{y11}\cdot\cos\theta_1+R_{y12}\cdot\cos\theta_2+R_{y21}+R_{y22}-R_{x11}\cdot\sin\theta_1- R_{x12}\cdot\sin\theta_2)\cdot(h_g+Z)\}\ /\ J_x \qquad (3)$$

$$\frac{d^2\gamma}{dt^2}=\{(R_{y11}\cdot\cos\theta_1+R_{y12}\cdot\cos\theta_2-R_{x11}\cdot\sin\theta_1-R_{x12}\cdot\sin\theta_2)\cdot a+(R_{y12}\cdot\sin\theta_2-R_{y11}\cdot\sin\theta_1+R_{x12}\cdot\cos\theta_2- R_{x11}\cdot\cos\theta_1)\cdot S_1/2-(R_{y21}+R_{y22})\cdot b+(R_{x22}-R_{x21})\cdot S_2/2\}\ /\ J_z \qquad (4)$$

where: - J_x, J_y, J_z – axial moments of inertia relative to the x, y, z axes of the sprung mass (body);
- a, b – the distance from the centers of the front and rear axles to the center of mass;
- S_1, S_2 – tracks of the front and rear wheels;
- h_g – the distance from the center of mass to the road surface.

The forces in the elastic elements of the suspension and in the shock absorbers were determined through the deformation rates of the elastic elements. The deformation rates of each elastic element are represented by equations taking into account the angular speed of the roll and deferent with respect to x and y axes, as well as the speed of movement of the center of mass along the Z axis:

$$V_{11}=\frac{d\beta}{dt}\cdot\frac{S_1}{2}+\frac{d\alpha}{dt}\cdot a-\frac{dz}{dt}$$

$$V_{12}=-\frac{d\beta}{dt}\cdot\frac{S_1}{2}+\frac{d\alpha}{dt}\cdot a-\frac{dz}{dt} \qquad (5)$$

$$V_{21}=\frac{d\beta}{dt}\cdot\frac{S_1}{2}-\frac{d\alpha}{dt}\cdot b-\frac{dz}{dt}$$

$$V_{22}=-\frac{d\beta}{dt}\cdot\frac{S_1}{2}-\frac{d\alpha}{dt}\cdot b-\frac{dz}{dt}$$

The deformations of the elastic elements of the suspension were determined by equation:

$$Z_{ijt}=Z_{ijt-1}+V_{ij}\cdot dt \qquad (6)$$

where: Z_{ijt} - deformations of elastic elements at a time t;

Z_{ijt-1} - deformations of elastic elements at a previous time $t-1$;

i - vehicle axis index: 1 - front, 2 - rear;

j - wheel axis index: 1 - front, 2 - rear;

$dt = 0,001\ c$ - step of integrating differential equations.

The forces in the shock absorbers and elastic elements of the suspension were determined taking into account the speed rates and deformations of the suspension and equations (5) and (6):

$$F_{Zij}=Z_{ij}\cdot C_{ij}+K_{ij}\cdot V_{ij} \qquad (7)$$

where: C_{ij} - stiffness of the elastic elements of the car suspension;

K_{ij} - shock absorber damping coefficients; they change their values depending on the suspension path (compression K_{cij} or rebound K_{oij}).

The moving (X, Y, Z) and fixed (X′, Y′, Z′) coordinate systems are connected by equations

$$V_x=\dot{x}\cdot\cos\omega_z-\dot{y}\cdot\sin\omega_z$$
$$V_y=\dot{x}\cdot\sin\omega_z-\dot{y}\cdot\cos\omega_z \qquad (8)$$

where: V_x - absolute speed of the center of mass relative to the X′ axis;

V_y - absolute velocity of the center of mass relative to the Y′ axis.

The wheel slip angles were determined by equations (9). The front wheel slip angles were determined taking into account the rotation angles of the wheels θ_1 and θ_2.

$$\delta_{11}=-\mathrm{arctg}\left[\frac{\dot{y}+\omega_z\cdot a}{\dot{x}+\omega_z\cdot S_1\cdot/2}\right]+\theta_1$$

$$\delta_{12}=-\mathrm{arctg}\left[\frac{\dot{y}+\omega_z\cdot a}{\dot{x}-\omega_z\cdot S_1/2}\right]+\theta_2 \qquad (9)$$

$$\delta_{21}=-\mathrm{arctg}\left[\frac{\dot{y}-\omega_z\cdot b}{\dot{x}+\omega_z\cdot S_2/2}\right]$$

$$\delta_{22}=-\mathrm{arctg}\left[\frac{\dot{y}-\omega_z\cdot b}{\dot{x}-\omega_z\cdot S_2/2}\right]$$

When choosing a mathematical model of the elastic tire, its peculiarity was taken into account - the shift (during braking) of the maximum of the longitudinal reaction Rx with an increase in the slip angle δ to the greater slippage S (Fig. 6). The operation of the anti-lock system was simulated. The ABS searches for maximum braking forces during the critical tire slippage and maintains the braking process at their level. Figure 6 shows the shift of critical slip S_{cr} at which the flat spot is disrupted, if the wheel moves with a slip angle. It can vary from 0.1 to 1.0.

Therefore, to describe the output characteristics of the elastic tire, the mathematical model developed by A.B. Dick [4,5] was applied. In the projection model of R_x and R_y, the longitudinal reaction of the braking wheel was determined using the normed slip function [4,5]:

$$f(s)=\sin\{a_1\cdot \mathrm{arctg}(b_1\cdot S)\} \qquad (10)$$

The normed function accurately describes *stationary characteristics* (Fig. 5) of car tires determined by processing the experimental grip characteristics.

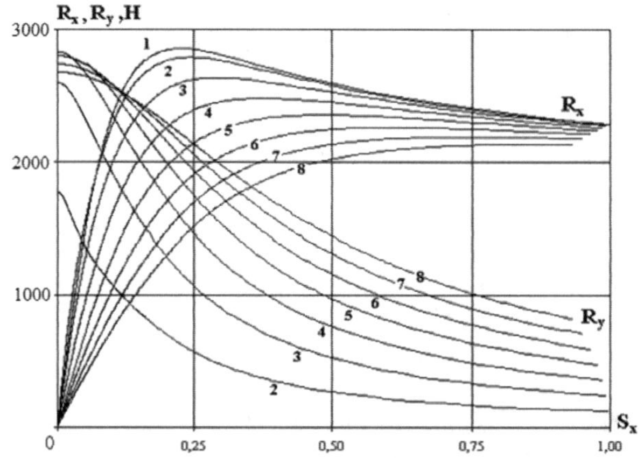

Figure 6. Dick's graphs of stationery characteristics of the 185/65 R14 TOYO STUDLESS GARIT G4 tire (R_z = 3000H), 1 – slip angle δ = 0°; 2 – slip angle δ = 5°; 3 – slip angle δ = 10°; 4 – slip angle δ = 15°; 5 – slip angle δ = 20°; 6 – slip angle δ = 25°; 7 – slip angle δ = 30°; 8 – slip angle δ = 35°

The power radius of the wheel (the rolling radius in the driven mode) r_w was determined taking into account the normal reaction R_z acting on the tire and the radius r_c of the unloaded wheel [4]:

$$R_w = r_c - (R_z \cdot C_2 + \sqrt{Rz \cdot C1}), \tag{11}$$

where C_1 and C_2 – coefficients adjusting the force radius function.

The total longitudinal reaction of the brake wheel in the contact spot was calculated by formula [4]:

$$R_\Sigma = R_z \cdot \varphi_{max} \cdot \boldsymbol{f(s)} \tag{12}$$

The projections of the longitudinal reaction of R_x и R_y on the x and y axes were determined as [4]:

$$R_x = R_\Sigma \cdot m_{\beta x} \cdot S_x / S$$

$$R_y = R_\Sigma \cdot m_{\beta y} \cdot S_y / S \tag{13}$$

The angular acceleration of the wheel was determined from the dynamic equilibrium

$$\text{equation: } \frac{d\omega_k}{dt} = \frac{Rx \cdot r_{ko} - M_T - M_f}{J_k} \tag{14}$$

where: $M_т$ - the braking torque;

M_f - the moment of rolling resistance;

J_k - the moment of inertia.

To describe the operation of the brake mechanism, we used a mathematical model that takes into account

- hysteresis losses in the brake;
- its inertia (*phase delay operation*);
- deadband of the brake depending on the gap in the friction pair;
- changes in the braking torque depending on the relative speed of movement of the elements of the friction pair.

The model is simple and accurate.

The model describes the inertia (*phase response delays*) of the brake using the first order dynamic link equation:

$$T_{mм} \cdot \Delta P_{mм} = P_m - P_{mм}, \tag{15}$$

where: $T_{mм}$ – a disc brake time constant;

$P_{mм}$ – fluid pressure in the brake;

P_m – fluid pressure at the brake inlet;

$\Delta P_{mм}$ – the rate of changes in fluid pressure pressure in the brake.

IOP Conf. Series: Materials Science and Engineering 1061 (2021) 012016 doi:10.1088/1757-899X/1061/1/012016

The nonlinear characteristics (dependences $M_t = f(P_{tm})$) were described by the piecewise linear approximation method. It takes into account the zone of insensitivity of the brake mechanism to fluid pressure pressure and the braking torque decreased to the grip moment when the wheel is blocked.

$$M_t = \begin{cases} 0, & \text{if } \Delta P_{тм} > 0 \text{ and } \Delta P_{тм} > \Delta_0 \text{ or } \Delta P_{тм} \le 0 \\ K_1 \cdot (P_{тм} - \Delta_0), & \text{if } \Delta P_{тм} > 0 \text{ and } K_1 \cdot (P_{тм} - \Delta_0) - M_{тр} > 0 \\ K_2 \cdot P_{тм}, & \text{if } \Delta P_{тм} \le 0 \text{ and } K_2 \cdot P_{тм} - M_{тр} < 0 \\ M_{тр}, & \text{if } \Delta P_{тм} > 0 \text{ and } K_1 \cdot (P_{тм} - \Delta_0) - M_{тр} \le 0 \\ M_{тз}, & \text{if } \Delta P_{тм} \le 0 \text{ and } K_2 \cdot P_{тм} - M_{тр} \ge 0 \\ R_z \cdot \varphi_{\max} \cdot f_б \cdot r_{к0}, & \text{if } \omega_к \le 0 \end{cases} \quad (16)$$

where Δ_0 – the zone of insensitivity of the brake mechanism to the increasing fluid pressure;

$M_{тз}$ and $M_{тр}$ – the braking torque during changes in a sign of the first pressure derivative in the actuator;

K_1 and K_2 – the rates of increase and decrease in the braking torque.

Rates K_1 and K_2 of increase and decrease in the braking torque are adjusted by changing the speed of the relative movement of the friction pair, or depending on the angular speed of wheel rotation:

$$K_1 = K_{10} - K_{\omega_1} \cdot \omega_к, \quad (17)$$

$$K_2 = K_{20} - K_{\omega_2} \cdot \omega_к, \quad (18)$$

where K_{10} and K_{20} – the coefficients of increase and decrease in the rate of changes in the braking torque at $\omega_к \to 0$;

K_{ω_1} and K_{ω_2} – the coefficients of adjustment of the rate of changes in the braking torque from the angular velocity of the wheel $\omega_к$.

The mathematical model simulates executive commands to trigger the ABS pressure modulator, depending on slip S and angular acceleration $\dot{\omega}_к$ of the elastic tire of the braking wheel.

The electronic control unit (ECU) opens and closes the solenoid shutdown and drain valves of the modulator using commands A and B. This changes fluid pressure pressure in the brake drive. Initially, the pressure shutdown valve A is open, and the drain valve B is closed.

With a three-phase ABS operation algorithm, the mathematical description of the ECU operation is as follows:

an increase in pressure: $$\frac{dP_{тм}}{dt} > 0 \text{ at} \begin{Bmatrix} A = 0 \\ Б = 0 \end{Bmatrix}, \text{ if } S < S_y \text{ and } \dot{\omega}_к < 0, \quad (19)$$

a decrease in pressure: $$\frac{dP_{тм}}{dt} < 0 \text{ at} \begin{Bmatrix} A = 1 \\ Б = 1 \end{Bmatrix}, \text{ if } S \ge S_y \text{ and } \dot{\omega}_к < 0, \quad (20)$$

pressure holding: $$\frac{dP_{тм}}{dt} = 0$$

$$\text{at} \begin{Bmatrix} A = 1 \\ Б = 0 \end{Bmatrix}, \text{ if } S \ge S_y \; \dot{\omega}_к \ge 0, \text{ or if } S < S_y \text{ and } \dot{\omega}_к \ge 0, \quad (21)$$

where $\dot{\omega}_к$ – angular acceleration.

The EC determines values S, S_y, $\dot{\omega}_К$ with errors and time delays whose values depend on the type of sensors, characteristics of the ABS computer, etc. All these time delays take into account the mathematical description of the modulator. At the same time, the operation of the computer was modeled without delays.

Thus, the mathematical model of fluid pressure pressure in the braking mechanism drive calculates pressure changes with time delays. These delays have a significant impact on the efficiency of the braking process of a car equipped with a running ABS.

The time delays are calculated by equations

$$\tau_m = \tau_m' + \tau_m'',\quad \tau_c = \tau_c' + \tau_c'',\quad \tau_в = \tau_в' + \tau_в'', \tag{22}$$

where τ_m, τ_c, $\tau_в$ – total time delays of the computer and the modulator for an increase, reset and pressure stabilization;

τ_m', τ_c', $\tau_в'$ – ECU-related delays;

τ_m'', τ_c'', $\tau_в''$ – modulator-related delays.

Figure 7 is a diagram illustrating the functioning of fluid pressure modulator by ABS commands and pressure delays in after the computer commands have been issued. The mathematical description of the pressure modulator was developed.

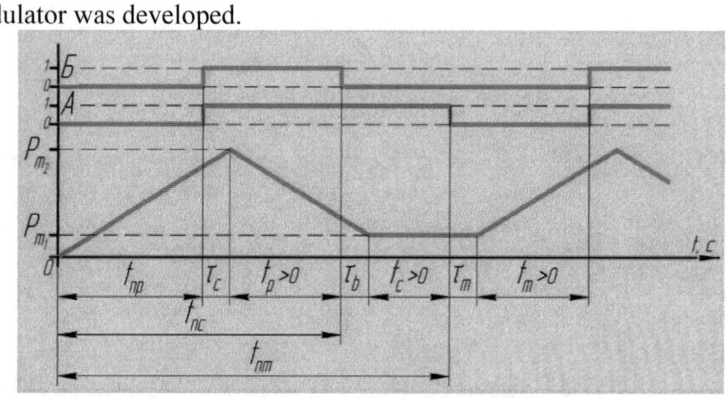

Figure 7. The pressure modulator schedule for three-phase ECU commands

An increase in K_t and a pressure release K_c were taken constant. They can be changed in the initial conditions of the calculation program. The mathematical description of fluid pressure pressure modulator is as follows:

$$P_m = \begin{cases} P_{m_1} + K_m \cdot t_m, & \text{if } A=0,\ Б=0,\ t_m>0 \\ P_{m_2} - K_c \cdot t_c, & \text{if } A=1,\ Б=1,\ t_p>0 \\ P_{m_1}, & \text{if } A=1,\ Б=0,\ t_c>0 \\ P_{m_{\max}}, & \text{if } P_m \ge P_{m_{\max}} \\ P_{m_{\min}}, & \text{if } P_m \le P_{m_{\min}} \end{cases}, \tag{23}$$

$$t_m = t_i - t_{nm} - \tau_m, \tag{24}$$

$$t_p = t_i - t_{np} - \tau_c, \tag{25}$$

$$t_c = t_i - t_{nc} - \tau_b, \tag{26}$$

where P_{m_1} and P_{m_2} – pressure values when changing the sign of its first derivative;

$P_{m_{\max}}$ and $P_{m_{\min}}$ – maximum and minimum pressure values in the actuator when $\frac{dP_{мм}}{dt} = 0$;

K_m and K_c –the rates of pressure increase and release;

t_m, t_p, t_c – current values of time of pressure increase, release and holding;

t_i – the time coordinate of changes in pressure in the fluid drive of the braking mechanism;

t_{nm}, t_{np}, t_{nc} – moments of time for issuing computer control commands to dump, lift and stabilize pressure.

The mathematical description of the braking process for a car equipped with ABS has a large number of nonlinear differential equations solved by computer numerical methods. The method of numerical integration of Euler differential equations was chosen.

At the next stage, the experimental results of the pedestrian visibility on night roads with contaminated headlights were processed in Microsoft Excel.

Figure 8 shows graphs of dependence of the pedestrian visibility distance on the headlight contamination level in dark winter conditions.

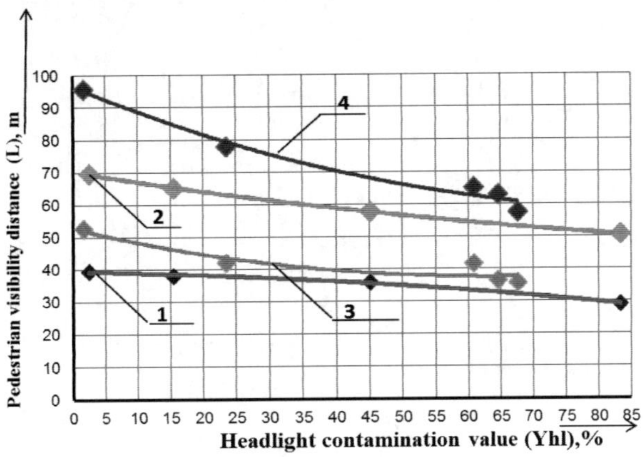

Figure 8. Graphs of dependence of the pedestrian visibility distance *(L_{pv})* on the headlight contamination level (Y_{hl}) for *TOYOTA FUN CARGO* and *NISSAN QASHQAI*.
1-the low beam mode of TOYOTA FUN CARGO; 2- the high beam mode of TOYOTA FUN CARGO; 3- the low beam mode of NISSAN QASHQAI; 4- the high beam mode of NISSAN QASHQAI.

The graph shows how the pedestrian visibility can be reduced if headlights are contaminated with CAIM.

The change in the TOYOTA FUN CARGO headlight contamination level from 2.5% to 83.5% decreases the pedestrian visibility:

- in the low beam mode: from 39.5 m. to 29 m.;
- in the high beam mode: from 69.5 m. to 50 m.

NISSAN QASHQAI headlight contamination reduces the pedestrian visibility from 1.7% to 67.6%:

- in the low beam mode: from 52,6 m to 35,6 m;
- in the high beam mode: from 95 m до 57,5 m

The largest braking distance on the packed snow surface for a car equipped with a running ABS at an initial braking speed of 60 km / h was 45 meters (Figure 2) [8].

According to formula (27), the stopping distance was calculated at a grip coefficient of 0.3 [4].

$$S_{stop} = (t_1+t_2+0.5\cdot t_3)\cdot \frac{V_a}{3.6} + \frac{V_a^2}{26\cdot g\cdot \varphi} \quad (27)$$

The grip coefficient of 0.3 is characteristic of the winter roads of Siberia covered with CAIM. The results of the experimental and analytical studies of the stopping distance are presented in Table 1.

Table 1. The results of the experimental and analytical studies of the stopping distance

Speed [km / h]	The stopping distance, [m] at a grip coefficient of 0.3				
	Calculated by formula (27)	According to the road tests		According to the mathematical model	
		With ABS	Without ABS	With ABS	Without ABS
0	0	0	0	0	0
10	5,19	5,3	5,3	5,1	5
20	13,0	13,0	12,51	12,2	12,12
30	23,4	21,9	22,79	19,7	23,31
40	36,4	33,1	36, 2	30,2	34,82
50	52,1	46,2	50,83	43,5	48,31
60	70,4	67,3	68,71	63,2	66,93
70	91,2	82,9	90,12	79,1	87,3
80	114,7	106,7	113,1	101,3	110,1
90	140,8	128,1	138,5	122,2	134,3

The results presented in Table 1 made it possible to build graphs of dependence of the stopping distance on the speed rate for packed snow roads (Fig. 9-11).

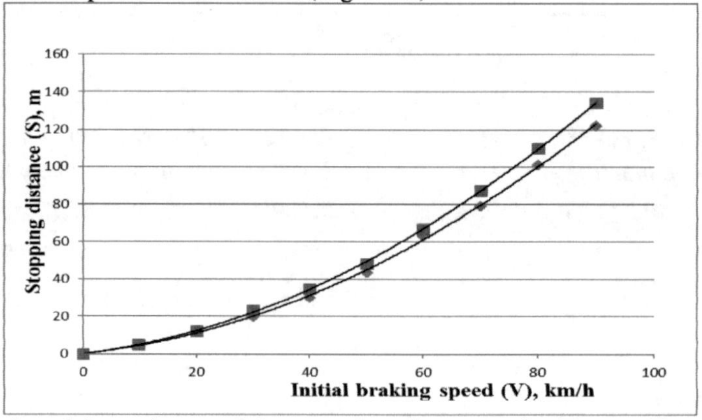

Figure 9. The graphs of dependence of the stopping distance of a Toyota Fun Cargo with winter 185/65 R14 TOYO STUDLESS GARIT G4 tires at the initial braking speed on the packed snow road (experiment): The upper graph - braking without an ABS; the bottom graph - braking with ABS

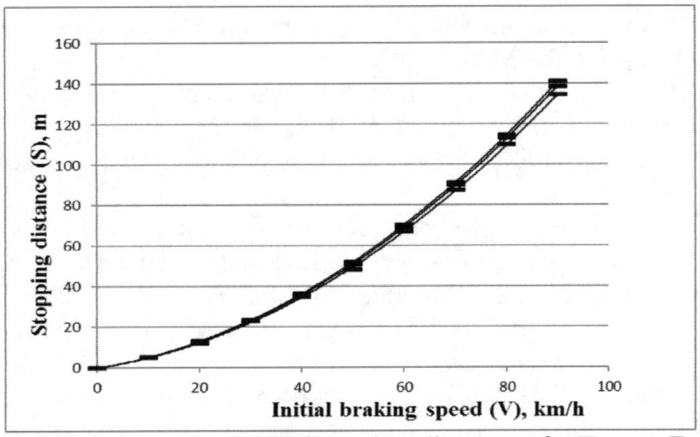

Figure 10. The graphs of dependence of the stopping distance of a Toyota Fun Cargo with winter 185/65 R14 TOYO STUDLESS GARIT G4 tires at the initial braking speed on the packed snow road (*braking without ABS*):

The upper graph was built using the mathematical model; the middle graph was built using formula (27); the bottom graph was built using the experimental results

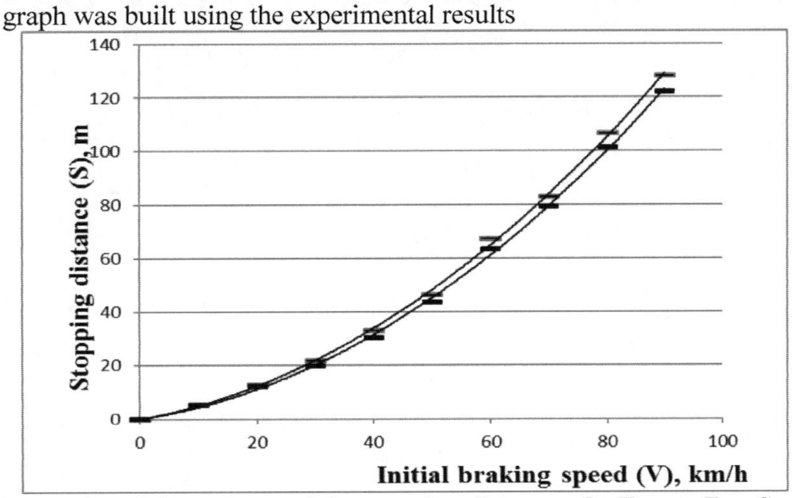

Figure 11. The graphs of dependence of the stopping distance of a Toyota Fun Cargo with winter 185/65 R14 TOYO STUDLESS GARIT G4 tires at the initial braking speed on packed snow road (*braking with ABS*): The upper graph is based on the experimental results; the bottom graph is based on the calculation results.

3. Conclusion

The results show that the mathematical model can calculate the stopping and braking distances for winter roads with an error not exceeding 3 - 5%. It can analyze the braking process of the vehicle equipped with ABS on winter roads.

The calculation results allow us to identify patterns of the influence of the initial braking speed on the stopping and braking distances of a car equipped with ABS on winter roads with a low grip coefficient.

The results make it possible to determine a safe speed rate of the vehicle under limited pedestrian visibility conditions due to the headlight contamination with chemical anti-icing materials.

References

[1] *Information on road safety indicators* 07.09.2017 available online: http://stat.gibdd.ru/
[2] IRC requirements for anti-icing materials Informavtodor 2003 (Moscow) p 72

[3] Vetrova V V 2006 Influence of anti-icing reagents on road conditions and traffic safety on the highway *Thesis: 05.23.11* MADI Moscow 136
[4] Dick A B 1985 Description of brake slip characteristics *Collection of scientific papers "Reliability and safety of vehicles*" MAMI pp 205–216
[5] Dick A B 1988 Calculation of stationary and non-stationary characteristics of the braking wheel *Thesis* Omsk p 228
[6] Balakin V D 2015 Expertize of the road accidents *manual Omsk SibADI* p 136
[7] Fedotov A I, Gromalova V O and Gergenov S M 2019 Influence of Headlight Contamination on the Stopping Distance and Maximum Allowable Speed on Roads Covered with Chemical Anti-Icing Materials Advances in Engineering Research, *International Conference on Aviamechanical Engineering and Transport* Volume 188
[8] Fedotov A I, Gromalova V O 2013 The effect of ABS on the braking performance and stability of the car *Bulletin of ISTU Issue* No. 8 Irkutsk pp 130–133

IOP Conf. Series: Materials Science and Engineering 1061 (2021) 012017 doi:10.1088/1757-899X/1061/1/012017

Ensuring priority of public urban passenger transport on urban road network

E V Fomin, K V Baklanova, E V Voevodin, A M Askhabov

Siberian Federal University, Krasnoyarsk, Russia

E-mail: 2325337@mail.ru

Abstract. Recently, in large and medium-sized cities, the increase in a load of highways is aggravated by the growth of motorization and the lack of development of the road network. An increase in the traffic intensity of vehicles to a critical level leads to an excess of the throughput of highways. The consequence is an increase in the level of road traffic delay. One of the most vulnerable groups is public urban mass passenger transport. This article discusses the methodology for determining the need to allocate a separate lane for the movement of public urban passenger transport on a given section of the road network. The purpose of this work is to identify patterns between the parameters of traffic flows and the parameters of the transportation program for public urban passenger transport, based on which criteria will be formulated for the need to allocate a separate lane for the movement of urban land transport on each separate section of the route network. For quality improvement of passenger transportation, a mathematical model has been developed, which is based on such indicators as the level of traffic delay and the share of the passenger flow in the total flow of traffic participants. Necessary conditions have been formulated, the strict implementation of which determines the need to ensure the priority of urban mass public transport on the considered section of the city's road network. The resulting dependencies allow determining the need to ensure the priority of urban transport. The proposed model makes it possible to determine the rationality of the allocation of individual lanes, traffic for public transport following the level of traffic in the city's road network. This study can become the basis for an algorithm for the operation of interactive signs that regulate traffic on a given stretch of the city's road network.

1. Introduction

For unload the city's road network, it is necessary to reduce the level of car traffic. One of the effective measures aimed at solving this problem is the task of increasing the attractiveness of urban mass passenger transport. That is, this task is the task of ensuring a sufficient level of passenger service, including the one consisting of a high speed of communication, which can be increased by giving priority to the movement of buses.

One of the promising directions for prioritizing public transport is the arrangement of separate lanes or streets for the movement of only the rolling stock of urban transport. In Krasnoyarsk, the first dedicated lanes for the movement of urban public transport appeared at the end of the nineties of the last century. Corresponding markings and road signs appeared on the central streets of the city. However, the above activities did not bring the desired effect. The reason for this is the massive violation of traffic rules by drivers of vehicles belonging to individual owners and taxi companies.

Most of the carriageway allocated for the movement of public urban passenger transport was operated in a parking space mode.

Consequently, bus drivers were forced to move in other lanes. Also, the problem was aggravated by the lack of control by the city administration over the regularity of urban passenger transport. "In pursuit of the ruble," drivers of urban passenger transport were forced to violate the speed limit established by traffic rules, endangering passengers.

With the advent of video recording complexes for traffic violations and complexes for monitoring the regularity of the rolling stock of public urban passenger transport, the situation has changed radically. The lanes allocated for the movement of urban transport began to be used for their intended purpose. Due to the lack of criteria according to which it is necessary to allocate separate lanes for the movement of urban transport rolling stock, they began to appear in the overwhelming majority on the main streets of the city. A large number of city highways appeared, used for the operation of only one route for the movement of urban public transport, with the organization of dedicated lanes on them. The need for such organizational measures is questionable; therefore, on some streets of the city of Krasnoyarsk, the allocated lanes for the movement of urban passenger transport are periodically canceled. Therefore, we need criteria that substantiate the need to allocate a separate lane for urban passenger transport on a given stretch of the city's route network.

Thus, the purpose of this work is to identify patterns between the parameters of traffic flows and the parameters of the transportation program for public urban passenger transport, based on which criteria will be formulated for the need to allocate a separate lane for the movement of urban land transport on each separately taken section of the route network.

2. Short review

There are active and passive methods of priority admission of urban mass passenger transport. The active methods include the methods of passing through the direct impact on the traffic light object [1].

This paper substantiates the feasibility of allocating lanes for the movement of urban mass passenger transport in general use.

The first step in domestic practice on the way of justifying the expediency of allocating lanes was the "Instruction on the organization of priority movement of public transport vehicles" approved by the USSR Ministry of Internal Affairs and the RSFSR Ministry of Automobile Transport in 1983[1].

According to this document, it is advisable to carry out priority traffic in the form of dedicated lanes if the following conditions are met:

- traffic intensity of public transport is not less than 40 units / hour;
- the traffic intensity of other vehicles (per one lane) is not less than 400 reduced units / hour;
- there are at least three lanes in one direction;
- the capacity of the road as a result of the allocation of a lane for public transport will be sufficient for the passage of other vehicles; traffic safety is not reduced, and an acceptable amount of delays is provided.

To date, the practical application of the instruction is impractical, since the given normative values of the motion parameters are outdated.

In the work of O.V. Popov, the main criterion for the expediency of allocating priority movement of urban public transport along the allocated lanes is the minimum total time spent on the movement of participants in the transport process. The author proposes to compare the capacity of a section of the road network with the traffic intensity of vehicles, excluding public transport [2].

Yu.D. Shelkov considers the following indicators to be the main factors for deciding on the possibility of allocating lanes intended for the movement of urban mass passenger transport[2]:

[1] Instructions on the organization of priority traffic of public transport vehicles. Approved by the USSR Ministry of Internal Affairs on 06/30/83, by the RSFSR Ministry of Transport on 06/28/83. Moscow: Transport, 1984, 32p.

- traffic intensity of public transport;
- traffic intensity of non-priority modes of transport;
- the number of lanes in the direction in question.

As the primary criterion for making a decision on the possibility of organizing dedicated bands in the studies of S.I. Smirnov advocates a reduction in the value of the total cost of delays of various types of vehicles. A mathematical description of the total cost of delays is given in the works of F.V. Akopov, N.O. Bludyan, R.S. Airieva, P.I. Kheifits and M.R. Yakimov [3–8].

Zyryanov V.V., Mironchuk A.A. in their study [9-11], they consider the development of methods for organizing priority traffic on the road network, namely, the concept of priority dedicated discontinuous lanes (PDDL). J. Vigas first described this concept in 1996. The PDDL system algorithm is carried out in the following stages:

- in the absence of buses on the road network section, all lanes are open for individual transport;
- when the bus approaches the PDDL at the calculated distance, the PDDL system is activated;
- activation is carried out by turning on controlled road signs;
- by the time the bus approaches, the priority lane is freed from individual transport; vehicles entering the PDDL section are prohibited from entering and changing lanes to the priority lane;
- the bus passes through the site in priority conditions;
- after the bus leaves the active section, the PDDL lane again becomes available for individual transport.

Assessment of the efficiency of allocated lanes for urban public transport is considered in the work of A.A. Fadyushina, D.S. Karmanov [12]. The assessment is carried out by determining the optimal parameters of the lane for public vehicles without widening the carriageway, and it is taken into account that for private vehicles, the traffic conditions should not worsen.

I.I.Shlippe, L.Yu. Chernobaeva, A.V. Akhterov consider in their work an approach to analyzing the effectiveness of creating dedicated lanes for city bus routes [13]. The assessment is made based on the economic assessment of the free time of the consumer of the transport service on the city public transport. The disadvantages of allocating a lane for public transport and damage to individual traffic are also considered.

The substantiation of the possibility of allocating lanes for the movement of urban passenger transport based on an integral indicator that takes into account the level of traffic on the road network and the level of passenger flow of the section under consideration was proposed in the studies of A.M. Belova [14-16].

In the works of scientists from the USA, Great Britain, South Korea and other countries, the following indicators are highlighted as the main factors for deciding on the possibility of allocating lanes intended for the movement of urban mass passenger transport [17]:

- the minimum value of the traffic intensity of priority types of transport;
- passenger traffic on the considered section of the road network.

In foreign practice, there are such network criteria as buffer index and buffer time or time buffer [18-22]. These parameters have become widely used for assessing the quality of traffic management in many countries.

Travel time index *TTI* is the ratio of the time taken by a vehicle to pass a section of the road network during peak periods to travel time under free-flow conditions:

$$TTI = \frac{T_{PP}}{T_{FF}} , \quad (1)$$

where TPP is the time taken by the vehicle to pass the section in rush hour conditions, min;

[2] Organization of traffic in cities: Methodological manual; Yu.D. Shelkova (ed.) / Research Center of the State Automobile Inspection of the Ministry of Internal Affairs of Russia. Moscow, 1995, 143 p.

TFF is the time taken by the vehicle to pass the section in free traffic conditions, min.

The time index TTI is mathematically quite simple to determine and allows assessing the impact of high loads on traffic conditions on urban and suburban roads.

The scope of the temporary index:

- assessment of the impact of high load on traffic conditions on segments of city streets and roads;
- assessment of the influence of the transport network load on time spent on moving along this network;
- comparative analysis of road conditions and the traffic quality management (TQM) in different cities or different areas of the city;
- assessment of the impact of high load on traffic conditions on country roads (access to cities, bypassing cities, main roads).

Table 1 contains data that allows evaluating the quality of traffic management on highways.

Table 1. Evaluating traffic conditions on city street and road segments

Service level	Time index value *TTI*	Driving conditions
A	<1,2	There is no deterioration in driving conditions during peak periods, excellent driving conditions.
B	1,21 – 1,3	During peak periods, there is a slight deterioration in traffic conditions.
C	1,31 – 1,5	During peak periods, there is a deterioration in traffic conditions.
D	1,51 – 2	During peak periods, there is a significant deterioration in traffic conditions, satisfactory traffic conditions.
E	>2,1	During peak periods, the segment does not function reliably. There may be congestion, poor traffic conditions.

The buffer time is estimated as the additional time T_b required to achieve the goal of movement with given reliability, for example, with a reliability of 90% or 95%. Accordingly, T_b is defined as the difference

$$T_b = T_{90\%(95\%)} - \overline{T}, \quad (2)$$

where $T_{90\%(95\%)}$ is the duration of movement with 90% or 95% availability;

$\overline{T}$ is the average travel time.

The indicator under consideration characterizes the reliability of the functioning of the urban road network or road network. In this case, the buffer time Tb can be used (using the cost of a passenger-hour, car-hour, and other factors) for estimating the economic costs that the user (driver or passenger) must bear in the form of additional time costs as a result of the unreliable functioning of the transport system.

The existing methods of justifying the need to allocate lanes for urban passenger transport are mainly based on indicators of traffic intensity and traffic capacity, as well as the values of passenger flow on a given section of the road network.

The considered methods do not fully characterize the level of existing delays. That is, these methods do not take into account the additional costs of waiting time associated with the regulation of traffic on a given section of the city's road network, which combines not only indicators of intensity and throughput, but also the parameters of traffic light regulation and the length of the queue. Also, the above methods do not take into account the share of passenger traffic in the total number of road users.

3. Materials and methods

To substantiate the need to allocate lanes for the movement of urban passenger transport, let us consider the section of the route network, along which vehicles move freely (without having priority), including the rolling stock of urban passenger transport.

It is not technically possible to designate a lane for urban traffic on roads with less than two lanes in one direction; therefore, the following condition must be met:

$$n \geq 2 \tag{3}$$

where n is the number of lanes in one direction on the stretch under consideration.

The movement of the rolling stock of urban passenger transport on the considered section of the route network in the general flow is considered. The movement of transport is associated with the limitation of the value of its throughput. The level of the throughput of urban passenger transport, which is not related to the technical operations of embarking and disembarking passengers, is mainly limited by the throughput of intersections of the city's road network and, as a rule, is associated with the indicator of traffic regulation delay. The delay indicator combines such indicators as the traffic intensity and throughput of the considered section of the road network, the length of the queue and the parameters of traffic light regulation. To ensure high speed of communication and increase the attractiveness of urban mass passenger transport, it is necessary to ensure an acceptable minimum level of delay.

The parameters of the movement of vehicles through the regulated intersection of the road network entirely depend on the traffic light regulation mode. That is, they depend on the duration of the regulation cycle, as well as on the duration of its constituent cycles, phases and the order of their alternation. In this case, the main movement of vehicles through the intersection is carried out during the period of the burning of the permitting traffic signal.

It should be noted that the beginning of the movement of vehicles at the moment when the permission signal of the traffic light comes on occurs with a certain delay (start delay) associated with the acceleration of vehicles and the necessary response time of the driver to the change of traffic signals. In this case, the traffic intensity of the flow of vehicles gradually increases to a value equal to the throughput of the considered direction [23,24,25,26].

At the moment when the prohibiting traffic light comes on, vehicles that do not have the technical ability to stop at the stop line continue their movement through the intersection, forming a "breakthrough" time.

Thus, the movement of vehicles through the regulated intersection of the road network begins somewhat later than the beginning of the burning of the permitting signal of the traffic light and ends during the period of the burning of the prohibited signal of the traffic light. The time of the actual implementation of the movement of vehicles through the intersection of the road network can be called the effective duration of the phase.

Consequently, the maximum (minimum) acceptable level of delay time for urban passenger transport, including when driving in free conditions, should be equal to the duration of the regulation cycle minus the time of the effective duration of the phase

$$Z = C - g_e \tag{4}$$

where: Z is delay time of urban passenger transport, sec;
C is the duration of the regulation cycle, sec;
g_e is effective phase duration, sec.

As practice shows, there are sections of the route network, traffic delays on which are acceptable or become acceptable at certain hours of the day. It is not uncommon for such tracks to have dedicated lanes intended for the movement of urban passenger transport. The presence of restriction (4) casts doubt on the need to allocate such lanes on the considered sections of the city's route network.

In the city of Krasnoyarsk, some stretches are part of only one route of urban passenger transport, where dedicated lanes operate to ensure its priority. The efficiency of work and the need to identify such bands are questionable. Ensuring the priority of urban passenger transport should be inextricably linked with the size of the passenger flow moving on the considered section of the road network. In this case, the size of the passenger flow should be commensurate with the number of participants moving in a free (without prioritization) flow along the maximum loaded traffic lane and moving in the same direction. Therefore, the following inequality is real:

$$Q \geq v_{cn} \cdot q_{mc}, \tag{5}$$

where: Q – the number of passengers traveling along the section under consideration per unit of time (for example, per hour), pass / hour;

v_{cn} – the intensity of the free flow per unit of time along the maximum loaded traffic lane (for example, per hour), given units / hour;

q_{mc} – the average number of passengers in a vehicle from a free flow of traffic (excluding urban passenger transport), pass.

Thus, taking into account the above, it is possible to formulate the following conditions, the presence of which determine the objective need to ensure the priority of public urban passenger transport on the considered section of the city's road network:

$$\begin{cases} n \geq 2; \\ Z \succ C - g_e; \\ Q \geq v_{cn} \cdot q_{mc}. \end{cases} \tag{6}$$

In the city of Krasnoyarsk, the route network spans were examined for their compliance with the conditions of restrictions (6). On the hauls of the route network with more than two lanes in one direction, at different times of the day, the following was recorded:

1. The level of transport delay, the duration of the regulation cycle and its components. For determine the level of delay, it is necessary to take into account vehicles arriving at the intersection of carriageways, but not having time to cross it because of the need to stop, and the total number of vehicles arriving at the intersection.
2. The power of the passenger flow was determined by the eye method directly on the stretch of the route network.
3. Traffic intensity and a number of passengers in the total traffic.

The examination revealed:

1. About 35% of the routes in the route network have dedicated lanes for the movement of urban passenger transport. The main number of spans with dedicated lanes appeared in preparation for the Universiade.
2. About 10% of the routes of the route network with dedicated lanes for the movement of route vehicles do not meet the requirements of restrictions (4). Consequently, the need to identify such bands is questionable.
3. About 5% of the routes in the route network with dedicated lanes for the movement of route vehicles do not meet the requirements of restrictions (5).

4. In between-peak and evening hours, the number of routes in the route network with dedicated lanes for the movement of route vehicles can be reduced (up to 45%).

5. About 6% of the runs of the city's route network, at peak hours, meet the conditions of restrictions (6). Therefore, they require the device of dedicated lanes for the movement of urban passenger transport.

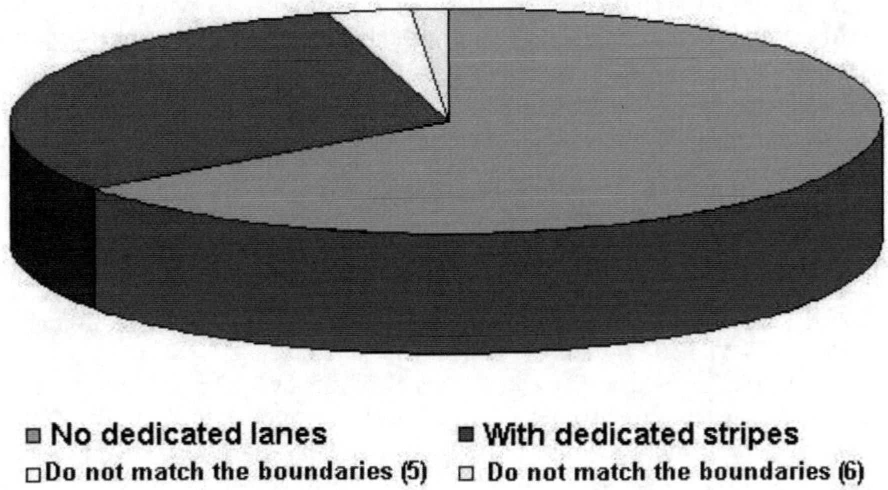

Figure 1. Results of the survey of the route network of Krasnoyarsk

Thus, the results of the survey carried out confirm the need to take into account conditions (6) when making decisions on the need to allocate separate lanes for the movement of urban passenger transport.

4. Conclusion

The throughput of urban passenger transport lines, not related to the process of embarking and disembarking passengers, is mostly limited by the throughput of intersections of the city's road network, which is significantly influenced by the parameters of random processes of vehicle arrival.

The proposed mathematical model forms the necessary conditions, the strict fulfillment of which determines the need to ensure the priority of public urban passenger transport on the considered section of the city's road network. The necessary conditions are based on such indicators as the level of traffic delay and the share of the passenger flow in the total flow of traffic participants.

The traffic delay level combines such indicators as traffic intensity and capacity of the considered section of the road network, queue length and regulation parameters. The amount of passenger flow is compared with the amount of the flow of participants moving in free mode.

Thus, the proposed model makes it possible to determine the rationality of the allocation of separate lanes, traffic for public transport under the level of traffic in the city's road network. This study, in turn, can become the basis for the algorithm for the operation of interactive signs that regulate traffic on a given stretch of the city's road network.

References

[1] Lytkina A A and Mikhailov A Yu 2011 The efficiency of using the priority of urban passenger transport at regulated intersections *Bulletin of the Irkutsk State Technical University* **7** (54) pp 60–65

[2] Popova O V, Gorev A E and Filimonova A M 2010 Improving the efficiency of using public transport through dedicated lanes *Motor transport enterprise* **8** pp 10–12

[3] Akopov F V, Bludyan N O and Ayriev R S 2014 An integrated approach to the organization of priority movement of ground passenger transport *Avtotransportnoe predpriyatie* **11** pp 5–9
[4] Akopov F V, Bludyan N O, Ayriev R S and Kheifits P I 2014 Demand regulation and integrated management of agglomeration passenger transportation *Automation and control in technical systems* **3** (11) pp 66–82
[5] Akopov F V, Bludyan N O, Ayriev R S and Real S 2015 Modern model of organizing multimodal interregional passenger traffic in the conditions of agglomeration *In the world of scientific discoveries* **6** (66) pp 233–242
[6] Yakimov M R 2011 Methodology for justifying the feasibility of allocating separate lanes for public transport on the road network of a large city *Bulletin of the Moscow Automobile and Road State Technical University (MADI)* 2 (25) pp 90–95
[7] Yakimov M R 2014 Quality indicators of the organization of road traffic in cities and methods of their assessment *Transport of the Urals* 2 (41) pp 90–92
[8] Yakimov M R 2011 Methods for analyzing infrastructural constraints of the public urban passenger transport system *Motor transport enterprise* 2011 pp 46–48
[9] Zyryanov V V and Mironchuk A A 2012 Priority movement of public transport: development of methods of organization *Transport of the Russian Federation* **3-4** (40-41) pp 22–25
[10] Mironchuk A A 2014 Development of the concept of priority lines of discontinuous action *Scientific Review* **7–3** pp 1050–1052
[11] Mironchuk A A 2013 Investigation of the influence of the level of development of the intelligent transport system on the effectiveness of the use of priority lines of discontinuous action *Engineering Bulletin of the Don* **3** (26)
[12] Fadyushin A A and Karmanov D S 2015 Features of the organization of road traffic in the central part of the city of Tyumen *Transport. Transport facilities. Ecology* pp 102–113
[13] Shlippe I I, Chernobaeva L Yu and Akhterov A V 2015 Analysis of the efficiency of creating dedicated lanes for city bus routes *Automation and control in technical systems (AUTS)* **3** pp 181–189
[14] Gorev A E, Popova O V and Filimonova A M 2010 Improving the efficiency of using public transport through dedicated lanes *Motor Transport Enterprise* pp 10–12
[15] Belova A M 2012 Basics of planning the organization of dedicated lanes for public transport *Bulletin of Civil Engineers* **6** (35) pp 123–129
[16] Belova A 2012 Technique bases of planning the organization of allocated strips for public transport movement *Bulletin of Civil Engineers* Vol 6 **35**
[17] Lytkina A A and Mikhailov A Yu 2011 Efficiency of application of the priority of urban passenger transport at regulated intersections *Bulletin of the Irkutsk State Technical University* **7** (54) pp 60–65
[18] Rumyantsev E A 2012 Improvement of methods for assessing the conditions of traffic flows on the urban street-road network *Bulletin of Irkutsk State Technical University* **9** (68) pp 148–152
[19] Sharov M I, Mikhailov A Yu and Kovaleva T S 2012 Assessment of the reliability of urban passenger transport in Irkutsk *Bulletin of the Irkutsk State Technical University* 9 (68) pp 174–178
[20] Sharov M I and Mikhailov A Yu 2014 On the issue of the development of a modern system of criteria for assessing the quality of the functioning of public passenger transport *Bulletin of the Volgograd State Technical University* Vol 9 **19** (146) pp 64–66
[21] Herman R and Ardekani S A 1984 Characterizing traffic conditions in urban areas *Transportation Science* **18** (2) pp 101–140
[22] Trofimov A V 2015 Time index - a criterion for assessing the impact of the traffic network load on the quality of its functioning *Bulletin of Irkutsk State Technical University* **10** (105) pp 181–185
[23] Levashev A G and Mikhailov A Yu 2004 The main parameters for assessing the throughput of controlled intersections *VNITI* **3** pp 21–26

[24] Akcelik R 1981 Traffic signals: capacity and timing analysis *Australian Road Research Board Research Report, ARR* **123**

[25] Olszewski P 1990 Modeling of Queue Probability Distribution at Traffic Signals *Transportation and Traffic Flow Theory, Elsevier Science Publishing Co., Inc., M. Koshi, Ed.* pp 569–588

IOP Conf. Series: Materials Science and Engineering 1061 (2021) 012018 doi:10.1088/1757-899X/1061/1/012018

Evaluation of the efficiency of the road network functioning

R N Gorbunov[1], Z V Gorbunova[1], L N Gorbunova[2], V S Kolchin[1], A I Karlina[1]

[1] Irkutsk National Research Technical University, 83 Lermontov St., Irkutsk, 664074, Russian Federation
[2] Higher School of Economics, 20 Myasnitskaya St., Moscow, 101000, Russian Federation

E-mail: gorbunow@list.ru

Abstract. Road transport has an impact to one degree or another on the life of every person. The article discusses approaches to assessing the costs caused by the deterioration of traffic conditions, which are subdivided into individual and social ones. Assessment of total costs should be carried out when adopting traffic management measures. The proposed approaches made it possible to estimate the individual costs caused by traffic management measures using the city of Irkutsk as an example.

1. Introduction.

Traffic management, its analysis and assessment are relevant research topics for the past years [1-9]. Intensive growth in the level of motorization in the context of the little-changing capacity of the existing road network, especially in the areas of historical development, leads to a decrease in the speed of communication in certain sections of road network, which is associated with the emergence of the concept of "transport congestion".

The term Congestion itself is borrowed from medicine and is used mainly in foreign specialized literature on the problems of urban road network functioning. Moreover, it is used in relation to both transport and the elements of transport infrastructure.

From the point of view of transport issues, the translation of the term Congestion into Russian is complicated, since this term refers to the set of states of the road network. The Joint Transport Research Center [10] provides the following definitions of the term "transport congestion" as:

1. The situation of excess demands for road space over supply.
2. Interference caused by vehicles against each other, in accordance with the dependence "speed-intensity" in conditions when the load on the road network approaches the value of its throughput.
3. The difference between the expected quality of the road network functioning and its real state.

Given the many formulations of the term, we can summarize them to the fact that the transport congestion is a load state of the road network, in which the volume of traffic approaches or exceeds the traffic capacity. [11]

The issue of the economic assessment of transport congestion is very important, since such an assessment will help in deciding on the need to introduce changes in the traffic organization on the city air traffic control and their consequences.

The term "transport congestion" describes the situation if the number of cars exceeds the capacity of the road. However, deterioration is possible not only in the case of such an excess, but also at a lower traffic intensity, as well as in case of changing the parameters of the road network functioning (changes in speed growth regime, placement of artificial irregularities, etc.). In the framework of the study, not only transport congestion was considered, but generally the deterioration of traffic conditions and the costs due to them was considered.

2. Materials and methods.

The costs incurred because of a decrease in the speed of communication are the sum of the costs of traveling in free conditions and the additional costs caused by the deterioration of traffic conditions. The costs in this study are understood as additional costs. Costs are borne by both road users and other people, companies and the environment.

Costs are divided into individual costs incurred by the participants in the movement, and social costs or negative impacts incurred by enterprises, the environment and other persons not involved in road traffic. The cost structure is given in table 1. It should be understood that some types of impacts can not always and not in full be compensated. Costs (financial losses) can also include lost revenue, for example, transport and logistics companies, taxi services and other organizations whose performance is reduced due to the deterioration of traffic conditions in urban road network.

Table 1. The structure of individual and public costs

Individual expenses	Public expenses
1. The time spent on movement;	1. The cost of maintaining roads;
2. Direct travel costs (fuel, etc.);	2. Environmental impact;
3. Indirect travel expenses (wear of parts, deterioration of clothing, etc.);	3. Impact on the psychophysical state of others (noise, dust, dirt, etc.);
4. Psychophysical costs;	4. Probability costs (for example, the risk of an accident, etc.);
5. Probable costs (for example, the risk of being late, the risk of pickpocketing in public transport, etc.);	5. Financial losses.
6. Financial losses.	

Cost estimation can be carried out both for each participant in the movement, and for road network as a whole, and the total individual and social costs can be determined according to (1, 2).

$$C_{ind} = C_{time} + C_{dir} + C_{indir} + C_{indpph} + R_{ind} + LR_{ind} \quad (1)$$

$$C_{pub} = C_{road} + C_{eco} + C_{pubpph} + R_{pub} + LR_{pub} \quad (2)$$

Each of the presented indicators can be determined on the basis of different approaches and have its individual dimension, for example, time costs are more easily estimated in seconds (minutes, hours), while direct costs are in rubles, and environmental impact can be estimated in m^3 of pollutants.

The calculation of total costs is possible in case of their economic evaluation by bringing all indicators to a single dimension – rubles.

Since in this study, costs are understood as additional costs caused by the deterioration of traffic conditions, such additional costs are the difference between the actual costs incurred and the potential costs that could be incurred in case of better conditions for the operation of the road network, for example, in the case of cars moving in free conditions .

$$C_{add} = C_{fact} - C_{ff} \quad (3)$$

For example, the time spent on traveling in the investigated road network section, caused by the deterioration of traffic conditions, is the difference between the actual duration of movement and the duration of movement in free conditions, can be graphically represented by a hatched figure, Figure 1.

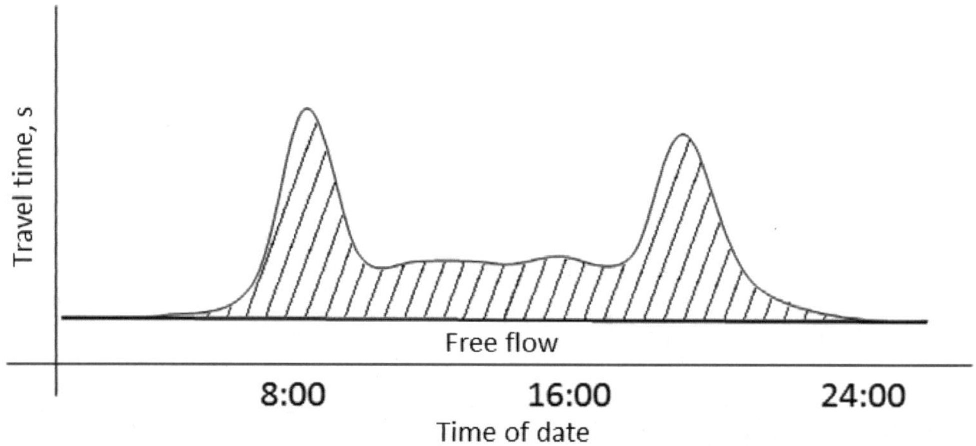

Figure 1. Time costs of traffic participants caused by difficulty in traffic conditions

Despite the apparent simplicity, the assessment of each of the indicators is very labor-intensive, since it requires the collection and processing of a large amount of data, individual indicators may be excluded from the estimated total costs, for example, the collection of data that is impossible or time-consuming and can be carried out in terms of aggregate and in terms of individual indicators.

3.Results and discussion.

The total costs incurred when traveling by road network by all participants for each type of cost represent the sum of the costs incurred by each participant in the road traffic and the amount of costs attributable to social impact, and can be represented graphically (Figure 2), where N is the traffic intensity vehicles in the road network section during the day T, and the Itching is the unit cost per vehicle, and it is assumed that the peak unit costs occur during peak periods of intensity.

Since the process of moving cars is continuous, the total costs will be determined by the volume of the figure shown in Figure 2, which is defined as:

$$C_{srn} = \iiint dCdNdT \tag{4}$$

In discrete form, total costs can be defined as a sum of individual and public costs (5). Such data can be obtained on the basis of geographic information received from road users.

$$C_{srn} = \sum_i C_{ind_i} + \sum_i C_{pub_i} \tag{5}$$

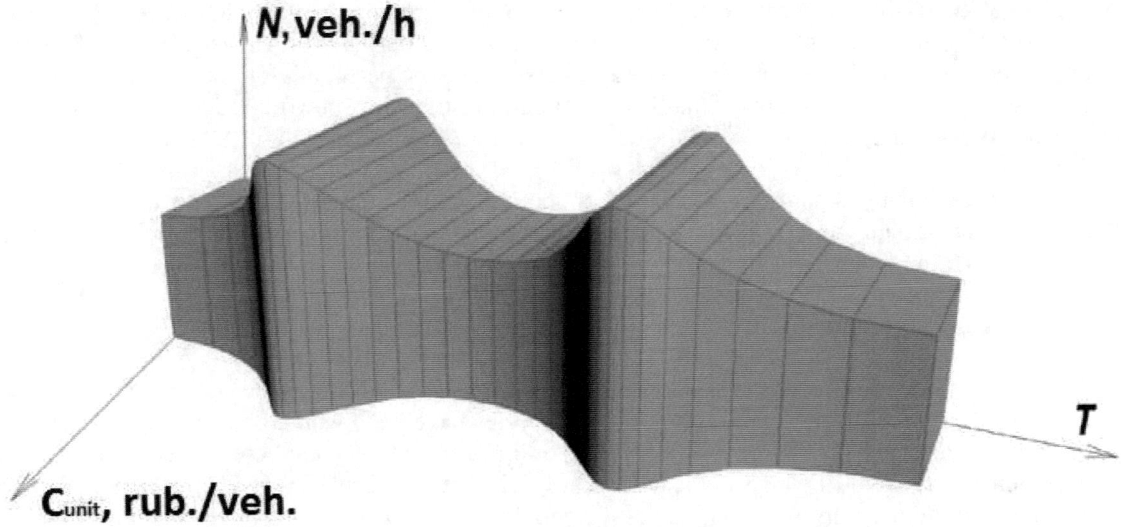

Figure 2. The total cost of movement on the road network

If it is necessary to calculate the total costs attributable to one road user, a number of questions arises when dividing public costs, since each participant has a different effect. It implies that the procedure for calculating the share of public expenses per 1 road user, requires a detailed analysis based on research and may be the subject of a separate study.

The assessment of individual costs is intuitively made by each participant in the road traffic and affects the choice of time for the start of movement, and the assessment of public costs in existing sections of the CDS should be carried out by state and municipal authorities responsible for the SDE, and design organizations at the stage of designing new sections of the CDS.

An integrated assessment of some indicators can be carried out analytically based on regulatory or tabular data. Actual assessment can be applied on the basis of geoinformation data received from participants in the movement, for example, on the basis of data from the Avtodata platform project.

The deterioration of the conditions of movements, first of all, affects the time costs, their monetary value is complex and can be carried out on the basis of average specific earnings.

$$C_{time_i} = \Delta T_i \cdot Sal_{s.a._i} \quad (6)$$

In addition, it is possible to assess the impact, both in the context of persons involved in road traffic, and in the context of vehicles, while the total costs of each road user can vary even in a single vehicle. When assessing costs, you should also pay attention to the fact that the value of some costs can take zero or negative values, for example, a passenger who rested during the trip will have a negative value of the psychophysical costs of C_{pph}, and those who did not pay for travel in public transport will not have direct costs.

Reducing specific public costs can be achieved using public transport. Public expenses are partially redistributed to the participants in the movement and are compensated by the state as taxes and other fees paid. Nevertheless, the psychophysical and environmental impact on the population remains uncompensated, so people who are not involved in road traffic are forced to bear the costs of road traffic on their own.

Since most of the movement to and from work falls on peak hours, a significant impact on the psychophysical state of people, as well as on the duration of rest, affects the duration of the trip.

Considering the described approach, it is possible to carry out an integrated assessment of the costs of movement, given below.

The effect of increasing the duration of movement on the mode of work and rest

Assuming an 8-hour working day with lunch, lasting 1 hour, as well as sleep, lasting 8 hours, the employee has 7 hours (420 minutes) per day to rest, carry out personal affairs and other activities. An increase in the duration of movement of only 10 minutes to and from work reduces the duration of the specified personal time by 4.7%, and the annual costs will be 4940 minutes or 82.3 hours, i.e. more than 3 days.

Using a similar valuation approach to commercial transportation, taking into account an 8-hour working day (480 minutes), reducing the duration of movement by 5 minutes will increase the efficiency of using working time by 1%.

Effect of 1 second delay

The traffic intensity in the sections of the road network changes during the day, for a more effective functioning, the traffic cycle of the traffic light object should be adjusted depending on the traffic intensity. Not on all traffic light objects, various cycles have been worked out in detail, depending on the time of day and the day of the week. In addition, in order to increase the efficiency of traffic management, it should be possible to temporarily change the regulation cycles in connection with a change in traffic intensity, for example, for the period of repair work, or in connection with a significant decrease in the intensity of both automobile and pedestrian traffic, for example the period of self-isolation of citizens for the reason caused by COVID-19.

The incorrect operation cycles of traffic lights can have a negligible effect on every road user. However, even a slight effect can be significant in cities. The population of the city of Irkutsk is about 623.5 thousand people. Daily average additional travel costs of only 1 second per city resident lead to a total time cost of more than 173 hours a day or more than 63 thousand hours (7.2 years) per year.

Environmental impact assessment

In accordance with GOST R 56162-2014 [12], in calculating of the mileage emissions of pollutants by road, a coefficient is used that takes into account the change in the amount of pollutants emitted depending on the average speed. When driving speed decreases from 60 to 30 km/h, this coefficient increases from 0.3 to 1, that is, emissions (excluding NxO) increase by 3.3 times, that is, there is more than 3-fold increase in environmental load per account of pollutants emitted, and when the speed decreases to 5 km/h from 0.3 to 1.4, that is, more than 4.6 times.

Thus, congestion not only increases the duration of movement, but also significantly increases the load on the environment.

Direct cost estimate

Direct costs incurred by road users can be divided into fare (taxi or public transport) and their individual direct costs in the case of using your own car. The change in travel costs in the case of a fixed fare, calculated not according to the duration of travel, does not change in the case of an increase in travel time, however, these costs will be borne by the carrier and either already taken into account in the fare, or their cost will be transferred to the fare in future. A consolidated estimate of direct costs may be based on fuel consumption.

Fuel consumption largely depends on the type and condition of the engines of vehicles involved in the movement. At the first stage, in order to calculate the averaged minimum additional costs, the specific idle fuel consumption can be used. As shown in [13] the average fuel consumption on cars with a maximum carrying capacity of up to 3.5 tons is about 0.85-1.2 liters/hour or an average of 2.86 · 10-4 liters/second. At a cost of 1 liter of AI-92 fuel in the city of Irkutsk, about 42 rubles per liter, the minimum cost per car will be 0.012 rubles per second or 43.2 rubles/hour.

To calculate the unit costs per person it is necessary to determine the average number of people per car. In many respects, this will depend on the availability of public transport in the road network surveyed area. For example, in Moscow, on average, from 1 to 1.7 people fall on 1 passenger car traveling along the road network, while taking into account the developed metro network and public transport in Moscow, which suggests that in cities where public transport is less developed, the number of people per 1 car traveling on CDS will be higher.

In addition, in some cases, for example, when calculating time costs, it is necessary to consider passengers of public transport. For our calculations of direct costs, we take the average value of 1.5 people per car. Thus, the unit cost per person will be 0.008 rubles per second, or 28.8 rubles/hour.

Impact assessment calming traffic on roads is used to increase safety

The adoption of measures to decelerate the movement is of a serious ethical nature, since it is assumed that, for the sake of the life and health of several people, others must sacrifice their time and health. For this reason, the choice of such measures must be approached reasonably. Measures of sedation include the installation of artificial irregularities and the reduction of speed limits on the sections of the road network, an assessment of the costs incurred by which will be given below.

Temporary loss on artificial bumps

In accordance with GOST R 52605-2006 [14], the speed of movement on artificial bumps should be reduced to 20 km/h. Assuming the maximum permissible speed V_0 = 60 km/h and acceleration during braking and acceleration a = 1.5 m/s2, the travel time during braking to V = 20 km/h will be:

$$t = \frac{V - V_0}{a} = \frac{\frac{20}{3.6} - \frac{20}{3.6}}{1.5} = 7.4s \tag{7}$$

A similar time will be required to accelerate to 60 km/h with the same acceleration, that is, the total cost of time will be about 14.8 seconds. In this case, the path l_1 traveled with a decrease in speed and the path l_2 traveled during its set will be 164.5 m:

$$l_1 = V_0 \cdot t + \frac{a \cdot t^2}{2} = \frac{60}{3.6} \cdot 7.4 + \frac{-1.5 \cdot 7.4^2}{2} = 82.3m \tag{8}$$

$$l_1 = V_0 \cdot t + \frac{a \cdot t^2}{2} = \frac{20}{3.6} \cdot 7.4 + \frac{1.5 \cdot 7.4^2}{2} = 82.2m \tag{9}$$

The same section of the track, in the absence of artificial irregularities, can be covered at a speed of 60 km/h in 9.9 seconds:

$$t_{60} = \frac{l_1 + l_2}{v_0} = \frac{82.3 + 82.2}{\frac{60}{3.6}} = 9.9s \tag{10}$$

Thus, the difference in the duration of the passage of the same section of passenger cars will be 4.9 seconds or about 0.5 hours per year. In the case of installing two artificial irregularities in a row, for example, as is done at pedestrian crossings with two-way traffic, in accordance with [14], the duration of movement will be longer.

Exceeding the maximum allowed speed

The maximum permitted speed on the roads of the Russian Federation is set in accordance with the traffic laws. Nevertheless, the technical capabilities of modern cars allow you to move at speeds higher than the established ones, and liability arises only when the permissible speed is exceeded by more than 20 km/h (non-fines), which many road users use, exceeding the maximum permitted speed is up to 20 km/h.

In connection with the discussed possibility of reducing the non-fined threshold, we can evaluate the effect of such a decrease on the duration of movement. A decrease in the non-fine amount of speeding will have a greater effect on the duration of movement in areas with a lower maximum permissible speed, Figure 3. Data on the change in the duration of movement for the most frequently set speeds on urban road network (40 and 60 km/h) are given in table 2.

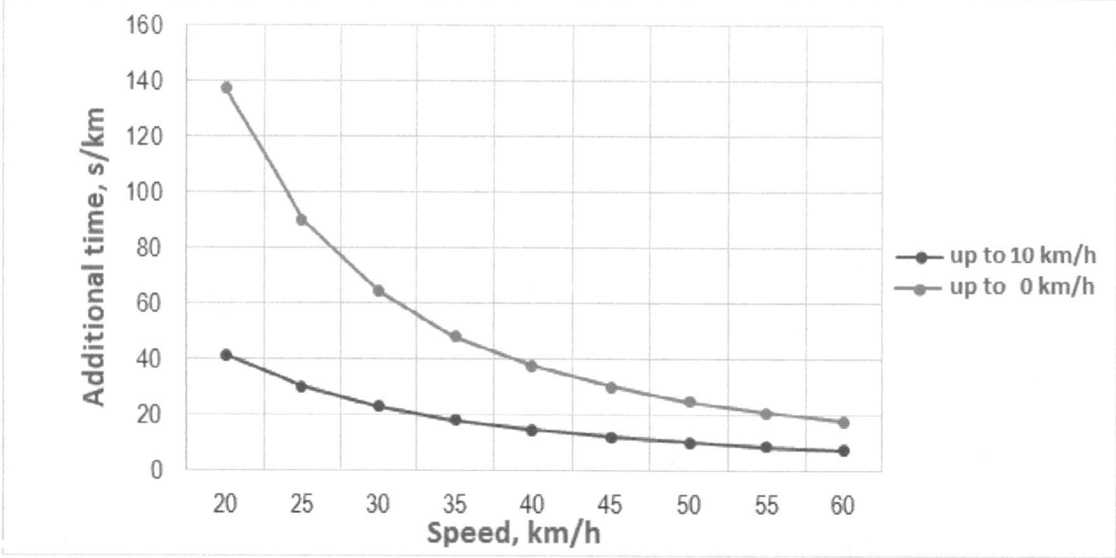

Figure 3. The dependence of the additional time spent on reducing non-fines over speed

Table 2. The effect of changes in non-fine speeding on the duration of movement

Maximum permissible speed, km/h	Non-standard excess, km/h	Actual speed of movement, km/h	Duration of movement in a section with a length of 1 km, s	Increase in duration of movement, s
40	20	55	65,5	0
	10	45	80	14,5
	0	35	102,9	37,4
60	20	75	48	0
	10	65	55,4	7,4
	0	55	65,5	17,5

Lowering the non-fine threshold to 10 km/h will increase the duration of travel at a maximum speed of 40 km/h by 14.5 s/km, and at a speed of 60 km/h by 7.4 s/km.

Setting the maximum allowed speed

Another way to calm the movement is to reduce the maximum permitted speed in individual sections of the road network. Speed reduction is carried out, as a rule, in the emergency sections of the road network to increase safety. An assessment of such a regulatory impact is possible at the planning stage. As an example, we can evaluate the changes in the speed regime adopted in the fall of 2019 in the city of Irkutsk at the dam of a hydroelectric power station with a length of 2.5 km, where the maximum permitted speed was reduced from 60 to 40 km/h.

The traffic intensity at the dam of a hydroelectric power station can exceed 4,000 aut./h in peak periods, table 3.

Table 3. Traffic intensity on the dam of the hydroelectric power station in the city of Irkutsk, vehicle/hour

Measurement Period	Direction		General Intensity
	to st. Baikalskaya	from the street Baikal	
in the morning rush hour	2370	864	3234
during the inter-peak period	1699	1037	2736
in the evening rush hour	2124	1950	4074

By extrapolating the data on the intensity, it is possible to obtain estimated data on the daily intensity of movement at the dam of the hydroelectric station, table 4.

Table 4. Estimated data of the daily dynamics of traffic intensity at the hydroelectric station dam, in the city of Irkutsk

Day period	0:00-6:00	6:00-7:00	7:00-8:00	8:00-9:00	9:00-10:00	10:00-17:00	17:00-18:00	18:00-19:00	19:00-20:00
Intensity auto/h	684	1492	2985	3234	2985	2736	3405	4074	3405

There are no artificial irregularities, unregulated pedestrian crossings, or other objects affecting the speed of vehicles at the dam. Speed control is carried out using cameras that take into account both temporal and spatial speeds, which gives reason to believe that most of the participants in the movement observe the speed regime, taking into account the excess of speed by a non-punishable value.

Assuming that the previously mentioned speed reduction from 60 to 40 km/h will affect only 60% of cars, taking into account the average number of people per car in the amount of at least 3 people, since there is active public transport on the dam of the hydroelectric power station, we get the average daily amount people P, which were affected by a decrease in speed in an amount of more than 90 thousand people/day:

$$P = 50308 \cdot 60\% \cdot = 90554\, people\, /\, day \tag{11}$$

The change in the duration of movement at a given speed and the total loss of time are shown in table 5.

Table 5. Estimation of time costs caused by a decrease in speed at the dam in Irkutsk

Indicators	Duration of movement, sec	Including excess by a non-fine amount, sec
At a speed of 60 km/h	150	120
At a speed of 40 km/h	225	164
Deviation	75	54
Quantity of people	90554	
	Additional expenses	
sec/day	6791550	4889916
h/day	1886,5	1358,3
day per day	78,6	56,6

The analysis showed that a decrease in speed from 60 to 40 km/h leads to additional time costs incurred by road users in the amount of more than 56.6 years each year, which is comparable to the duration of a human life. In addition, a decrease in the speed of movement, as has been shown earlier, significantly increases the environmental impact.

IOP Conf. Series: Materials Science and Engineering 1061 (2021) 012018 doi:10.1088/1757-899X/1061/1/012018

The obtained values of additional time costs can be estimated in monetary terms. One of the options for such an assessment is the assessment based on average earnings.

The average salary in Irkutsk is about 45 thousand rubles, while in 2020, according to the production calendar, there will be 1971 working hours at a 40-hour working week, which gives an average cost of 1 hour of work of about 274 rubles/hour (excluding holidays, taxes, contributions and other expenses related to remuneration) or 7.61 kopecks per second.

Estimated time and direct costs in monetary terms are given in tables 6.7.

Table 6. Estimation of time costs in monetary terms

Estimated indicator	Time costs		Unit cost, rub.	Cash Valuation of Time Costs	
	Unit of measurement	Value		Unit of measurement	Value
The impact of an additional 10 minutes of travel costs per person	hour/year	82,3	274	rubles/year	22 550
The effect of 1 second delay in Irkutsk	hour/year	63145	274	rubles/year	17 301 730
Losses on artificial irregularities (1 car passage)	seconds	4,9	0,0761	rubles/artificial roughness	0,3729
Reduction of non-fines to 10km/h at 40 km/h	seconds/km	14,5	0,0761	rubles/km	1,1
Reduction of non-fines to 10km/h at 60 km/h	seconds/km	7,4	0,0761	rubles/km	0,56
Decrease in the maximum allowed speed at Dam in Irkutsk	hour/year	495 816	274	rubles/year	135 853 584

Table 7. Assessment of direct costs in monetary terms

Estimated indicator	Time costs		Unit cost, rub.	Cash valuation direct costs	
	Unit of measurement	Value		Unit of measurement	Value
The impact of an additional 10 minutes of travel costs per person	hour/year	82,3	28,8	rubles/year	2 370
The effect of 1 second delay in Irkutsk	hour/year	63145	28,8	rubles/year	1 818 576
Losses on artificial irregularities (1 car passage)	seconds	4,9	0,008	rubles/artificial roughness	0,04
Reduction of non-fines to 10km/h at 40 km/h	seconds/km	14,5	0,008	rubles/km	0,12
Reduction of non-fines to 10km/h at 60 km/h	seconds/km	7,4	0,008	rubles/km	0,06
Decrease in the maximum allowed speed at Dam in Irkutsk	hour/year	495 816	28,8	rubles/year	14 279 501

As can be seen from tables 6.7, direct costs in value terms amount to 10% of the time. An enlarged economic assessment shows that an increase in the duration of movement to and from work by 10 minutes is comparable to costs (temporary + direct) of about 25 thousand rubles per year for each person, which corresponds to half the average monthly wage, which will lead to colossal costs, because, as shown in the study, a one-second delay per day for each resident of Irkutsk corresponds to costs of more than 19 million rubles per year.

A single passage of a car of a single artificial roughness increases the time by 4.9 s, which corresponds to costs of about 41.2 kopecks. In the case of daily traffic intensity of 10,000 cars/day, the additional costs caused by the daily passage of artificial bumps will exceed 2.25 million rubles per year.

A decrease in the size of the non-fine amount of speeding from 20 to 10 km/h means an increase in costs of 62 to 122 kopecks/km, which corresponds to 6-11% of the unit cost of operating a car, countries will lead to annual costs of billions of rubles.

An insignificant amount of costs, measured in kopecks, gives impressive amounts when converted to all participants in the movement. As it was shown, the decrease in the maximum permitted speed at the dam of the hydroelectric power station in the city of Irkutsk from 60 to 40 km/h corresponds to the cumulative increase in costs for the movement of participants in more than 56 years/year, which in monetary terms corresponds to more than 150 million. rubles/year.

The similar amounts of additional costs, calculated in the hundreds of millions of rubles on the road network section of a small length (about 2.5 km), confirm the need for an economic assessment when making decisions on the organization of road traffic.

Accounting for other types of costs indicated in table 1 will give an even greater total value of additional costs.

4. Conclusion.

Despite the enormous importance of economic costs, it should be emphasized that the issue of safety is a key issue in the organization of traffic. However, the most economically efficient improvements in road safety that can be achieved by various methods should be made taking into account the additional costs incurred, both individual and public.

Further research should be aimed at studying each of the types of costs, both individual and public, approaches to their assessment, as well as the development of a multifactor model of economic assessment of travel difficulties (congestion).

References

[1] Bennecke A, Friedrich B, Friedrich M and Lohmiller J 2011 Time-dependent Service Quality of Network Sections *Proc. of. the 6th Int. Symp.on Highway Capacity and Quality of Service* (Stockholm) pp 364-373

[2] Chang J S 2010 Assessing Travel Time Reliability in Transport Appraisal *Journal of Transport Geography*, Vol 18 pp 419-425

[3] Lyman K and Bertini R L 2008 Using Travel Time Reliability Measures to Improve Regional Transportation Planning and Operations *Transportation Research Record* Vol 2046 pp 1-10

[4] Mikhailov A Yu 2004 Integral criterion for assessing the quality of functioning of road networks *Izvestiya IGEA* (Irkutsk) Vol 2 (39)

[5] Pu W 2011 Analytic Relationships Between Travel Time Reliability Measures *Transportation Research Record* (Washington) Vol 2254 pp 122-130

[6] Rumyantsev E A and Mikhailov A Yu 2010 On the effectiveness of criteria for assessing the conditions of movement of traffic flows *Sat. Report 9th Int. Conf. Traffic Safety Organization* (St Petersburg) pp 121-123

[7] *Travel Time Reliability: Making It There On Time, All The Time,* United States Department of Transportation, Federal Highway Administration, available at: https://ops.fhwa.dot.gov/publications/tt_reliability/TTR_Reporthtm

[8] Yazici M A, Kamga C and Mousko K 2012 Analysis of Travel Time Reliability in New York

City Based on Day-of-Week Time-of-Day Periods *Transportation Research Record* (Washington) Vol 2308 pp 83–95

[9] *Travel time index,* Bureau of Transportation Statistics, available at: https://www.bts.gov/content/travel-time-index

[10] *Managing urban traffic congestion* Transportation research centre, availabled at: https://www.itf-oecd.org/sites/default/files/docs/07congestion.pdf

[11] Gorbunov R N, Mikhailov A Yu and Pirov Zh T 2017 Service level assessment based on reliability criteria *Bulletin of Irkutsk State Technical University* (Irkutsk) Vol 10 p 188-194

[12] GOST R 56162-2014 Air pollution emissions. Method of calculating vehicles emissions under summary computational modeling for urban settlements

[13] Zeer V A and Martynov A A 2008 Technique for rating the fuel consumption of vehicles with disconnected engine cylinders *Izvestia TPU* Vol 4 pp 45-48

[14] GOST R 52605-2006 Traffic control devices. Road bumps and road humps. General technical requirements. Application rules

Features of grinding surfaces of double curvature with cylindrical flap wheels

V P Koltsov[1], Le Tri Vinh[1], V B Rakitskaya[1], E V Tardybaeva[2]

[1] Irkutsk National Research Technical University, 83, Lermontov St., Irkutsk, 664074, Russia
[2]Baikal State University, 11, Lenin St., Irkutsk, 664003, Russia

E-mail: kolcov@istu.edu

Abstract. In the manufacture of the panels and sheaths of outer-forming parts, after milling operations or shot peen forming, grinding the surface with flap wheels is widely used. Surface grinding is often performed with flap wheels, which may have a cylindrical or profiled surface. Since profiled flap wheels are designed to handle surfaces with a specific curvature, cylindrical flap wheels are more universal and more often used. The main task of technological operation of grinding the complex surfaces is to ensure a same removal over the entire part surface. In this paper, an analytical study of the interaction features of the flaps of the cylindrical flap wheel along the width of the flap wheel when grinding the surfaces of double curvature is carried out. The analysis of possibilities for leveling the degree of influence of the flaps of the cylindrical flap wheel on the treated surface by re-overlapping the processing surface to ensure uniformity of grinding is carried out. A variant of grinding the surfaces of outer-forming parts with a cylindrical flap wheel with overlapping passages, which provides uniform treatment of the surface of double curvature, is proposed.

1. Introduction

In the manufacture of the panels and sheaths of outer-forming parts, after milling operations or shot peen forming, grinding the surface with flap wheels is widely used [2-12].The results of numerous experiments have shown that the material removal linearly depends on the deformation value (the bend value of the flaps from pressing the flap wheel to the treated surface) of the flaps of flap wheel within the studied limits [1, 6, 7]. Therefore, in order to obtain a uniform (same) material removal from the treated surface of the panel and sheaths, it is necessary to ensure a constant value of deformation while maintaining the remaining operating parameters. However, due to the curvature of the panel and the cylindrical shape of the flap wheel, the same deformation is not possible across the entire width of the contact of flaps with the treated surface. Thus, it is necessary to find a compromise solution that will minimize the uneven removal along the width of the flap wheel. The way out in this case is to re-pass (partially overlapping) the trace of the previous processing at the next passage. In the repeated (following) passage, the amount of overlapping is chosen so that the sum of flap deformation in the overlapping zone of the two passages was approximate equal to the maximum flap deformation in the central zone of the flaps of each passage that will minimize the deviation of flap deformation across the width of the processing and thereby to provide more same material removal from the machined surface (fig.1a).

IOP Conf. Series: Materials Science and Engineering 1061 (2021) 012019 doi:10.1088/1757-899X/1061/1/012019

2. Formation of flap deformation during grinding panels of double curvature with straight profile flap wheels

The Let's consider a general case of grinding with the flap wheels with overlapping of the flap wheel passages for a panel that has a variable curvature in transverse direction. In this case, we assume that the largest amount of deformation *T* (approximately in the center of the flap wheel) is constant for each passage.

Fig. 1 shows the model of flap deformation when grinding the panel with different radius of curvature, where after the first passage of grinding the section with a radius of curvature R_1 with the deformation T (position 2 in fig. 1a) the flap wheel proceeds to perform the second passage of the other section with a radius of curvature R_2 with the same deformation T without changing other operating parameters (position 3 in fig. 1a) with an overlapping of width *bb*.

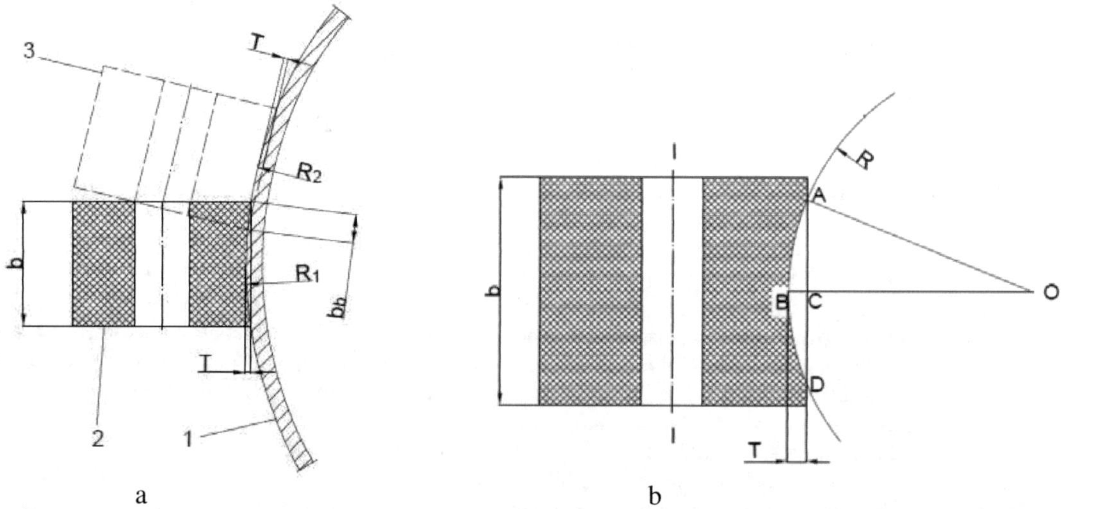

Figure 1. Grinding the aircraft panel with flap wheel, where: a) the position of the flap wheel during grinding, b) to determine the width of the processing, 1 - panel, 2 - flap wheel with width of b, 3 - the position of the flap wheel in the next passage with an overlapping with width of b_b.

From Fig. 1b it is not difficult to notice that the width of processing b0 for one passage of the flap wheel at the flap deformation value of T is determined by the following expression:

$$b_0 = AD = 2\sqrt{R^2 - (R-T)^2} = 2\sqrt{2RT - T^2} \quad , \qquad (1)$$

where: *T* is the maximum deformation of the flaps in the central zone of the flap wheel, *R* is the radius of curvature in the panel grinding zone of this passage.

Fig. 2 shows the model of the overlaying of deformation curvature of the flaps of the second passage L_2 on the deformation curvature of the flaps L_1 obtained in the first passage, since the radius of curvature (more 4m) is much higher than the value of flap deformation (≤4mm) and during grinding the flap wheel is directed normal to the panel surface in the contact zone, so for the convenience of analysis of the interaction of the flap wheel with the processing surface the model of overlapping passes conditionally constructed in a straight line AH (Fig.2).

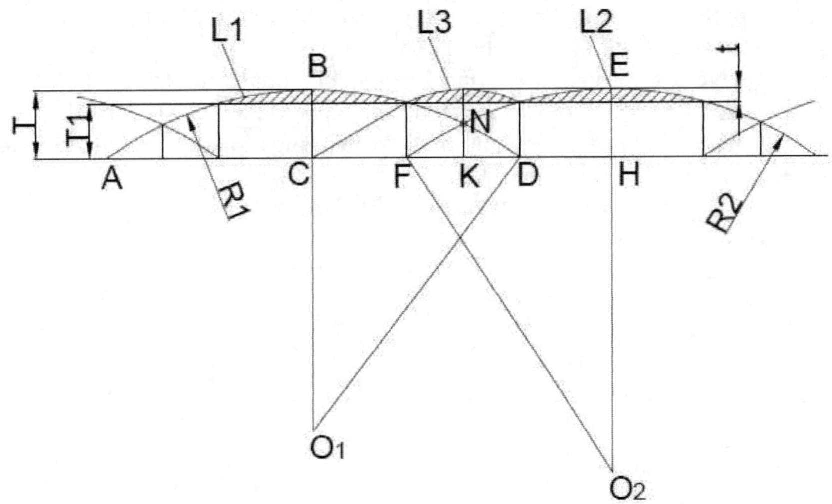

Figure 2. Formation of flap deformation during grinding with overlapping passages.

Let's assume that after the 1st passage of the flap wheel with the deformation value of *T*, a processing stripe with the width *AD* is obtained (Fig. 2). It is obvious that in the center of the flap at the highest *T*, the impact of the flap wheel on the surface of the part is greatest, and when moving away from the flap center to the edges, it decreases to a minimum and the value of material removal decreases accordingly. To align the removal of the next passage of the flap wheel, the grinding should be carried out with the overlapping of the traces of processing of the previous passage, so that the degree of impact on the treated surface was with the smallest deviations. Since the amount of removal during grinding is proportional to the flap deformation [1,6], we assign a point N (the intersection of the circles R_1 and R_2) at a distance of *T/2* from the line *AD* (see Fig. 2). In this case, the total amount of deformation at point *N* will be equal to *T*. When you move from the line *NK* to points *F* and *D*, the sum of the deformation values will decrease, approaching the value *T1* and forms the line *L3*, which is the arc of the sector *L3*. Accordingly, arcs R_1 and R_2 with a line drawn at a distance *T1* parallel to *AD* also form sectors (shaded). The distances of the arcs *L1*, *L2*, and *L3* from this straight line represent fluctuations in the amount of deformation *t*, which in accordance with [1,6] are proportional to the amount of removal and characterize the uniformity of grinding.

The width of the overlapping for the subsequent passage in accordance with Fig. 2 is defined as follows:

$$b_{ov} = FD = FK + KD = (FH\text{-}KH) + (CD\text{-}CK)$$

From Fig. 2 it follows that:

$$b_{ov} = FD = \left(\sqrt{{R_1}^2 - (R_1 - T)^2} - \sqrt{{R_1}^2 - \left(R_1 - \frac{T}{2}\right)^2}\right) + \left(\sqrt{{R_2}^2 - (R_2 - T)^2} - \sqrt{{R_2}^2 - \left(R_2 - \frac{T}{2}\right)^2}\right) = \sqrt{2R_1T - T^2} + \sqrt{2R_2T - T^2} - \sqrt{R_1T - \frac{T^2}{4}} - \sqrt{R_2T - \frac{T^2}{4}} \quad (2)$$

Then the width of the main stripe (the stripe processed without overlapping the stripe of first passage), taking into account the radius of curvature R_1 and the smooth change of the radii of curvature to R_2, can be determined by the following expression:

$$b_{o6} = 2(CD - FD) = 2\left(\sqrt{R_1T - \frac{T^2}{4}} + \sqrt{R_2T - \frac{T^2}{4}} - \sqrt{2R_2T - T^2}\right) \quad (3)$$

Similarly, you can calculate the width of the main processed stripe (without overlapping) for the second and subsequent passages with different curvature radii.

Thus, due to an overlapping in point with flap bend value equated to 1/2 of the deformation T for subsequent passages, we obtain a more smoothed treated surface with a profile bounded by the curvature curves of the panel and curves in the overlapping zones with a maximum deviation proportional to t (see the shaded areas in Fig. 2).

The maximum amount of deformation deviation t (Fig. 2) is defined as follows:

$$t = T - T_1$$

At the same time:

$$R_1 - T + T_1 = \sqrt{R_1{}^2 - CF^2},$$

so:

$$t = T - T_1 = R_1 - \sqrt{R_1{}^2 - CF^2} = R_1 - \sqrt{R_1{}^2 - (CD - FD)^2} = R_1 - \sqrt{R_1{}^2 - \left(\sqrt{2R_1T - T^2} - \sqrt{2R_1T - T^2} - \sqrt{2R_2T - T^2} + \sqrt{R_1T - \frac{T^2}{4}} + \sqrt{R_2T - \frac{T^2}{4}}\right)^2},$$

It means that:

$$t = R_1 - \sqrt{R_1{}^2 - \left(\sqrt{R_1T - \frac{T^2}{4}} + \sqrt{R_2T - \frac{T^2}{4}} - \sqrt{2R_2T - T^2}\right)^2}. \quad (4)$$

It is easy to see that the curves *L1*, *L2*, and *L3* (Fig. 2) change in shape close to the sinusoidal law, then the height of the segment of these curves is expressed by the following general formula:

$$t_i = t \sin \propto,$$

where: t is the maximum deviation, $\alpha \in (0 \div \pi)$

The average deviation t_{av} in this case can be determined as follows:

$$t_{av} = \frac{1}{\pi}\int_0^{\pi} t \cdot \sin\alpha = -\frac{1}{\pi} t \cdot (\cos\pi - \cos 0) = \frac{t}{\pi} \cdot 2 = 0{,}636 \cdot t \quad (5)$$

For the second passage, the flap wheel must move to the next distance s (Fig.2):

$$s = KC + KH = \sqrt{R_1T - \frac{T^2}{4}} + \sqrt{R_2T - \frac{T^2}{4}} \quad (6)$$

Fig. 3, 4, 5 and 6 showed the graphs of width change of main processed stripe, overlapping stripe width and offset of the flap wheel to complete the following passage depending on the radius of panel curvature and the magnitudes of flap deformation of flap wheel in the case of $R_1 = R_2$.

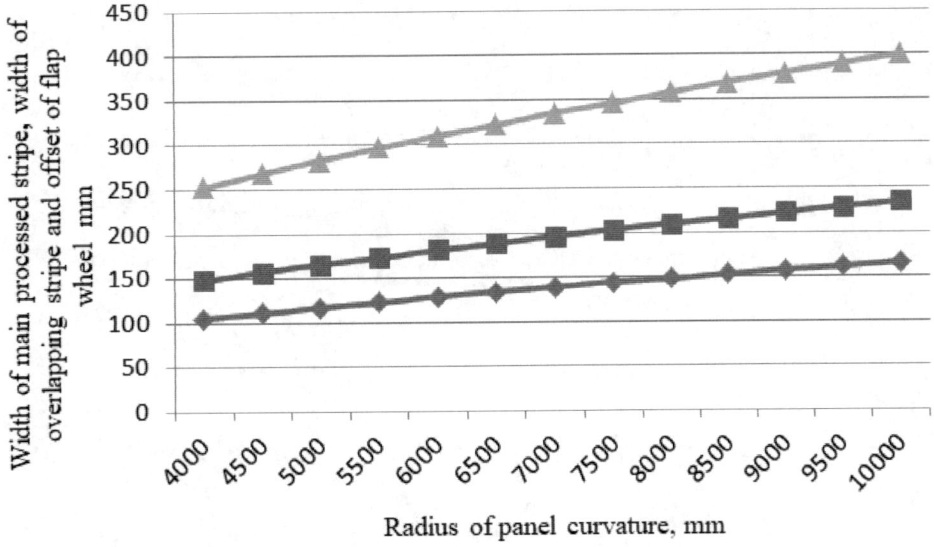

Figure 3. Dependence of width of overlapping stripe, width of main processed stripe and displacement of flap wheel depending on curvature radius at deformation value of 4mm, where: bottom curve - width of overlapping stripe, middle curve – width of main processed stripe, top curve- offset of flap wheel for following grinding passage.

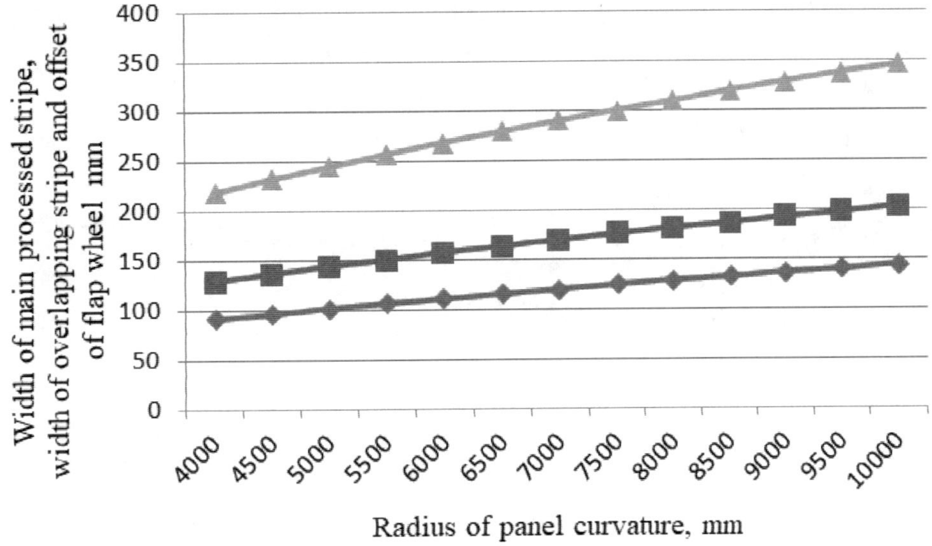

Figure 4. Dependence of width of overlapping stripe, width of main processed stripe and displacement of flap wheel depending on radius of curvature at deformation value of 3mm, where: bottom curve - width of overlapping stripe, middle curve – width of main processed stripe, top curve- offset of flap wheel for following grinding passage.

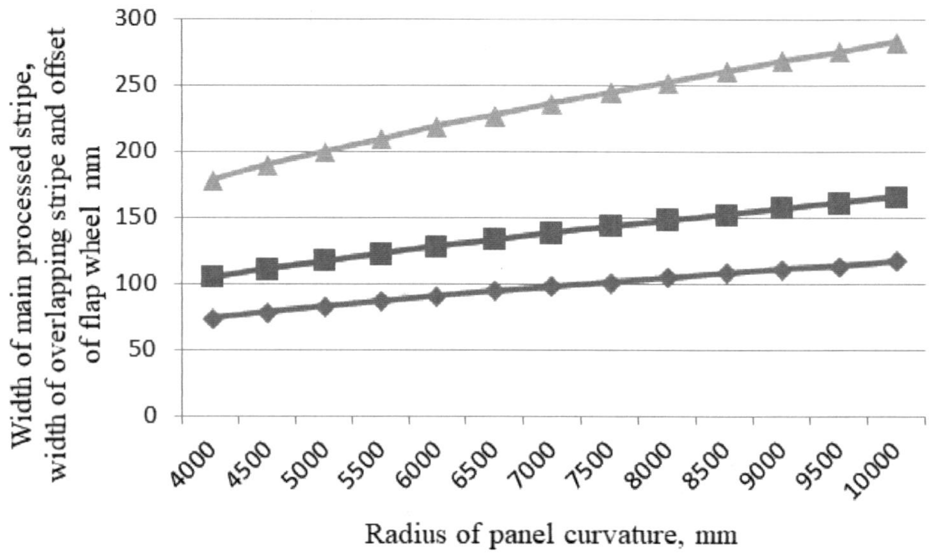

Figure 5. Dependence of width of overlapping stripe, width of main processed stripe and displacement of flap wheel depending on radius of curvature at deformation value of 2mm, where: bottom curve - width of overlapping stripe, middle curve – width of main processed stripe, top curve- offset of flap wheel for following grinding passage.

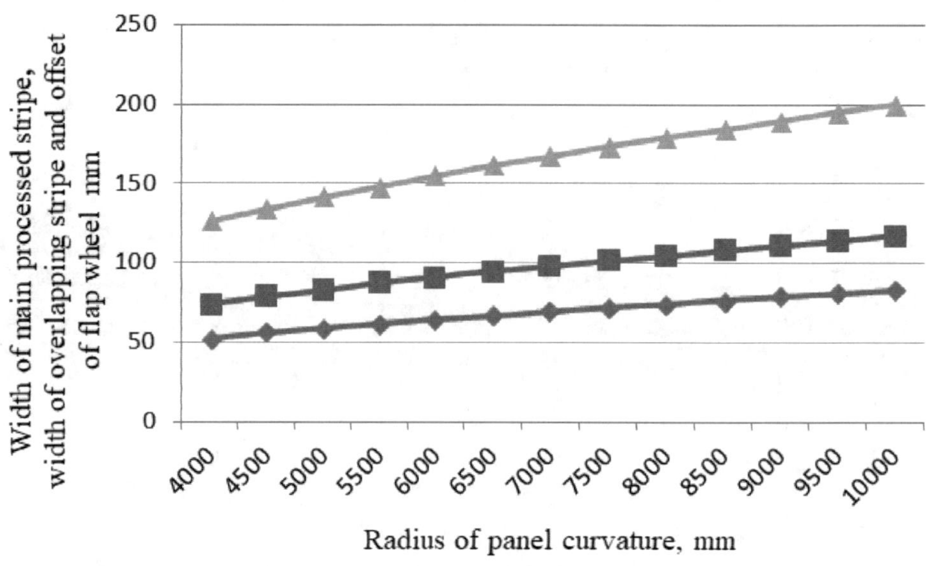

Figure 6. Dependence of width of overlapping stripe, width of main processed stripe and displacement of flap wheel depending on the radius of curvature at deformation value of 1mm, where: bottom curve - width of overlapping stripe, middle curve – width of main processed stripe, top curve- offset of flap wheel for following grinding passage.

From Fig. 3, 4, 5 and 6 it should be noted that for processing the panel in areas with a small radius of curvature, it is recommended to use a wide flap wheel, since in this case the width of the flap wheel provides the maximum width of the main processed stripe per passage. This will reduce the number of passages when processing the entire part and, consequently, reduce the processing time. The specific width of the flaps required for processing in this way must be preliminarily calculated, taking into account the condition that the maximum deviation of the value of the flap deformation in the zone of contact with the curved surface of the panel T in the transverse direction must be greater than or equal to the amount of needed deformation (according to the processing mode) set by the technologist.

Fig. 7 shows the dependence of the average deviation of the resultant deformation value on the radius of curvature for processing modes with a deformation of 1, 2, 3 and 4 mm during grinding with overlapping passages.

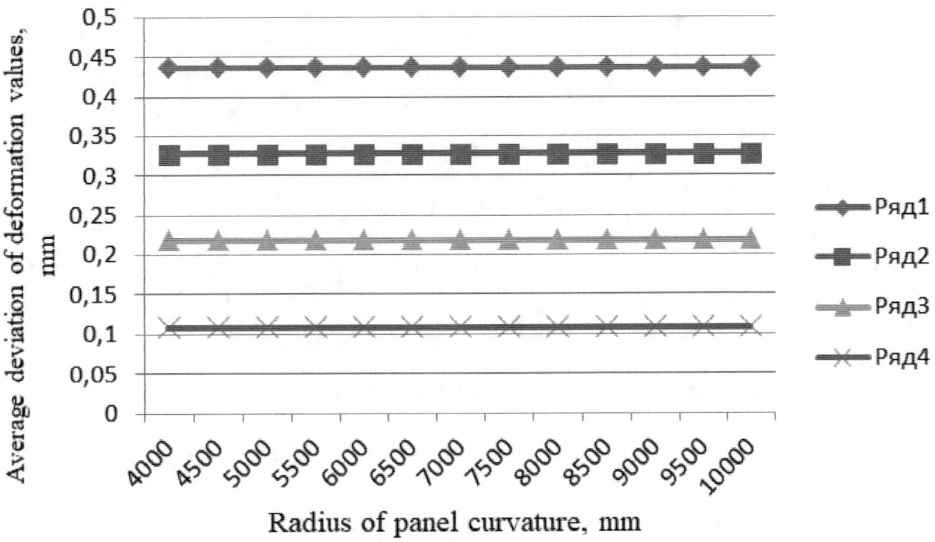

Figure 7. Dependence Average deviation of resultant deformation during grinding panel with different processing modes, where: bottom curve (curve 1) – for grinding with a deformation of 1 mm, curve 2 (above the curve 1) – for grinding with a deformation of 2 mm , curve 3 (above the curve 2) - for grinding with a deformation of 3 mm, top curve (curve 4) - for grinding with a deformation of 4 mm.

The results of the study of deviations of the resultant deformation during grinding with passage overlapping at a point with flap deformation equal to 1/2T showed that its deviation does not exceed 11% of the specified technological deformation of the central part of flap, which is expected to significantly improve the uniformity of material removal from the machining surface. Thus, the deviation of the surface shape should not exceed 11% of the thickness of the removed material, respectively.

3. Conclusion

To ensure a more uniform removal of material during grinding the double curvature panel with flap wheels in areas with a small radius of curvature, it is recommended to perform grinding with overlapping passages at the intersection of passages with the flap deformation equal to ½ of the flap deformation in central contact zone.

The developed method for calculating the width of the overlapping stripe, the width of the main processed stripe, and the necessary displacement of the flap wheel to perform the subsequent passage can minimize the uneven material removal from the surface of the double curvature panel for sections of the panel with a small transverse radius.

References

[1] Shmatkova A V 2006 *Optimization of the process of machining parts by the periphery of flat flap wheels* Author's abstract of the candidate's dissertation of technical Sciences: 05.02.08, 05.03.01//Shmatkova Anna Viktorovna (Irkutsk) p 19

[2] Pashkov A E 2005 *Technological connections in the process of manufacturing long sheet parts (*Irkutsk: Publishing house of Irstu) p 138
[3] Pashkov A E Veprev A A 2012 *Complex technology of the aircraft panels forming* (IJAMFO vol. 4 No. 1) pp 39-54
[4] Likhachev A A, Gerasimov V V, Pashkov A A 2015 Implementation of shot peen forming control system on contact type installations. *Vestnik of ISTU.* 12(107) 42-47
[5] Dimov U V , Poashev D B 2015 Cutting forces in machining by elastic abrasive wheels. *Vestnik of ISTU.* 7(102) 47-55
[6] Koltsov V P, Starodubtseva D A, Kozyreva M V 2015 *Analysis of the dependences of removal and surface roughness of the part during processing with flap wheels based on the results of a factor experiment* (Vestnik of ISTU No. 1 (96)) pp 32-41
[7] Koltsov V P, Starodubtseva D A, Chapyshev A P 2017 *To determine the value of the allowance for grinding the surface of panels and sheaths with a flap wheels after shot peen forming* (Bulletin of the Kazan state technical University. A. N. Tupolev T. 73) pp 25-30
[8] Dimov U V , Poashev D B 2015 Cutting zone temperature under elastic fap disc machining. *Vestnik of ISTU.* 2(97) 38-42
[9] Jerald V S 2008 Understanding residual stress effects and corrective shot peening action for die casting tools. *Die casting engineer Part 2* 2-6
[10] Diak A U 2015 Promising methods to determine surface coverage under shot peening. *Vestnik of ISTU.* 12(107) 19-25
[11] Diak A U 2014 Promising methods to determine surface coverage under shot peening. *Vestnik of ISTU.* 7(90) 12-17
[12] Pashkov A A 2015 Automation of large scale panels shootblasting process morphogenesis on contact type installations. *Vestnik of P. A. Solovyov Rybinsk State Aviation Technical University.* 4 34-39

The influence of flank lands of end mills on machined surface in milling in the context of chatter

S I Dyadya[1], Ye B Kozlova[1], V I Tretiak[1], E V Kondratyuk[2]

[1] National University "Zaporizhzhya Polytechnic", 64, Zhukovsky str., Zaporizhzhia, 69064, Ukraine
[2] Zaporizhzhia Machine-Building Design Bureau "Ivchenko-Progress", 2, Ivanov str., Zaporizhzhia, 69061, Ukraine

E-mail: kozlova@zntu.edu.ua

Abstract. Reliability and cost-effectiveness are the main requirements for aircraft materials and structure components used in manufacturing of aviation engines. This defines the requirements for them. The majority of components are thin-walled and made of hard-to-machine materials: nickel and titanium alloys, heat-resistant steels. The use of such materials places restrictions on the use of certain cutting modes. During rough milling and semi-finish milling with end-mill cutters chatter occur. The vibrations deteriorate the machined surface of engine parts, tool life decreases and machine units are worn out. Different process technologies and constructive features of tools are used to decrease the intensity of chatter. The use of end mills having lands has yielded mixed results. The influence of land usage on surface condition and decrease of chatter in up-milling and down milling has been studied. The mechanism of influence on machined surface and chatter in up-milling and down milling has been determined. At the experimental stand studies were carried out on its effect on changes in the laws of motion of detail during the cutting process. It has been shown that in up-milling and down milling, a land $f_l = 0.05$ mm wide on the rear surface along the cutting tooth reduces the intensity of chatter has been shown. The results can be used to select the feed direction to reduce the intensity of chatter when milling thin-walled parts with end mills with a land on the rear surface along the cutting edges.

1. Introduction

Light weight and durability are the main requirements to machine parts and components in aviation industry. Therefore, the parts are made of hard-to-process materials and contain thin-walled elements.

During end milling operations tooth entry is accompanied with shock causing vibrations. Their intensity (A) depends on stiffness of a work piece (k) and kinetic energy (E) at tooth entry [1]. Eq. 1.

$$A = \sqrt{\frac{2E}{k}} \tag{1}$$

While cutting speed grows the intensity of vibration increases. It should be mentioned, there are different modes of vibration at different cutting speeds [2]. Therefore, all cutting speeds are divided into five zones. Each zone has its own values of constituent equation (CE) between cutting time – t_{cut}

and period of chatter – $T_{chatter}$ of a work piece ($CE = t_{cut}/T_{chatter}$) [2]. Methodology of calculations of cutting time for different types of end mills can be seen there [2]. Free vibration period is determined depending on stiffness of a work piece [1].

Division into speed zones gives the possibility to reveal unfavorable modes with chatter intensity more than 0.02 mm. Unfavorable modes should be detected at pre-production stage. These modes are common for rough end milling and semi-finish end milling of hard-processed materials in the third speed zone of vibrations ($1 < CE < 7$). Different process technologies and constructive features of tools are used to decrease the intensity of chatter in that zone.

The primary goal is to eliminate wave regeneration on cutting surface and change the conditions of rubbing between flank face and machined surface.

In the first case use cutters with variable step [3 – 8], different angles of flute helix, use modulation of cutting speeds.

Having analyzed the results of milling with alternating pitch cutters [9 – 13] it was determined tooth-to-tooth variation (pitch variation) is to be equal to a half wavelength of chatter. Under this condition the suppression of chatter takes place.

In cutter designs regulated by GOST 17025–71, 17024–71, 18372–73, 15086-69 alternating pitch is considered to be standard, therefore, the cutters made in that way are able to suppress chatter of a certain period. The period limitedness terminates the use of such cutters. Alternate gash mill cutters are able to provide necessary cutting speed in a broad vibration-free range [16 – 19].

Modulated cutting speed results in continuous phase shift between vibration trace and in-process vibrations of the system [14]. At CNC machine tools spindle speed is a programmed parameter. Spindle movement options influence speed modulation. However, the applicability of this approach of suppression of chatter is limited. It is caused by time lag of spindle and the necessity of choosing modulation depth and modulation rate for every type of processed material.

The method of suppression of chatter due to rubbing between flank and machined surface does work properly [2]. There is a flat (relieved) land (GOST 18372–73) or a cylindrical land (GOST 17025–71, 17026–71) with a width no more than $f_l = 0.05$ mm in end mills designs. Possible geometry of flank sharpening is shown in fig. 1.

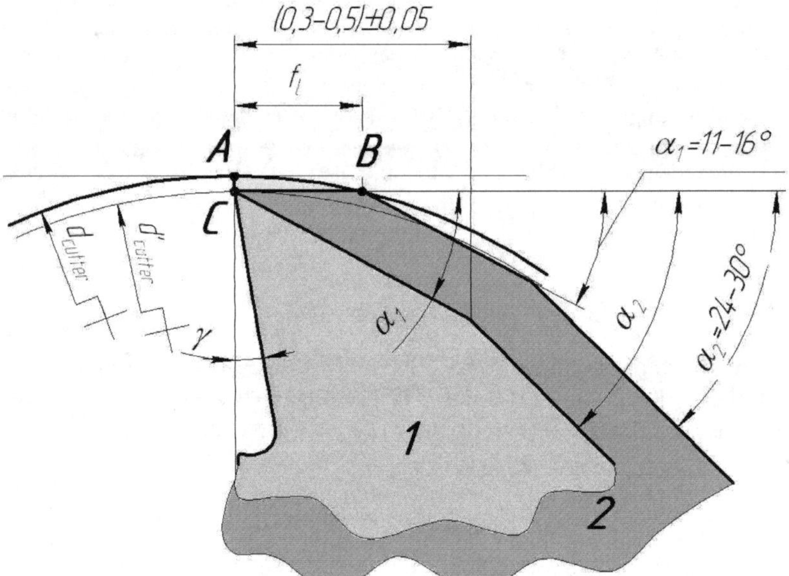

Figure 1. Geometry of flank sharpening: 1. Flat relieved tooth. 2. A tooth with a land f_l: *AB* – a cylindrical land, *CB* – a flat (relieved) land

The characteristics of the mill tooth are the following: the tooth is plain with double-angle point $\alpha_1 = 11 - 16°$ and $\alpha_2 = 24 - 30°$. Tooth 1 is conventional, flat relieved, while tooth 2 has a land f_l. Depending on the method of sharpening a land can be either cylindrical – *AB* or flat (relieved) – *CB*. The shape of the land determines conditions of tooth surface rubbing with the machined surface.

When the tip of tooth *A* travels on the circumference – d_{cutter} a cylindrical land touches the tip of tooth, so relief angle in every point is $\alpha = 0°$. When the tip of tooth *C* travels on the circumference – d'_{cutter} relief angle of flat land *CB* becomes negative. At point *B* flank can have plastic contact with the machined surface.

Both decrease of vibrations intensity and good level surface are equally important in milling processes. It is evidently that level surface after up-milling using a tool with a land has practically the same waviness as after milling using a conventional flat-relieved tool.

Consequently, the research of mechanism of using end mills tools with lands to decrease vibrations intensity and refine level surface of machined parts is of primary importance. Equally important is to determine the parameters of such influence in both up-milling and down milling.

2. Materials and methods

A special testing device was used for cutting a thin-walled work piece with an end mill during the experiments [2]. The design of the testing device gave the opportunity to separately investigate three factors influencing vibration processes. These factors are: characteristics of elastic system, properties of the machined material and the choice of cutting modes. The testing device with a specimen on it was firmly clamped to the table of vertical milling machine. A milling cutter of a special structure was fixed in the spindle [2].

Cutters were mounted sequentially in the cutter holder. A conventional flat relieved tool ($f_l = 0$) was mounted first, while a tool with a land on flank face ($f_l = 0.05$ mm) was mounted after that. The machining was being done in the third speed zone both in up-milling and down milling.

Vibrations of elastic system of workpieces in milling with the use of conventional tools and tools with lands of 0.05mm wide were being recorded in oscillograms.

Profilograms of machined workpieces were recorded after milling. The values of waviness height and waviness step were measured. Oscillograms were divided into basic fragments. They helped to determine range of chatter – R_2 and period of chatter – $T_{chatter}$, deviation of first wave from position of elastic equilibrium in up-milling and deviation of last wave – $\Delta_{d\ prof}$ in down milling.

These parameters of vibratory motion influenced by the availability of the land are linked with formation of machined surface profile. Amplitude of chatter is a measure of their intensity. The period of chatter is linked with waviness step of the machined surface. Deviations of first wave from elastic equilibrium in up-milling and deviations of last wave in down milling determine the position of profile forming cavity on machined surface after each cut of the tooth [20].

The experiments were done under the following conditions.

- Cutting modes: spindle rotation rate – $n = 280$ rev/min; radial depth of cut $a_e = 0.5$ mm; axial depth of cut $a_p = 3.4$ mm; feed per tooth $S_z = 0.1$ mm; feed path – up-milling and down milling; free cutting.
- Cutter tool: cutter diameter $d_{cutter} = 50$ mm; teeth number $z = 1$; cutting edge inclination $\omega = 0°$; material of cutting edges – ВК8; free vibration frequency $f_{fvf} = 833$ Hz.
- Elastic system of the work piece: free vibration frequency – $f_{fvf} = 439$ Hz.
- Specimen material in St. 3 GOST 380-2005.

An electrical contact device was used for making oscillograph charts [2]. Tooth-in-cut time – t_{cut}, i.e. the time span from the tooth entry into a work piece till the tooth exit was clocked and recorded.

3. Presentation of the main material and analysis of the results

Figure 2 shows profilograms of machined surfaces after both up-milling and down milling with a conventional tool and a tool with a land $f_l = 0.05$ mm.

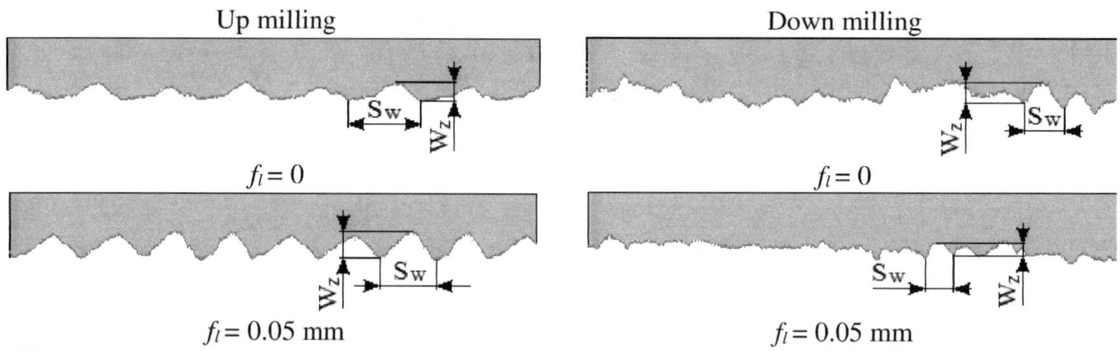

Figure 2. Profilograms of machined surfaces after both up-milling and down milling with a conventional tool and a tool with a land f_l

Profilograms show waviness height – W_z and waviness step – S_w. Analysis of values got after up-milling and down-milling shows that characteristics are steadier after up-milling. Average numerical values are done in table 1. The obtained data prove the tool having the flank land influences surface forming, first of all decreasing waviness step. As for waviness height, it considerably decreases in down milling while practically does not change in up milling.

Chatter during cutting process influence machined surfaces. The changes on the surfaces are caused by changes in vibration motion and peculiarities of up-milling and down milling. Up-milling implies minimum chip thickness at the very beginning, so chatter manifests themselves immediately after the tooth entry. Down milling starts with maximum chip thickness, so chatter appear when chip thickness decreases reaching definite values.

Table 1. Average values of waviness step and waviness height of machined surfaces

Feed path	Up-milling		Down milling	
	Conventional tool ($f_l = 0$)	With land ($f_l = 0.05$ mm)	Conventional tool ($f_l = 0$)	With land ($f_l = 0.05$ mm)
Waviness step, S_w, (mm)	2.65	2.33	1.09	0.87
Waviness height, W_z, (mm)	0.086	0.099	0.111	0.051

Fragments of oscillograms of vibrations of work pieces in up-milling and down milling with the use of a conventional tool and a tool with a land of $f_l = 0.05$ mm wide are shown in fig. 3. Tool cutting time – t_{cut}, range of chatter – R_2 and period of chatter – $T_{chatter}$, deviations from the position of elastic equilibrium – *PEE* in the zone of surface formation – $\Delta_{up\ prof}$ and $\Delta_{d\ prof}$ are shown in fig. 3.

Table 2 shows values of amplitude and period of chatter in up-milling and down milling using a flat-relieved tooth and a tooth having a land.

The presence of land $f_l = 0.05$ mm on flank reduces intensity of chatter caused by rubbing on machined surface in down milling by a factor of 1.86. The same process under the same conditions in up-milling results in decrease of vibration by a factor of 1.34. The period of vibration proportionally decreases. Waviness step on the machined surface depends on the period of vibration [20]. This statement is confirmed with measurements shown in table 1.

It should be emphasized that the use of land $f_l = 0.05$ mm has no influence on the results of up-milling in the zone of machined surface profiling. Fig. 3 clearly proves this provision.

Deviations of first wave from elastic equilibrium have close numerical values both in up-milling and in down milling. The choice of a flat-relieved tool or a tool having a land of $f_l = 0.05$ mm wide practically does not influence the resulting values. The influence of land is manifested in profiling

IOP Conf. Series: Materials Science and Engineering 1061 (2021) 012020 doi:10.1088/1757-899X/1061/1/012020

zone, in particular, it results in decreasing both chatter and their period. All these considerations explain why waviness step decreases while waviness height does not change.

Maximum chip thickness at the beginning of down milling process dampens chatter. When a flat relieved tooth travels deeper the chip thickness minimizes, so the intensity of vibration increases. With the use of a tool having land of $f_l = 0.05$ mm wide the intensity of vibration increases only marginally due to the land contact with machined surface. In down milling process the surface is formed at the tooth exit. Therefore, decrease in intensity of chatter due to the use of land of $f_l = 0.05$ mm wide contracts the period of vibrations and reduces the value of $\Delta_{d\ prof}$ (fig. 3). Moreover, waviness step and waviness height on machined surface become less (fig. 2)

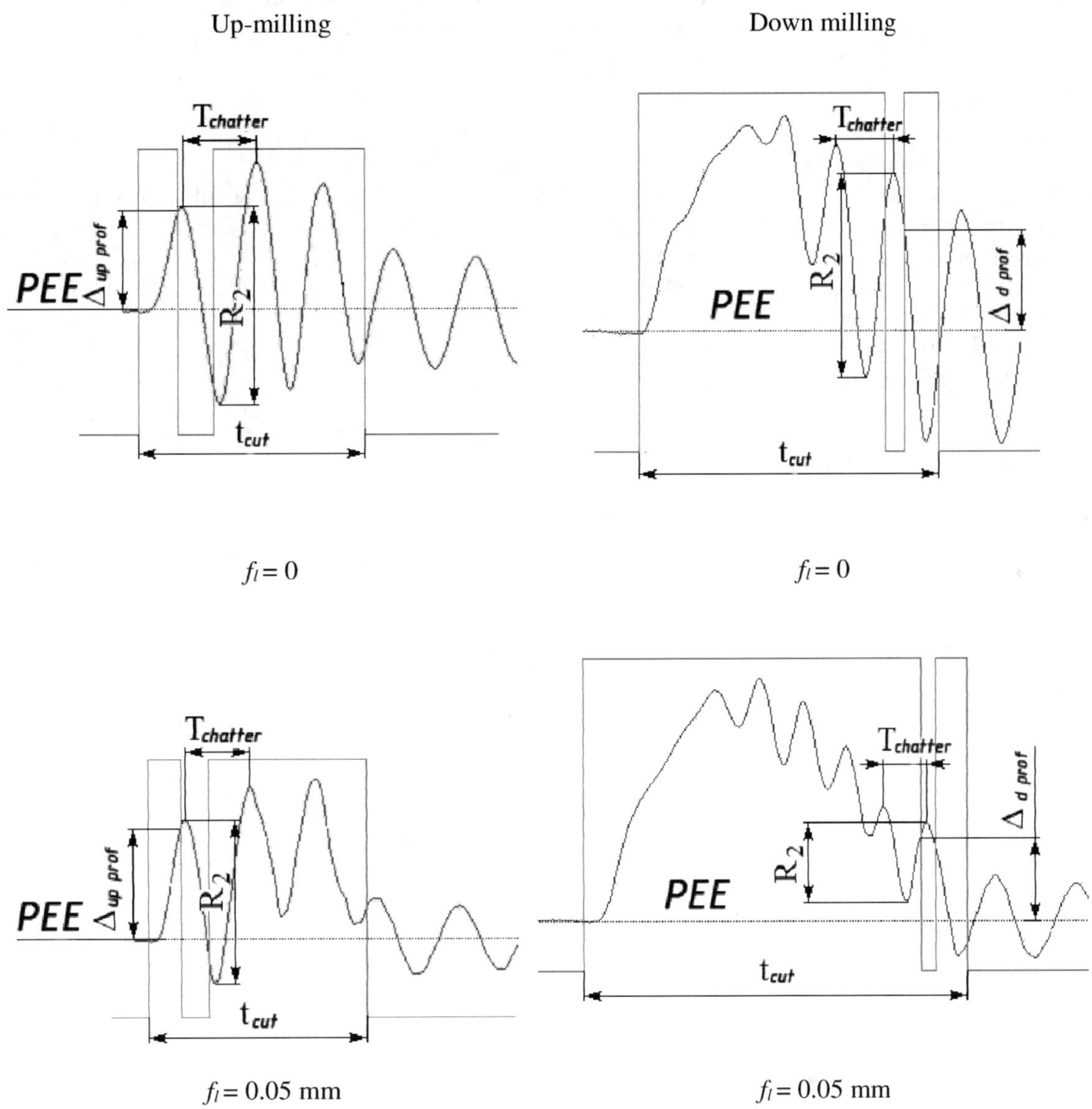

Figure 3. Fragments of oscillograms of work piece vibration in up-milling and down milling with the use of a conventional tool and a tool with a land of $f_l = 0.05$ mm wide

Table 2. Values of amplitude and period of chatter in up-milling and down milling using a flat-relieved tooth and a tooth having a land

Feed path	Up-milling		Down milling	
Cutting edge	Conventional tool ($f_l = 0$)	With land, ($f_l = 0.05$ mm)	Conventional tool ($f_l = 0$)	With land, ($f_l = 0.05$ mm)
Range of chatter, R_2, (mm)	0.130	0.097	0.126	0.069
Period of chatter, $T_{chatter}$, (sec)	0.00205	0.00191	0.00153	0.0013

4. Conclusion

On the basis of the conducted research, the influence of flank lands on the rear surface of the end mills on the formation of the machined surface have been established.

1. The presence of lands of $f_l = 0.05$ mm wide on flanks of end mills positively influences on suppression of chatter due to changes in rubbing of tool flank and machined surface both in up-milling and down milling.
2. Both in up-milling and down milling the use of lands of $f_l = 0.05$ mm wide decreases the period of chatter and therefore causes reducing of waviness step on machined surface.
3. In up-milling chatter occur at the beginning of cutting process and impact the profiling zone until the land contacts the surface being cut.
4. In up-milling the use of land of $f_l = 0.05$ mm wide does not influence waviness height of machined surface. The influence of the land on suppression of chatter is manifested only after profiling zone is machined.
5. In down milling the land of $f_l = 0.05$ mm wide has stronger effect on suppression of chatter because forming occurs at the exit of the tool, therefore the land has contact with the surface being cut.
6. The best results of using the land $f_l = 0.05$ mm wide on the flank of end mills in the third speed zone are achieved in down milling.

References

[1] Kuzmichev V 1987 Laws and formulas of physics Kiev: Scientific opinion 864

[2] Vnukov Yu, Dyadya S, Kozlova Ye, Logominov V and Chernovol N 2017 Chatter in milling thin-walled elements of parts Zaporozhye: ZNTU 208

[3] Svinin V M and Savilov A V 2017 Application of variable teeth pitch face mill as chatter suppression method for non-rigid technological system *Conf. Series: Materials Science and Engineering* 327

[4] Budak E 2017 An Analytical Design Method for Milling Cutters with Nonconstant Pitch to Increase Stability, *Part 1: Theory; Part 2: Application, ASME J. Manuf. Sci. Eng.* **125** 29–38

[5] Slavicek J 1965 The Effect of Irregular Tooth Pitch on Stability of Milling *Proc. 6th MTDR Conference* (London: Pergamon Press) 15–22

[6] Opitz H, Dregger E U and Roese H 1966 Improvement of the Dynamic Stability of the Milling Process by Irregular Tooth Pitch *Proc. 7th International MTDR Conference* (New York: Pergamon Press) 213–227

[7] Suzukia N, Kojimaa T, Hinoa R and Shamotoa A 2012 A Novel Design Method of Irregular Pitch Cutters to Attain Simultaneous Suppression of Multi-Mode Regenerations *3rd CIRP Conference on Process Machine Interactions* 98–102

[8] Vanherck P 1965 Increasing Milling Machine Productivity by Use of Cutters with Non-Constant Cutting Edge Pitch *8th MTDR Conference* (Manchester) 947–960

[9] Tlusty J, Ismail F and Zaton W 1983 Use of Special Milling Cutters Against Chatter *Proc. NAMRC* **11** 408–415

[10] Shalamov V 1991 Choice of tooth divergence of cutters, *Progressive technology of finishing and finishing processing coll. of articles* (Chelyabinsk) 14–22

[11] Altintas Y, Engin S and Budak E 1991 Analytical Stability Prediction and Design of Variable Pitch Cutters, *ASME J. Manuf. Sci. Eng.* **121**, **2**, 173–178

[12] Shamoto E and Saito A 2016 A novel deep groove machining method utilizing variable-pitch end mill with feed-directional thin support *Precision Engineering* **43** 277–284

[13] Sellmeier V and Denkena B 2011 Stable islands in the stability chart of milling processes due to unequal tooth pitch International *Journal of Machine Tools and Manufacture* **51** **(2)** 152–164

[14] Svinin V M, Savilov A V, Zarak T V 2018 Self-oscillation regeneration control by irregular tooth pitches. *Advances in Engineering Research: International Conference on Aviamechanical Engineering and Transport (AviaENT 2018)* **158** 399–405

[15] Sims N D, Mann B and Huyanan S 2008 Analytical prediction of chatter stability for variable pitch and variable helix milling tools, *Journal of Sound and Vibration* **317** **(3–5)** 664–686

[16] Wan M, Zhang W H, Dang J W and Yang Y 2010 A unified stability prediction method for milling process with multiple delays, *International Journal of Machine Tools and Manufacture* **50** **(1)** 29–41

[17] Yusoff A R and Sims N D 2011 Optimisation of variable helix tool geometry for regenerative chatter mitigation, *International Journal of Machine Tools and Manufacture* **51** **(2)** 133–141

[18] Dombovari Z and Stepan G 2012 The effect of helix angle variation on milling stability *Journal of Manufacturing Science and Engineering* **134** **(5)** 661–683

[19] Sims N D 2016 Fast chatter stability prediction for variable helix milling tools *Proceedings of the institution of mechanical engineers, part C, Journal of mechanical engineering science* **230** **(1)** 133–144 doi: 10.1177/ 0954406215585367

[20] Dyadya S 2017 Investigation of the formation of the machined surface of a thin-walled part element in end cylindrical milling with chatter *Modern Engineering Technologies* **12** 5–18

The optimization of the lateral motion control system of an unmanned aerial vehicle

V A Kramar[1], **A A Kabanov**[1]

[1] Sevastopol State University, 33, Universitetskayast., Sevastopol, 299053, Russia

E-mail: kramarv@mail.ru

Abstract. The article is devoted to the possibility of constructing an aircraft lateral motion control system only using the rudder; a mathematical model of such a system is obtained based on the structuring of an unmanned aerial vehicle model. The design of optimal lateral motion control of a UAV is carried out using the linear-quadratic optimization method in the complex domain for single-control systems. As an optimization criterion, a quadratic criterion in the form of restrictions on the control energy is used. Examples of system modeling are given.

1. Introduction

The unmanned aerial vehicle (UAV) in landing mode, especially in difficult conditions (above the sea surface), is very sensitive to control inputs, so special attention is paid to the modeling and design of its control systems. In this case, the task of automatic control of the lateral movement of UAV is very important [1-3]. The linear and non-linear mathematics models of UAV are used in the time and the complex domain, [4]. The deep mutual relations between the channels of the onboard control systems that affect the flight of the UAV lead to the complication of mathematical models that describe the dynamic processes of the UAV movement. In the synthesis of UAV (the aircraft type) control laws, the spatial motion of UAV is often separated using various approaches [5, 6]. A very clear representation of the system is in the Laplace domain, and therefore, its representation in the form of a structural diagram.

The robust and adaptive approaches to the construction of UAV control [7, 8] in particular, lateral movement, as well as modal control methods [6] are widely known. It is of considerable interest in the development of UAV control algorithms, including lateral movement, obtaining regulators that implement the optimal laws, which is carried out using linear-quadratic optimization (LQO) methods both in the time and in the complex domain. This task does not lose its relevance, especially concerning UAV, given its high maneuverability and high-speed characteristics.

2. The mathematical model of the lateral movement of the UAV

We consider a linearized model of the lateral movement of the UAV in the form:

$$\frac{dx(t)}{dt} = Ax(t) + Bu(t) \tag{1}$$

where $x = (\nu, \omega_y, \omega_x, \gamma, \psi, z)^T$; $\nu = -V_z = -V \sin\beta \approx -V\beta$ – side glide speed; $u(t)$ – the control vector; $u = (\delta_э, \delta_н)^T$, $\delta_э$ – the aileron deflection angle; $\delta_н$ – the rudder angle; γ – the roll angle; β – the glide angle; ψ – the yaw angle; ω_x – the angular velocity of rotation around the longitudinal axis (the roll angular velocity); ω_y – the horizontal angular rotation speed (the yaw rate); V – the UAV speed.

When structuring the mathematical model, the factors of the second order of smallness are discarded, i.e. is accepted $a_{13} = a_{23} = a_{32} = a_{42} = b_{12} = b_{21} = b_{32} = 0$. Applying the Laplace transform to simplified equations, we obtain:

$$\begin{cases} (s - a_{11})V(s) = a_{12}\omega_y(s) + a_{14}\gamma(s), \\ (s - a_{22})\omega_y(s) = a_{21}V(s) + b_{22}\delta_H(s), \\ (s - a_{33})\omega_x(s) = a_{31}V(s) + b_{31}\delta_э(s), \\ s\gamma(s) = \omega_x(s),\ s\psi(s) = \omega_y(s),\ sz(s) = V(s) + a_{65}\psi(s). \end{cases} \quad (2)$$

Adding the relationships of control laws that reflect the idea of the nested tracking loops,

$$\delta_э(s) = W_{p\gamma}(s)\varepsilon_\gamma(s),\ \ \varepsilon_\gamma(s) = \gamma_*(s) - \gamma(s),\ \gamma_*(s) = W_{pz\gamma}(s)\varepsilon_z(s), \quad (3)$$

$$\delta_н(s) = W_{p\psi}(s)\varepsilon_\psi(s),\ \varepsilon_\psi(s) = \psi_*(s) - \psi(s),\ \psi_*(s) = W_{pz\psi}(s)\varepsilon_z(s), \quad (4)$$

we get the system model in the form of the following structural diagram:

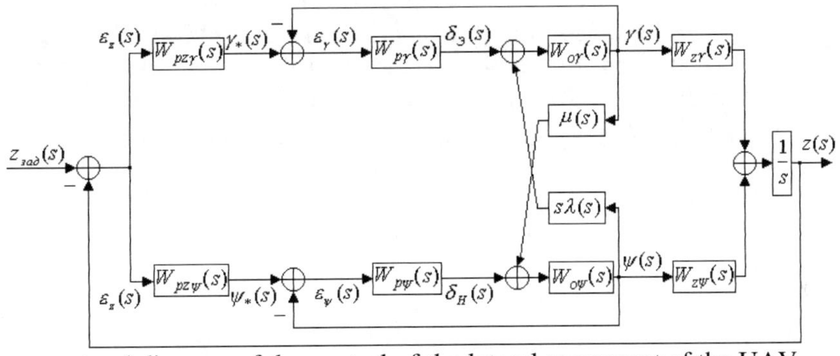

Figure 1. The structural diagram of the control of the lateral movement of the UAV

In the case of small values of a_{11} and a_{31}, we accept $a_{11} = 0, a_{31} = 0$. Then $\lambda(s) \equiv 0$ and the influence of yaw on the roll is excluded, and the lateral coordinate can be controlled only with the help of the roll. The yaw loop performs the functions of a stabilization loop (maintaining $\psi \approx 0$, i.e., suppressing disturbances in angular motion).

In the general case, the exclusion of the mutual influence of the circuits can be achieved by introducing the compensating relationships in the laws of control

$$\delta_э(s) = W_{p\gamma}(s)\varepsilon_\gamma(s) + [-s\lambda(s)\psi(s)],\ \delta_н(s) = W_{p\psi}(s)\varepsilon_\psi(s) + [-\mu(s)\gamma(s)]. \quad (5)$$

In the case when $a_{11} \neq 0$, the lateral force has a significant effect, so it is possible to consider the control of the lateral coordinate using the yaw control. Then the roll loop from the lateral displacement control can be excluded and perform only the functions of the roll stabilization loop (the process of maintaining $\gamma \approx 0$).

3. The statement of the problem and the basic theoretical principles

We will consider the problem of synthesis of regulators, i.e. transfer functions $W_{p*}(s)$ indicated in Figure 1, by the method of linear-quadratic optimization (LQO). After eliminating the mutual influence of the contours (due to the accepted assumptions or due to the introduction of compensating

relationships), it is possible to use the methods of one-dimensional optimization for each channel in two stages - optimization of the internal and then the external circuits.
For a one-dimensional system (Figure 2) there are the relations

$$\Phi(s)=\frac{W_p(s)W_o(s)}{1+W_p(s)W_o(s)},\ W_p(s)=\frac{\Phi(s)}{W_o(s)(1-\Phi(s))} \tag{6}$$

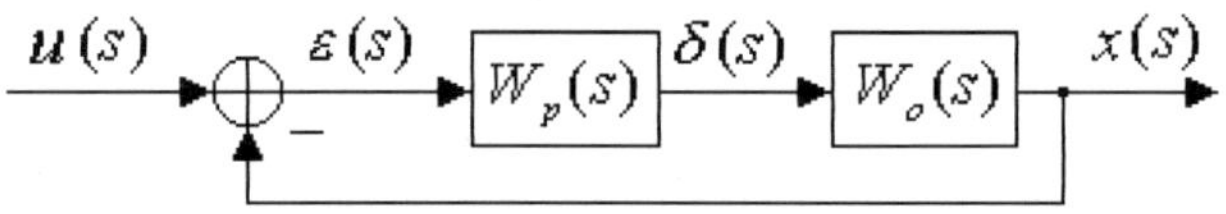

Figure 2. The structural diagram of a one-dimensional system

We introduce the optimality criteria:

$$I_{21}=\int_0^\infty \varepsilon^2(t)dt \to \min;\quad I_{22}=\int_0^\infty \delta^2(t)dt \le C\,, C\,>0 \tag{7}$$

where the parameter C determines the restrictions on the energy of the control.

Given the relation (6), the transfer function of the controller can be determined by the transfer function of the closed system $\Phi(s)$, which is optimal in the sense of criterion (7) and belongs to the class of fractional rational functions with the left poles. We assume that the input action also belongs to this class.

Following figure 2, we write the expression for the error and deviation of the control in the complex area:

$$\varepsilon(s)=[1-\Phi(s)]u(s);\ \delta(s)=\frac{\Phi(s)}{W_o(s)}u(s). \tag{8}$$

Let a stable system be considered and $\varepsilon(\infty)=0$.

Following (8), one can write the following expressions for the integral quality criteria:

$$I_{21}\{\Phi\}=\frac{1}{2\pi j}\int_{-j\infty}^{+j\infty}[1-\Phi(s)][1-\Phi(-s)]u(s)u(-s)ds; \tag{9}$$

$$I_{22}\{\Phi\}=\frac{1}{2\pi j}\int_{-j\infty}^{+j\infty}\frac{\Phi(s)\Phi(-s)}{W_o(s)W_o(-s)}u(s)u(-s)ds. \tag{10}$$

To find the optimal transfer function $\Phi(s)$, we use the following known position. Let $\Im$ means the class of fractional rational functions with left poles. Let:

1. $\Phi(s,\lambda)$ –the solution to the problem of unconditional extremum:

$$\Phi(s,\lambda)=\arg\min_{\Phi\in\Im}\left[I_{21}\{\hat{\Phi}\}+\lambda I_{22}\{\hat{\Phi}\}\right],\ \lambda\in R^+.$$

2. $\forall\hat{C}\le C\quad(\hat{C}>0)\quad\exists\hat{\lambda}:I_{22}\{\Phi(s,\hat{\lambda})\}\le\hat{C}$.

3. The area of valid parameter values is determined λ $\Lambda = \{\lambda \in R^+ : I_{22}\{\Phi(s,\lambda)\} \leq C\}$ and $\lambda_0 = \arg\min_{\lambda\in\Lambda} [I_{21}\{\Phi(s,\lambda)\}]$.

Then $\Phi(s) = \Phi(s,\lambda_0)$ – the optimum transfer function.

Let $R(s^2) = 1 + \dfrac{\lambda}{W_o(s)W_o(-s)}$ and $R(s) = Y^+(s)Y^-(s)$, where $Y^+(s)$, $Y^-(s)$ –the rational functions, with zeros and poles in the left and right half-planes, respectively.

Then, to solve the problem of unconditional extremum $\Phi(s,\lambda)$, we have

$$Y^+(s)u(s)\Phi(s,\lambda) = \left[\frac{u(s)}{Y^-(s)}\right]_+ ;\ u(s) \in \Im ,$$

where the "+" sign means the separation result, i.e. separation of the fractional rational function with the left poles.

The factorization $R(s^2) = \dfrac{A(s^2)}{B(s^2)}$ is performed to obtain $Y^+(s)$ and $Y^-(s)$ in the general case, by calculating the roots of the polynomials $A(x), B(x)$, $x = s^2$ with subsequent representations

$$A(x) = a_n(x - x_1^a)\cdot\ldots\cdot(x - x_n^a),\ B(x) = b_m(x - x_1^b)\cdot\ldots\cdot(x - x_m^b), \tag{11}$$

$$x - x_i = (s^2 - x_i) = (s + \sqrt{x_i})(s - \sqrt{x_i}),\ \operatorname{Re}\sqrt{x_i} > 0. \tag{12}$$

The separation of a fractional rational function is performed using its decomposition into simple fractions.

The solution to the whole problem, namely, the optimal transfer function of a closed system will be the function $\Phi(s) = \Phi(s,\lambda_0)$. To determine it based on the well-known methods for calculating the integrals

$$I_2 = \frac{1}{2\pi j}\int_{-j\infty}^{+j\infty} \frac{c(s^2)}{q(s)q(-s)} ds \tag{13}$$

the construction of dependencies of optimality criteria and $I_{21}(\lambda)$, $I_{22}(\lambda)$.

4. The simulation results

For the UAV model in landing configuration (indicated in dimensionless quantities)

$$A = \begin{bmatrix} -0.089 & -2.21 & 0.328 & 0.319 & 0 & 0 \\ 0.076 & -0.217 & -0.166 & 0 & 0 & 0 \\ -0.602 & 0.327 & -0.975 & 0 & 0 & 0 \\ 0 & 0.15 & 1 & 0 & 0 & 0 \\ 0 & 1 & 0 & 0 & 0 & 0 \\ 1 & 0 & 0 & 0 & 2.21 & 0 \end{bmatrix};\ B = \begin{bmatrix} 0 & -0.0327 \\ 0.0264 & 0.151 \\ 0.227 & -0.0636 \\ 0 & 0 \\ 0 & 0 \\ 0 & 0 \end{bmatrix}$$

Correspondences $I_{21}(\lambda)$ and $I_{22}(\lambda)$ with a single step effect are of the form:

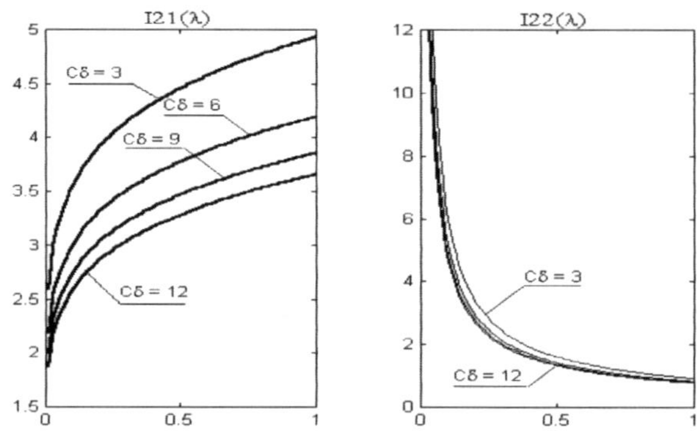

Figure 3. Dependencies $I_{21}(\lambda)$ and $I_{22}(\lambda)$ for the roll control

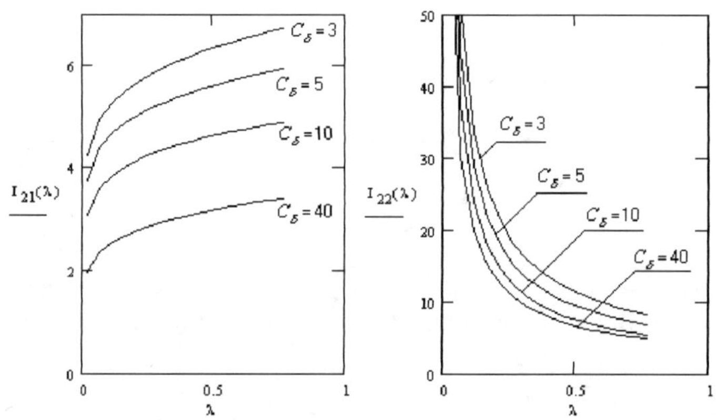

Figure 4. Dependencies $I_{21}(\lambda)$ and $I_{22}(\lambda)$ for the yaw control

Given the value of the restriction parameter C, the dependence $I_{22}(\lambda)$ determines the value of λ_0 the left boundary value of the set Λ.

The procedure for the synthesis of optimal contours at given levels of stepwise influences and the constraints of the integral quadratic deviations of the controls includes, in the general case, the calculation of the roots of polynomials, factorization and separation of rational functions, the calculation of integral quadratic functionals for different $\lambda \in R^+$, the calculation of the parameters of the optimal transfer functions and ends with modeling of optimal processes. As a result, tables of form 1, 2 can be calculated (the data given are calculated at inputs $1(t)$).

Table 1. The maximum value of the deviation of the ailerons and the angle of heel at various restriction parameters and C_δ и $C_{\gamma*}$

δ_{max} (deg)					γ_{max} (deg)				
C_δ \ $C_{\gamma*}$	**0.5**	**1**	**2**	**4**	C_δ \ $C_{\gamma*}$	**0.5**	**1**	**2**	**4**
3	0.6	0.8	1.26	1.9	**3**	0.09	0.11	0.15	0.19
6	1.1	1.68	2.6	4	**6**	0.12	0.17	0.23	0.27
9	1.6	2.4	3.85	5.95	**9**	0.15	0.2	0.26	0.35
12	2.1	3.2	5	7.8	**12**	0.17	0.22	0.3	0.4

Table 2. The maximum value of the deviation of the rudder and yaw angle at various parameters of restriction and C_δ и $C_{\psi *}$

δ_{max} (deg)					ψ_{max} (deg)				
$C_{\psi *}$ / C_δ	**15**	**20**	**30**	**40**	$C_{\psi *}$ / C_δ	**15**	**20**	**30**	**40**
5	0.64	0.77	1	1.19	**5**	0.15	0.17	0.18	0.2
10	1.7	2	2.6	3.1	**10**	0.23	0.25	0.27	0.3
15	3.9	4.7	6	7.2	**15**	0.3	0.34	0.38	0.42
20	5.5	6.6	8.6	10	**20**	0.36	0.4	0.45	0.5

The totality of such tables for different levels of exposure allows you to choose the parameters of the optimal controllers for the specified permissible deviations of the controls.

5. Conclusion

An approach to developing an optimal UAV lateral motion control system using only the rudder has been proposed in the paper. A mathematical model of such a system is obtained based on the structuring the initial UAV model. A linear-quadratic optimization method in the complex domain for systems with single control input was used as the optimization method. After eliminating the mutual influence of the yaw and roll control loops, it is possible to use one-dimensional optimization methods for each loop. The results indicate that optimization of the aircraft lateral motion control system using the rudder method using the LQO method of the integrated area for systems with one control body gives quite acceptable results and confirms the possibility of building a control system of the UAV lateral movement with separate control channels.

6. Acknowledgments

This work was financially supported by the Ministry of Science and Higher Education of the Russian Federation as part of the project "Fundamentals of mechanics, control and management systems for unmanned aerial systems with shape-building structures, deeply integrated with power plants, and unique properties not used today in manned aviation", No. FEFM-2020-0001.

References

[1] Oh J C, Yun E, Golkovski M G and Lee S 2003 *Mater. Sci. Eng. A* **351** pp 98–108

[2] Yilbas B S, Patel F and Karatas C 2015 *Laser Surface Engineering* (Waugh J.L.G. Woodhead Publishing)

[3] Stepanova N V and Razumakov A A 2013 *The 8 international forum on strategic technologies (IFOST 2013)* **1** pp 240–242

[4] Kimon P Valavanis and George J. Vachtsevanos 2015 *Handbook of Unmanned Aerial Vehicles* (Springer Science+Business Media Dordrecht) p 3015

[5] Nelson R C 1998 *Flight Stability and Automatic Control* 2nd Edition (Mc Graw Hill) p 452

[6] Zubov N E, Mikrin E A, Ryabchenko V N and Sorokin I V 2019 *Russ. Aeronaut.* **62** pp 22–31

[7] Zubov N E, Mikrin E A, Ryabchenko V N and Fomichev A V 2017 *Russ. Aeronaut.* **60** pp 64-73

[8] Chowdhary G and Johnson E N 2010 *The 49th IEEE Conference on Decision and Control* Atlanta pp 3674–3679

IOP Conf. Series: Materials Science and Engineering 1061 (2021) 012022 doi:10.1088/1757-899X/1061/1/012022

Variations in health of piezoelectric elements of the Common Rail electric hydraulic nozzles in the operating conditions

S N Krivtsov[1], T I Krivtsova[1].

[1] Iirkutsk National Research Technical University, 83, Lermontova ave., Irkutsk, 664074, Russia

E-mail: krivcov_sergei@mail.ru

Abstract. The article discusses operational reliability of piezoelectric elements of the Common Rail electric hydraulic nozzles. Attention is paid to the piezoelectric element's health and its influence on the output injector parameters. It was assumed that capacity of the piezoelectric packet can be used as a diagnostic parameter. The studies carried out in the operating conditions identified failures of some elements of the electro-hydraulic nozzles, patterns of variations in capacity of the piezoelectric packet exposed to short or long-lasting heat treatment. Based on the data obtained, conclusions about the frequency of monitoring of the piezoelectric element's health were drawn.

1. Introduction

One of the key problems of road transport is harmful emissions of exhaust gases, including carbon dioxide. To meet stringent environmental requirements for exhaust emissions, modern engines are equipped with electronic fuel control systems. In order to meet the Euro-6 emission standards, diesel engines are equipped with battery-powered fuel supply systems with a high injection pressure.

Currently, the leading manufacturers of fuel systems produce Common Rail systems with piezoelectric nozzles. According to the BOSCH Company, they can reduce fuel consumption by 3%, harmful emissions of exhaust gases by 20%, engine noise by 3 dB, and increase power by 7% [1]. In addition, the piezoelectric injector is characterized by stable multiphase fuel injections, minimal portions of preliminary injections, short intervals between preliminary and main injections and a compact design [2,3]. Compared to the electromagnetic nozzle, piezo nozzles consume less fuel and have higher efficiency [4].

BOSCH piezo injectors are the most common injectors. Their piezoelectric transducer consists of 350 quartz plates with a thickness of 90 microns. When DC voltage is applied, each plate is extended by up to 0.13 microns. The maximum (total) elongation of the piezoelectric packet can reach 45 μm [5]. The initial size of the piezoelectric element is restored when stress is removed from the plates.

During long-term operation, battery fuel supply system's health worsens. This violates the law of fuel supply and changes the environmental characteristics [4,5]. It is important to minimize emissions of harmful substances and carbon dioxide at the design and production stages and during the life cycle of the car. Currently, a lot of publications deal with the issue of optimization of the fuel supply systems and their injectors. The issue of operational reliability is understudied. In this regard, it is necessary to study the issue of reliability of electro-hydraulic nozzles with a piezo drive and search for methods of regular monitoring of health parameters.

IOP Conf. Series: Materials Science and Engineering 1061 (2021) 012022 doi:10.1088/1757-899X/1061/1/012022

2. Problem statement.

A block diagram of the Bosch injector with a piezoelectric element, which shows the elements affecting the fuel supply process (grouped by their functions), is presented in Figure 1.

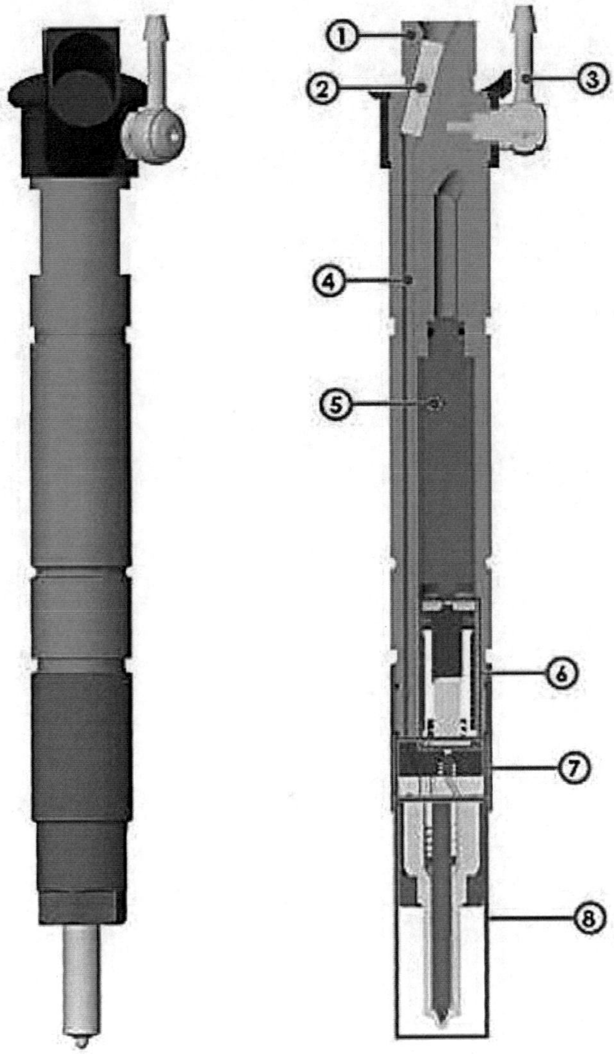

Figure 1. The block diagram of a piezo injector.

1 – high pressure connection; 2 – edge filter; 3 – back leak channel; 4 – feed channel; 5 – piezzo driver; 6 – hydraulic amplifier; 7 – control valve; 8 – nozzle

For the electro-hydraulic nozzle with a piezo driver, such elements (Fig. 1) are an atomizer, a control valve, and a piezo driver.

The most wearing part of the CR BOSCH piezo injector is a control valve [6-9]. During operation, cones of the valve plate and fungus wear out (functional node 7, Fig. 1). This violates tightness of the nozzle and changes operation parameters.

The second problem is nozzle shut-off cones (functional node 8, Figure 1), which create droplets when the nozzle works and is out of work, because the fuel is under pressure.

With long-term operation of the piezo nozzles, characteristics of the piezoelectric elements worsen, which changes (decreases) the value of movement of moving elements, including the sprayer needle. This affects the cyclic fuel supply.

3. Theoretical background.

When the piezoelectric element operates as a control valve actuator, voltage is applied to it [10,11]. The force developed by the piezo drive can be determined by the following expression:

$$F_{piezz} = \frac{d_{33} \cdot E \cdot A}{t} \cdot V - \frac{E \cdot A}{N \cdot t} \cdot x - b \cdot \dot{x} \quad (1)$$

where d_{33} - piezoelectric strain constant
E - modulus of elasticity of the material
V - driving voltage
A - cross sectional area of the piezo actuator
t - thickness of each individual layer
N - number of stack layers
b - damper configuration
x - tip stroke
The damper configuration is calculated by formula

$$b = \frac{\sqrt{km}}{Q} \quad (2)$$

where m - piezo weight
k - stiffness of the piezo element
Q - mechanical quality factor

$$Q = \frac{d_{33} \cdot E \cdot A}{t} \cdot x - C^S V \quad (3)$$

The relationship between the electromechanical coupling factor and the dielectric constant can be expressed as follows:

$$C^S = \frac{\varepsilon_T AN(1-K_{33}^2)}{t} \quad (4)$$

where ε_T is the dielectric constant
K_{33} is the electromechanical coupling factor calculated by formula

$$K_{33} = \sqrt{\frac{E}{\varepsilon_T}} \quad (5)$$

An analysis of expressions (1)-(5) allows us to conclude that capacity of the piezoelectric element can act as a parameter describing its health and properties.

4. Materials and methods.

The data on failures of the piezoelectric hydraulic nozzles were collected in 2017-2018 in Irkutsk. Mileage was recorded for each set of nozzles, after which they were checked on the test desk. In total, over 500 electro-hydraulic nozzles with a piezoelectric element produced by the BOSCH company were tested.

The nozzle was tested on the Dieselland CR-JET desk, where the cyclic supply and control flow were determined according to the test plan. The following parameters of the piezoelectric element were determined using the Cristina Piezo instrument (Figure 2):

Figure 2. 2 Diesel test – CRistina Piezo.

-piezoelectric resistance;
- insulation resistance;
- smoothness;
- capacity.

The temperature was determined using a pyrometer. The piezoelectric element and nozzle were heated up to 500C for 8 hours during 7 days. Mathematical data were processed in the MatLab environment (MathWorks).

5. Results and discussion.

After checking 500 nozzles, it was found that the injector life varies for one car and injector failures are caused by wear of the control valve, defects of the atomizer and piezoelectric elements. Not all failures are recorded by the self-diagnosis system. The processed experimental data allowed us to establish the share of worn control valves (38%), atomizers (23%), and piezoelectric elements (10%). (Fig. 3) The nozzle malfunction is due to the wear of one or several elements. About 29% of causes of wear were functional (see Fig. 3).

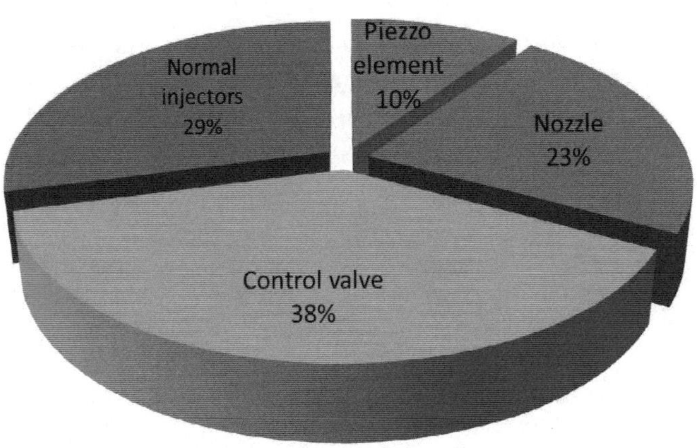

Figure 3. The diagram of correlation of failures of the electro-hydraulic piezo nozzle Common Rail.

Depending on the mileage, injector's failure probability distribution densities were built for the piezoelectric element, the atomizer and the control valve. The distribution laws and statistical characteristics were determined. These data are summarized in Table 1. Figure 4 shows the piezoelectric elements' failure probability distribution densitiy depending on the mileage of a vehicle equipped with the Common Rail power supply system with piezo-drive nozzles.

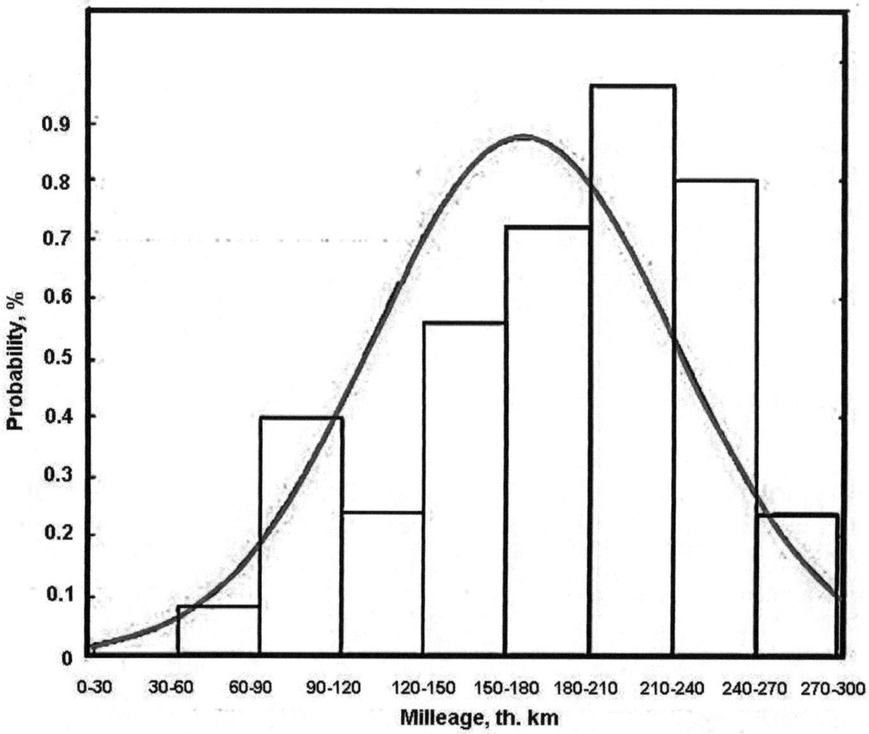

Figure 4. The diagram of failure probability distribution densitiy for the piezoelectric elements of electro-hydraulic nozzles depending on the vehicle mileage.

Table 1. Statistical characteristics of the failure distribution laws for the elements of electro-hydraulic nozzles.

		Statistical Characteristics			
Element	Distribution law	Math. expectation, thousand km	Mean-square deviation, thousand km	Variation ratio	Confidence interval, thousand km (p=0,95)
Valve	Norm	166	57,3	0,34	51,4…280,6
Nozzle	Norm	171	56	0,32	59…283
Piezo-element	Norm	171,7	55,3	0,32	61,1…282,3

To confirm the theoretical assumptions about the relationship between piezoelectric packet's capacity and health, malfunctioning injectors were checked. The nozzles were divided into two groups:

1. Malfunctioning nozzles which have no cyclic feed in maximum load conditions.

IOP Conf. Series: Materials Science and Engineering 1061 (2021) 012022 doi:10.1088/1757-899X/1061/1/012022

2. Technically sound (which may have deviations from the standards values of the test plan).
In order to determine the range of values corresponding to technically sound and malfunctioning conditions, statistical characteristics of the laws of distribution of capacity of the piezoelectric element were identified (Fig. 5). The data are summarized in Table 2

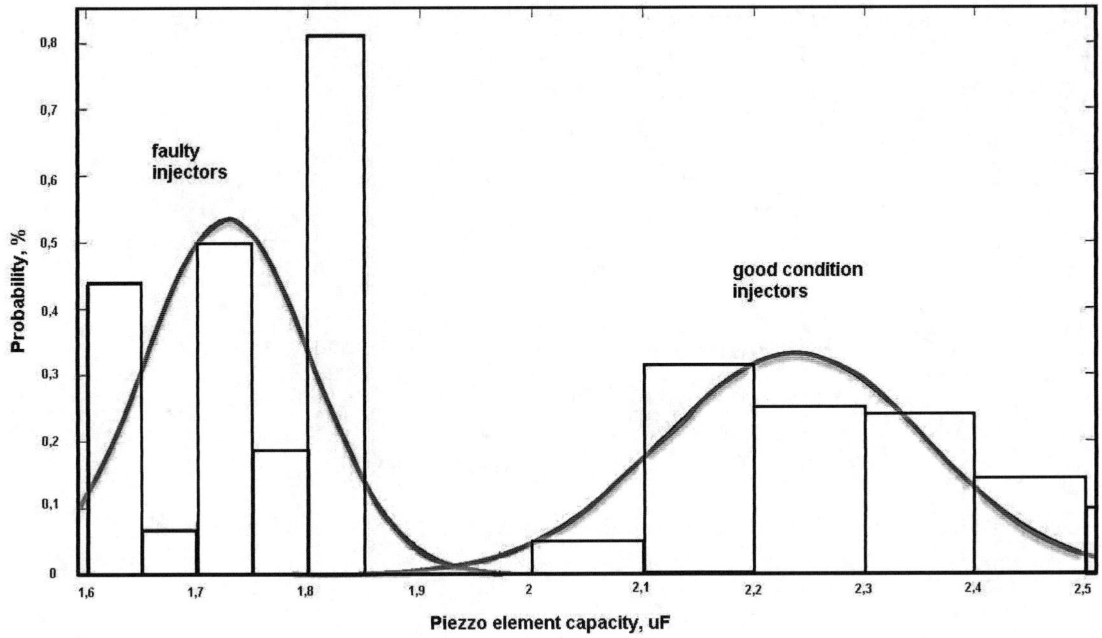

Figure 5. The diagram of probability density distribution for capacity of the piezoelectric element of technically sound (with deviations from the standards) and malfunctioning nozzles (without cyclic feed).

Table 2. Statistical characteristics of the failure distribution laws for the piezoelectric elements of electro-hydraulic nozzles.

Health	Statistical Characteristics		
	Distribution law	Math. expectation, uF	Mean-square deviation, uF
Technically sound (with deviations)	Norm	2,23	0,120
Malfunctioning	Norm	1,727	0,07

By heating various modifications of the piezoelectric injector, the dependence of capacity of the piezoelectric packet on its temperature was determined (Fig. 6).

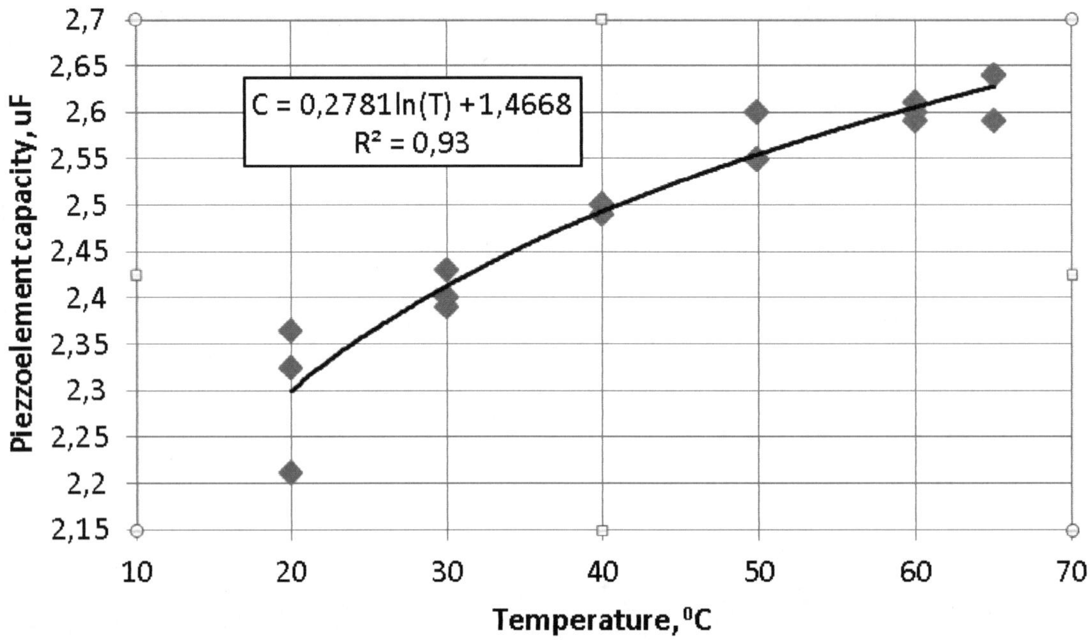

Figure 6. The diagram of dependence of nozzle piezo-packet's capacity on temperature.

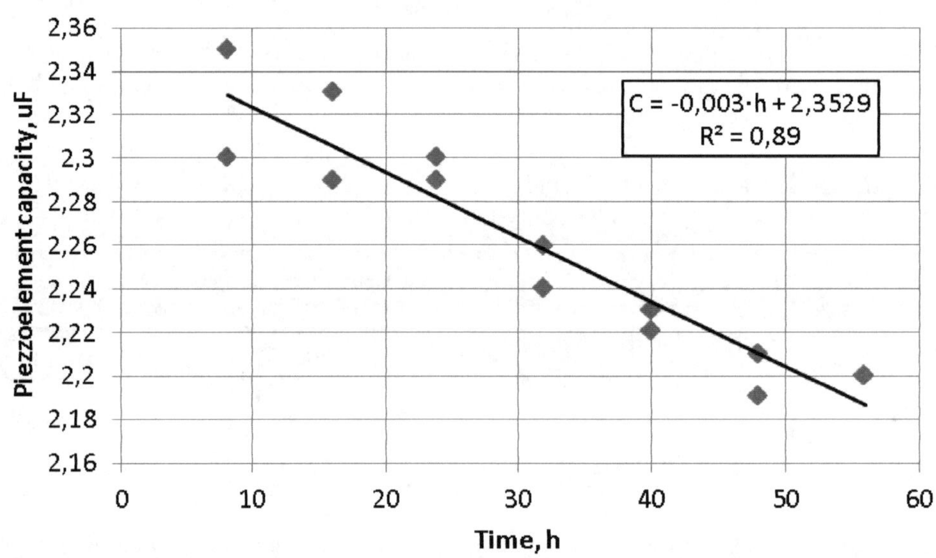

Figure 7. The diagram of dependence of capacity of the piezoelectric element on the duration of temperature exposure.

An analysis of the data shown in Figure 5 allows us to conclude that capacity of the piezoelectric element increases with an increase in temperature. Statistical data obtained after processing the test results of more than 50 nozzles "116" show that the numerical range of capacity of the technically sound nozzle is 2.1-2.4 uF at room temperature.

The nozzles were exposed to longer temperature effects. During 7 days, these nozzles were heated up to 50^{0}C. The temperature dependence is shown in Fig. 7.

6. Conclusion

1. Life cycle is different for different nozzles. Up to 30% of nozzles tested may be suitable for further operation during the intercontrol run.

2. The largest share of failures is due to wear of the control valve (38%) and abrasive wear of the control valve, the atomizer and the piezoelectric element.

3. Failures of the elements of piezoelectric injectors are governed by the normal distribution law and can occur jointly or separately. For each element, the statistical characteristics are similar.

4. The average life cycle of these injectors is 166 thousand km, the coefficient of variation is 0.32-0.34, the standard deviation is 55.3 thousand km.

5. The optimal frequency of monitoring is 50-60 thousand kilometers. It should be timed to the next scheduled maintenance.

6. Capacity of the piezoelectric packet is an informative diagnostic parameter of the electro-hydraulic nozzle's health; when developing the distribution laws, the error area was <3% of the total area of the distribution law. The mathematical expectation of capacities of the piezoelectric element: for the technically sound injector, the mathematical expectation is 2.23 uF with a standard deviation of 0.120 uF; for malfunctioning injectors (without cyclic feed) - 1.727 uF with a standard deviation of 0.07 uF.

7. With an increasing temperature, capacity of the piezoelectric element increases according to the logarithmic dependence

8. During operation, capacity of the piezoelectric packet decreases due to the degradation of properties of the piezoelectric packet.

References

[1] Reif K 2010 *Moderne Diesel-Einspritzsysteme* (Wiesbaden: Vieweg+Teubner Verlag)

[2] Palamondon E and Seers P 2014 Development of a simplified dynamic model for a piezoelectric injector using multiple injection strategies with biodiesel / diesel-fuel blends *Applied Energy* **131** pp 411–424

[3] Bianchi G M, Falfari S, Pelloni P, Kong S-C and Reitz R D 2002 Numerical Analysis of High Pressure Fast-Response C.R. Injector Dynamics *SAE Technical Paper Series 2002-01-0213*

[4] Xin-yuan Huang Zhen-ming Liu 2019 Research on Drive Control of Piezoelectric Injector May *DOI: 10.12783/dtcse/icaic2019/29464*

[5] Pogulyaev Y D, Baitimerov R M and Rozhdestvenskii Y V 2015 Detailed dynamic modeling of common rail piezo injector *Procedia Engineering. International Conference on Industrial Engineering, ICIE 2015* pp 93–98

[6] Yakimov I V, Krivtsov S N, Potapov A S and Svirbutovich O A 2019 Fuel flow and pressure in common return line as a diagnostic parameter of electro-hydraulic injectors technical state *IOP Conf. Series: Materials Science and Engineering* **632** *012058 IOP Publishing doi:10.1088/1757-899X/632/1/012058.*

[7] Stoeck T 2020 Methodology for Common Rail fuel injectors testing in case of non-typical faults *Diagnostyka* **21(2)** pp 25–30

[8] Yordanov N, Hadjiev K and Stankov E. 2018 Experimental simulation of common rail electromagnetic injectors wearing *International Scientific Journal "Machines, Technologies, Materials"* **12(4)** pp 208–211

[9] Krivtsov S N, Yakimov I V and Ozornin S P2018 Numerical Analysis and Experimental Studies on Solenoid Common Rail Diesel Injector with Worn Control Valve. *IOP Conf. Ser.: Mater.Sci. and Eng.* **327** *042057 doi:10.1088/1757-899X/327/4/042057*

[10] Byounggul Oh, Seungsuk Oh, Kangyoon Lee, Myoungho Sunwoo 2007 Development of an Electromechanical Model for Piezo Actuated Common Rail Injectors *SAE Technical paper 2007-01-3537*

[11] Zelong Zhang, Zhenming Liu, Guangyao Ouyang, Xiaofeng Li 2012 Study on Design of Drive Circuit for Piezoelectric Actuator *Advances in Electronic Engineering, Communication and Management* **Vol.1** *DOI: 10.1007/978-3-642-27287-5_58*

Mathematical analysis of the model of a low-melting metal coating on the surface of the guide

M A Mukutadze, E O Lagunova, N S Zadorozhnaya

Rostov State Transport University, Rostov-na-Donu, Russia

E-mail: murman1963@yandex.ru, elagun0va@yandex.ru, simon@sfedu.ru

Abstract. The work presents the asymptotic solution of the system of differential equations in the form of series in powers of the parameter due to the melt of the surface of the guide covered with a low-melting metal alloy and the rate of dissipation of mechanical energy, taking into account the dependence of the viscosity of the lubricant on pressure. Subsequently, a self-similar solution was found, as a result of which the fields of velocities, pressure and the main operating characteristics, load capacity and friction force were determined when the working gap was not completely filled with lubricant.

1. Introduction

The increase in loads on friction units, which affect their durability, efficiency and reliability, depends mainly on the design and quality of the bearings [1-9]. Therefore, when designing new machines and mechanisms, it becomes necessary to take into account additional factors. In the existing theoretical calculation methods [10-12] for plain bearings, one of the drawbacks in most cases is the fact that the incomplete (pre-emergency state) filling of the working gap with the lubricant is not taken into account at all, or is partially taken into account.

Thus, the development of a mathematical calculation model with an incomplete filling of the working gap with a lubricant when the surface of the guide is coated with a low-melting metal alloy is one of the promising areas of theoretical research in modern tribology. The latter determines the novelty and relevance of the results obtained.

2. Materials and Methods

A mathematical model is proposed that describes the motion of an incompressible true-viscous lubricant for a "thin layer" taking into account the dependence of viscosity and many additional factors in case of incomplete (pre-emergency state) filling of the working gap with lubricant from pressure. Also, a comparative analysis of the newly obtained results and the existing ones was carried out, which confirmed the approximation of the new model for solving practical problems.

A wedge-shaped sliding bearing is considered. It is assumed that the surface between the slider and the guide is not completely filled with lubricant, the slider is stationary, and the guide is covered with a low-melting metal alloy and moves around the axis Oy at speed u* (Fig. 1).

The following dependence gives the dependence of the viscosity of the lubricant on pressure:

$$\mu' = \mu_0 e^{\tilde{\alpha} p'}, \tag{1}$$

where μ_0 is the characteristic viscosity, μ' is the coefficient of dynamic viscosity of the lubricant, p' is the hydrodynamic pressure in the lubricating layer, $\tilde{\alpha}$ is a constant.

In the Cartesian coordinate system (Fig. 1), the equation of the linear slider contour and the guide molten surface are written as:

$$y' = h_0 + x' tg\alpha, \quad y' = -\Phi(x), \tag{2}$$

where h_0 is the thickness of the lubricating film in the initial section, Φ_0 is the thickness of the molten layer in the initial section, $\Phi(x)$ is the function characterizing the thickness of the melt of the guide surface, α is the angle of inclination of the linear contour of the slider to the axis Ox'.

The initial basic equations are the dimensionless equation of motion of a viscous incompressible fluid in the "thin layer" approximation, the continuity equation and the equation that determines, taking into account the expression for the rate of dissipation of mechanical energy, the molten contour of the guide. In the Cartesian coordinate system, the above system of equations with boundary conditions will be written in the form:

$$\frac{\partial v}{\partial y} = 0, \quad \frac{\partial^2 v}{\partial y^2} = e^{-\alpha p}\frac{dp}{dx}, \quad \frac{\partial v}{\partial x} + \frac{\partial u}{\partial y} = 0, \quad \frac{d\Phi(x)}{dx} = K\int_{-\Phi(x)}^{h(x)}\left(\frac{\partial v}{\partial y}\right)^2 dy, \tag{3}$$

$$v = 0, \quad u = 0 \quad at \quad y = h(x) = 1 + \eta x, \quad v = -1, \quad u = 0 \quad at \quad y = -\Phi(x),$$

$$\Phi = -\frac{h^*}{l} \quad at \quad x = 0, \quad p(x_1) = p(x_2) = 0, \quad \Phi(0) = 0. \tag{4}$$

where u, v are the components of the velocity vector of the lubricating medium, $\eta = \frac{l tg\alpha}{h_0}$, $K = \frac{2\mu u^*}{h_0 L'}$ are the parameters due to the melt and the rate of dissipation of mechanical energy, L' is the specific heat of fusion per unit volume, x_1, x_2 are the angular coordinates of the beginning and end of the free surface of the lubricant, respectively.

Dimensionless quantities are related to dimensional quantities by the following relationships:

$$x' = lx, \quad y' = h_0 y, \quad v_{x'} = u^* v, \quad v_{y'} = u^*\frac{h_0}{l}u, \quad p' = p^* p, \quad p^* = \frac{\mu_0 u^* l}{h_0^2}, \quad \mu' = \mu_0\mu, \quad \alpha = p^*\tilde{\alpha}.$$

(5)

Let us introduce a notation, let $z = e^{-\alpha p}$. Differentiation of both sides of the equality leads to the following form of equation (3):

$$\frac{\partial^2 v}{\partial y^2} = -\frac{1}{\alpha}\frac{dz}{dx}, \quad \frac{\partial v}{\partial x} + \frac{\partial u}{\partial y} = 0, \quad z\frac{d\Phi(x)}{dx} = K\int_0^{h(x)}\left(\frac{\partial v}{\partial y}\right)^2 dy, \tag{6}$$

with the boundary conditions:

$$v = 0, \quad u = 0 \quad at \quad y = h(x) = 1 + \eta x, \quad v = -1, \quad u = 0 \quad at \quad y = 0,$$

$$z(x_1) = z(x_2) = 1. \tag{7}$$

Write down the boundary conditions u and v for on the contour $y=-\Phi(x)$:

$$v(0-H(x))=v(0)-\left(\frac{\partial v}{\partial y}\right)\bigg|_{y=0}\cdot H(x)-\left(\frac{\partial^2 v}{\partial y^2}\right)\bigg|_{y=0}\cdot H^2(x)+...=-1$$

$$u(0-H(x))=u(0)-\left(\frac{\partial u}{\partial y}\right)\bigg|_{y=0}\cdot H(x)-\left(\frac{\partial^2 u}{\partial y^2}\right)\bigg|_{y=0}\cdot H^2(x)+...=0 \quad (8)$$

We seek the asymptotic solution of the system of differential equations (3) - (4) and (8) in the form of series in powers of the parameter K:

$$v(x,y)=v_0(x,y)+Kv_1(x,y)+K^2v_2(x,y)+...$$
$$u(x,y)=u_0(x,y)+Ku_1(x,y)+K^2u_2(x,y)+...$$
$$\Phi(x)=-\Phi_0(x)-K\Phi_1(x)-K^2\Phi_2(x)-...$$
$$z(x)=z_0(x)+Kz_1(x)+K^2z_2(x)+... \quad (9)$$

Substituting (9) into the system of differential equations (3) - (4), we obtain the equations:

– for zero approximation (not taking into account melting):

$$\frac{\partial^2 v_0}{\partial y^2}=-\frac{1}{\alpha}\frac{dz_0}{dx},\quad \frac{\partial v_0}{\partial x}+\frac{\partial u_0}{\partial y}=0,\quad \frac{d\Phi_0(x)}{dx}=0, \quad (10)$$

with the boundary conditions:

$$v_0=0,\quad u_0=0\quad at\quad y=1+\eta x,\quad v_0=-1,\quad u_0=0\quad at\quad y=0,$$
$$z_0(x_1)=z_0(x_2)=1,\quad \Phi_0=\frac{h_0^*}{l}. \quad (11)$$

– for the first approximation (taking into account the melting of the low-melting coating):

$$\frac{\partial^2 v_1}{\partial y^2}=-\frac{1}{\alpha}\frac{dz_1}{dx},\quad \frac{\partial v_1}{\partial x}+\frac{\partial u_1}{\partial y}=0,\quad -z_0\frac{d\Phi_1(x)}{dx}=\int_0^{h(x)}\left(\frac{\partial v_0}{\partial y}\right)^2 dy, \quad (12)$$

with the boundary conditions:

$$v_1=\left(\frac{\partial v_0}{\partial y}\right)\bigg|_{y=0}\cdot\Phi_1(x),\quad u_1=\left(\frac{\partial u_0}{\partial y}\right)\bigg|_{y=0}\cdot\Phi_1(x),$$
$$v_1=0,\quad u_1=0\quad at\quad h(x)=1+\eta x,\quad z_1(x_1)=z_2(x_2)=0. \quad (13)$$

The solution to the problem for the zero approximation (not taking into account melting) will be sought in the form:

$$v_0=\frac{\partial\psi_0}{\partial y}+V_0(x,y);\quad u_0=-\frac{\partial\psi_0}{\partial y}+U_0(x,y)$$
$$\psi_0(x,y)=\tilde{\psi}_0(\xi);\quad \xi=\frac{y}{h(x)};$$

$$V_0(x,y)=\tilde{v}_0(\xi);\quad U_0(x,y)=\tilde{u}_0(\xi)\cdot h'(x). \tag{14}$$

Substitution of (14) into the system of differential equations (10) - (11) allows obtaining the following system of differential equations:

$$\tilde{\psi}_0'''=\tilde{C}_2;\quad \tilde{v}_0''=\tilde{C}_1;\quad \tilde{u}_0'-\xi\tilde{v}_0'=0;\quad \frac{dz_0}{dx}=-\alpha\left(\frac{\tilde{C}_1}{h^2(x)}+\frac{\tilde{C}_2}{h^3(x)}\right); \tag{15}$$

with the boundary conditions:

$$\tilde{\psi}_0'(0)=0,\quad \tilde{\psi}_0'(1)=0;\quad \tilde{u}_0(1)=0,\quad \tilde{v}_0(1)=0;$$

$$\tilde{u}_0(0)=0,\quad \tilde{v}_0(0)=-1;\quad \int_0^1\tilde{v}_0(\xi)\,d\xi=0. \tag{16}$$

By direct integration, we get

$$\tilde{\psi}_0'(\xi)=\frac{\tilde{C}_2}{2}(\xi^2-\xi),\quad \tilde{v}_0(\xi)=\tilde{C}_1\frac{\xi}{2}+\left(1-\frac{\tilde{C}_1}{2}\right)\xi-1,\quad \tilde{C}_1=-6. \tag{17}$$

From the condition $z_0(x_1)=z_0(x_2)=1$ up to terms of the second-order of smallness $O(\eta^2)$ for $\tilde{C}_2$, we obtain the expression:

$$\tilde{C}_2=6\left(1+\frac{\eta}{2}(x_2-x_1)\right). \tag{18}$$

Now we can write:

$$z_0=3\eta\alpha\left((x-x_1)^2-(x-x_1)(x_2-x_1)\right)+1. \tag{19}$$

The definition $\Phi_1(x)$, taking into account equation (17), is reduced to the following equation:

$$\frac{d\Phi_1(x)}{dx}=\frac{h(x)}{z_0}\int_0^1\left(\frac{\tilde{\psi}_0''(\xi)}{h^2(x)}+\frac{\tilde{v}_0'(\xi)}{h(x)}\right)d\xi. \tag{20}$$

Integration of equation (20) leads to:

$$\Phi_1(x)=\left(\int_0^x\frac{\Delta_1dx}{h^3(x)}+\int_0^x\frac{\Delta_2dx}{h^2(x)}+\int_0^x\frac{\Delta_3dx}{h(x)}\right)\cdot 0{,}9998. \tag{21}$$

$$\Delta_1=\int_0^1\left(\Phi_0''(\xi)\right)^2d\xi=\frac{\tilde{C}_2^2}{12};\quad \Delta_2=\int_0^1 2\Phi_0''(\xi)\tilde{v}_0'(\xi)d\xi=\tilde{C}_2;\quad \Delta_3=\int_0^1\left(\tilde{v}_0'(\xi)\right)^2d\xi=4. \tag{22}$$

The solution of equations (21) – (22) up to terms of the second-order of smallness leads to:

$$\Phi_1(x)=0{,}9998\left(13x-\frac{25}{2}\eta x^2+9\eta(x_2-x_1)\right). \tag{23}$$

Taking into account (23), boundary conditions (13) have the following form:

$$v_1=0,\quad u_1=0\quad at\quad y=1+\eta x,\quad u_1=\left(\frac{\partial u_0}{\partial y}\right)\Bigg|_{y=0}=0,$$

$$v_1(x)=\left[1+5\eta x-3\eta(x_1+x_2)\right]\cdot\Phi_1(x),\qquad z_1(x_1)=z_2(x_2)=0. \tag{24}$$

The solution of the equation of motion of a viscous incompressible fluid for the first approximation (12) taking into account the boundary conditions (24) using the Leibniz formula on the parameters is as follows:

$\frac{\partial}{\partial x}\int\limits_0^{1+\eta x} v_1(x)dy = 0$, for z_1 we get:

$$z_1 = -\alpha\left[117\eta\left(x_2^3 - x^3\right) + \frac{117}{2}\eta\left(x_1 + x_2\right)\left(x_2^2 - x^3\right)\right]\cdot 0,9998, \tag{25}$$

then

$$z = z_0 + Kz_1 = 2,99\alpha\eta\left[\left(x - x_1\right)^2 - \left(x - x_1\right)\left(x_2 - x_1\right) - K\left(39\left(x_2^3 - x^3\right) + \frac{39}{2}\left(x_1 + x_2\right)\left(x_2^2 - x^2\right)\right)\right] + 1 \tag{26}$$

Applying the expansion of the function $e^{-\alpha p}$ up to terms inclusive $O\left(\alpha^2\right)$ and solving the resulting equation up to $O\left(\alpha^3\right)$, for the hydrodynamic pressure we obtain:

$$p = 2,99\eta\left[\left(x - x_1\right)^2 - \left(x - x_1\right)\left(x_2 - x_1\right) - K\left(39\left(x_2^3 - x^3\right) + \frac{39}{2}\left(x_1 + x_2\right)\left(x_2^2 - x^2\right)\right)\right]. \tag{27}$$

Determine the bearing capacity and friction force:

$$W = p^* l \int\limits_{x_1}^{x_2} p dx = \frac{\mu_0 u^* l^2}{h_0^2}\left[2,99\alpha\eta\left[\frac{x_2^3}{3} - \frac{x_1^3}{3} - x_1\left(x_2^2 - x_1^2\right) + x_1^2\left(x_2 - x_1\right) - \left(\frac{x_2^2}{2} - \frac{x_1^2}{2}\right)\left(x_2 - x_1\right) + \right.\right.$$

$$\left.\left. + x_1\left(x_2 - x_1\right)^2 - K\left[39\left(x_2^3\left(x_2 - x_1\right) - \frac{x_2^4}{4} - \frac{x_1^4}{4}\right) + \frac{39}{2}\left(x_2^2\left(x_2^2 - x_1^2\right) - \left(x_1 + x_2\right)\left(\frac{x_2^3}{3} - \frac{x_1^3}{3}\right)\right)\right]\right]\right],$$

$$L = \mu\int\limits_{x_1}^{x_2}\left[\left.\frac{\partial v_0}{\partial y}\right|_{y=0} + K\left.\frac{\partial v_1}{\partial y}\right|_{y=0}\right]dx = \frac{\mu_0\left(1 - \alpha p + \frac{\alpha^2 p^2}{2}\right)}{1,0018}\left[\left(x_2 - x_1\right) + \frac{5}{2}\eta\left(x_2^2 - x_1^2\right) - 3\eta\left(x_2^2 - x_1^2\right) + \right.$$

$$\left. + K\left[\left(x_2 - x_1\right) + 2\eta\left(x_2^2 - x_1^2\right) - \frac{117}{2}\left(x_2^3 - x_1^3\right) + \frac{117}{4}\eta\left(x_1 + x_2\right)\left(x_2^3 - x_1^3\right)\right]\right]. \tag{28}$$

In an experimental study, a wedge-shaped sliding bearing with a low-melting metal coating made of Wood's alloy was considered (Table 1). Based on the results of the experiments, the value of the friction coefficient was determined. The value makes it possible to judge the presence of a hydrodynamic friction regime both when the bearing is operating with a lubricant with Newtonian properties, and when the fusible coating of the guide surface is melted from Wood's alloy. The temperature regime and the transition of the hydrodynamic friction regime to boundary friction were also determined. The analysis of experimental studies shows that the melt of a low-melting coating made of Wood's alloy affects the friction coefficient 2.5 – 4 times more intensively than the rheological properties of the applied liquid lubricants. A set of experimental studies, which confirmed the reliability of the developed theoretical models and the data of their numerical analysis, in the considered range of design and operational parameters of wedge-shaped sliding bearings with low-melting metal coatings from Wood's alloy, as a result of satisfactory convergence of theoretical and experimental results.

Table 1. Comparison of theoretical and experimental research

	№	Theoretical research		Experimental research
		Thrust bearing without fusible coating	Wood alloy low-melting metal-coated thrust bearing	Wedge-shaped slide bearing with wood alloy low-melting metal coating
Coefficient of friction	1	0,0038	0,0019	0,0028
	2	0,0042	0,0021	0,0031
	3	0,0045	0,0022	0,0034
	4	0,0047	0,0023	0,0035
	5	0,0051	0,0024	0,0037

3. Results

Theoretical studies have shown that the bearing capacity with a low-melting metal coating of the guide surface, taking into account the dependence of the general rheological properties of the lubricant used and the melt of a low-melting metal coating with valid viscosity properties on pressure, increases ≈ by 5 – 8% with an increase in the parameter α, which characterizes the dependence of viscosity on pressure, and the length $(x_1 - x_2)$ of the loaded region, and the coefficient of friction, in this case, decreases ≈ by 8 – 9%.

Based on the calculated models obtained in the theoretical part, an experimental study was carried out, because of which the area of promising operation of the developed tribosystem was determined.

As a result of experimental studies, tribological characteristics have been determined, which make it possible to judge the presence of the duration of the hydrodynamic friction regime and the reliability of the developed theoretical calculation models and the data of their numerical analysis.

4. Conclusions

New multivariable expressions have been developed for the main operating characteristics (bearing capacity and friction force) of a wedge-shaped sliding bearing, taking into account the rheological properties of a true-viscous lubricant with an incomplete filling of the working gap, and also taking into account the melt of the guide surface covered with a low-melting metal alloy.

An assessment of the influence of the parameters of variable factors caused by the melt of the surface of the guide covered with a low-melting metal alloy and the dependence of the viscosity of the lubricant on pressure when the working gap is not completely filled is given.

The obtained refined design models of wedge-shaped sliding bearings allow adjusting the ratio of its bearing capacity and the coefficient of friction as a result of varying the coating from a low-melting metal coating on the surface of the guide.

Satisfactory convergence of theoretical and experimental studies was established in support of the theoretical conclusions.

Acknowledgments
The work has been carried out under the grant of OJSC RZD No. 2210370 / 22.12.2016 for the development of scientific and pedagogical schools in the field of railway transport.

References

[1] Kragelsky I V 1968 *Friction and wear* (Moscow: Mechanical engineering)
[2] Bowden F P and Tabor D 1968 *Friction and lubrication of solids* (Moscow: Mechanical engineering)
[3] Kropachev D Y, Grishin A A and Maslo A D 2012 Methods for the Rapid Measurement of the Temperature of Metal Melt for the Needs of Machine-Building Enterprises *Cast and Metallurgy* **3** (66) pp 126–127
[4] Levanov I, Doykin A, Zadorozhnaya E and Novikov R 2017 Investigation of Antiwear Properties of Lubricants with the Geo-Modifiers of Friction *Tribology in Industry* **39** (3) pp 302–306
[5] *Physical Quantities. Directory.* 1991 (Moscow: Energoatomizdat)
[6] Khavin V Y 1991 *Brief Reference Book of the Chemist* (Moscow: Khimiya)
[7] Perelman V I 1964 *Brief Reference Book of the Chemist* (Moscow, Leningrad: Khimiya)
[8] Prokopiev V N, Zadorozhnaya E A, Karavaev V G and Levanov I G 2010 Improvement of the Calculation Technique for Complex Loaded Sliding Bearings Lubricated With Non-Newtonian Oils *Problems of machine building and machine reliability* **1** pp 63–67
[9] Mukchortov I V, Pochkaylo K A and Zadorozhnaya E A 2016 The Influence of Anti-Wear Additives on the Bearings Hydro-Mechanical Characteristics *Procedia Engineering* **150** pp 607–611
[10] Prikhodko V M and Kotelnitskaya L I 2001 Mathematical Model of Hydrodynamic Lubricant during Melting of Bearing Surface of Radial Bearing *Friction and wear* **22** (6) pp 606–608
[11] Akhverdiev K S, Mukutadze M A, Lagunova E O and Vasilenko V V 2017 Hydrodynamic Calculation of a Radial Bearing Lubricated By a Melt of a Fusible Coating in Presence of a Lubricant *Bulletin of Rostov State University of Transport* **2** (66) 129–135
[12] Lagunova E O 2017 Simulation of Wedge-Shaped Sliding Supports Taking into Account the Rheological Properties of an Electrically Conductive Lubricant *Bulletin of Southern Ural State University. Series "Machine Building"* **4** pp 13–25 DOI: 10.14529/engin170402

Excavator implement for pipeline repair

A B Letopolsky[1], P A Korchagin[1], and I A Teterina[1]

[1] Siberian State Automobile and Highway University (SibADI), 644089, Omsk, Mira Ave., 5 Russia

Iateterina@mail.ru

Abstract. The article presents an option to improve the design of the equipment of an excavator of the 3 dimensional group. The equipment is intended for selective repair of the pipeline by an excavator, without involving other road construction equipment (for example, a pipeline). The proposed version of the design of the working body allows you to increase the speed of repair of the pipeline, by reducing the time required to change equipment, its relocation, which as a result reduces the cost of repair work. The main idea is to supplement the main equipment of the excavator with a structure consisting of three teeth, forming, in conjunction with the bucket, a gripping mechanism. To confirm the operability of the proposed design, theoretical studies were carried out aimed at determining the stability of the machine for the most dangerous design operating positions, and the permissible sizes (pipe sizes) were established, with which the stability of the machine is maintained.

1. Introduction

The creation of an effective energy infrastructure, including pipeline transport, occupies one of the priority places both in the long-term socio-economic development of the country and in the strategic activities of the Ministry of Energy of the Russian Federation. At present, 100% of the gas produced, 93% of the oil produced, and more than 20% of the refined products are transported via trunk pipelines [1,2].

During operation, pipelines and their elements wear out. The nature of wear can be very different and is determined by the operating conditions, the properties of the material from which the pipeline is made, its design features, insulation quality, etc [1].

Overhaul is a repair performed to restore serviceability and complete (or close to complete) restoration of the resource of an object with the replacement or restoration of its components, including base parts. Overhaul of the linear part is divided into the following types: with the replacement of pipes; with replacement of insulating coating; selective repair [3].

Selective repairs include: repair of the pipeline section with the elimination of defects in the pipe wall or with the installation of the coupling; repair of the pipeline section with the replacement of the "coil", pipe, assembly of linear fittings. [3].

It is possible to reduce the time and financial costs of performing operations on selective repair of the pipeline by improving the equipment of the excavator, which is the basic machine for performing selective repair of the main pipeline [4].

Researching

To improve the design of excavator equipment for selective repair of the pipeline, confirming its operability by checking the machine for stability and making the necessary calculations of the hydraulic system.

In order to increase the speed of repair and reduce the cost of work, the authors developed the design of the equipment of the excavator, which allows selective repair of the pipeline without involving additional equipment (Figure 1).

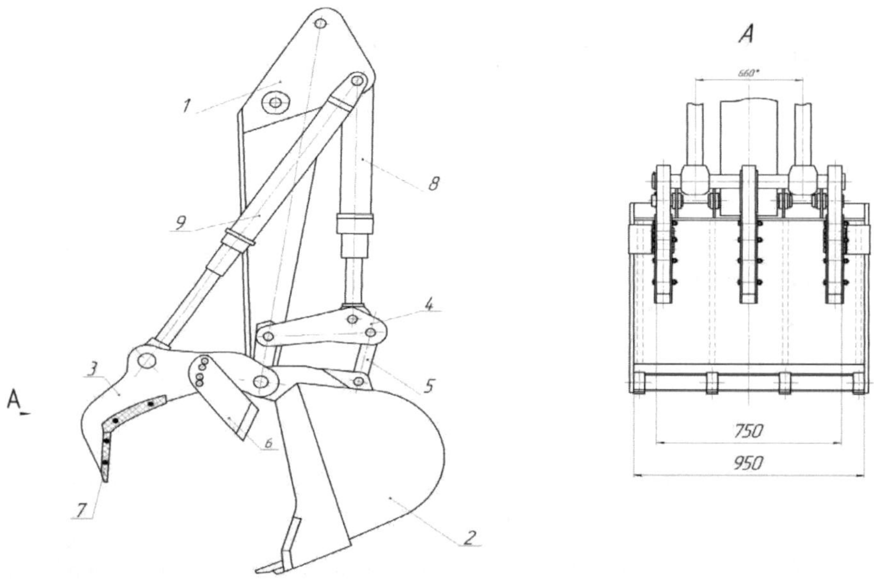

a. b.

Figure 1. General view of the equipment: a – main view; b – left view

The proposed design of the working equipment is: 1 – handle; 2 – bucket; 3 – capture; 4 – thrust; 5 – bucket traction; 6 – limiter; 7 – rubber pad; 8 – bucket hydraulic cylinder; 9 – capture cylinder. Welding of all elements of the working equipment is carried out in accordance with GOST 5264-80 [5].

Capture consists of three teeth located in parallel and fixed on one base. The edges of the teeth have a rubberized coating to prevent deformation of the pipe during the lifting [6]. Designed equipment allows you to develop soil in the places of repair of pipelines and to dismantle the opened pipeline using gripping devices. The working equipment allows selective repair of pipelines with a diameter of 720 ...1220 mm [7].

2. Theoretical studies of the design of equipment

Theoretical studies require two tasks:

1. To carry out calculations of the hydraulic system of the equipment;
2. Carry out calculations to determine the stability of the machine [8].

It is possible to determine the forces on the rod of the capture hydrocylinder by projecting onto the y axis equal $y_i = 0$.

$$-G_{c1} + F \cdot \sin 60° - G_{c2} - G_p - G_{k+t} - G_e = 0, \tag{1}$$

where F – force of pressing teeth to the surface of the pipeline, N; G_{c1} – weight of arrow 1; G_{c2} – weight of arrow 2; G_p – handle weight; G_{k+t} – weight of bucket and pipe; G_e – the weight of the excavator.

$$F = \frac{G_{c1} + G_{c2} + G_p + G_{k+t} + G_e}{\sin 60°}. \tag{2}$$

The net power of the hydrocylinder was determined by the formula [9,10]

$$N = F \cdot V, \tag{3}$$

where N is a useful power developed by a hydrocylinder, W; F – force on stock, N; V – stock speed, m/s.

The actual speed of the stock was determined by the formula [11]

$$V_D = \frac{Q_{fc}}{S_{ef}}, \tag{3}$$

where V_D – actual stock speed, m/s; Q_{fc} – fluid consumption, m^3/s; S_{ef} – is an effective piston area, m^2.

For the piston working cavity, the effective area of the piston was determined by the formula [12]

$$S_{ef} = \frac{\pi D^2}{4}, \tag{4}$$

where D is the piston diameter, m.

The payload power of the pump is determined by the formula [11]

$$N_{pp} = \Delta p_p \cdot Q_s , \tag{5}$$

where N_{pp} – useful power of the pump, W; $\Delta P_{н}$ – pressure drop across the pump, Pa; $\Delta p_{н} = p_{nom}$; Q_s – pressure supply, m^3/s.

To confirm the operability of a machine equipped with a modernized equipment, theoretical studies were conducted aimed at determining the stability of the machine and preventing its overturning [13]. This parameter was determined through the stability coefficient, which is determined by the formula [13].

$$K_y = \frac{M_y}{M_o}, \tag{6}$$

where M_y – the moment of all the forces holding the excavator from tipping over; M_o – the moment of all the forces contributing to the overturning of the excavator.

When determining the stability coefficient, the size group of the machine, the type of equipment, the working position of the machine and a number of other parameters are taken into account [2]. For different types of interchangeable equipment and different operating modes, its own limits of stability coefficient are recommended. Under normal conditions, $K_y = 1.1 ... 1.2$ [14]

The efficiency of the modernized equipment excavator was calculated to work with pipes with a diameter of 720, 1020 and 1220 mm.

3. Research results

The stability of the excavator is checked not only for the conditions under which the working process is performed, but also for the conditions of transportation of the unit, when changing equipment, etc. [15].

Theoretical studies were carried out for two design positions of the excavator, taking into account the variable shoulders of the gravity equipment and the machine, where the running gear is located perpendicularly and parallel to the trench (Figure 2).

Figure 3 shows the graphical dependencies constructed during theoretical studies. It can be seen from the graph that the stability of the excavator is fully preserved for both the first and second working position. The x axis shows that stability is maintained when working with pipes having a diameter of 720, 1020 and 1220 mm.

Also, as a result of research, it was found that the operation of the machine with a pipeline, the diameter of which is 1420 mm, with the use of equipment in combination with a gripper, leads to a rollover. In other words, the stability coefficient of the machine in position II is below the established coefficient for machines of this size category.

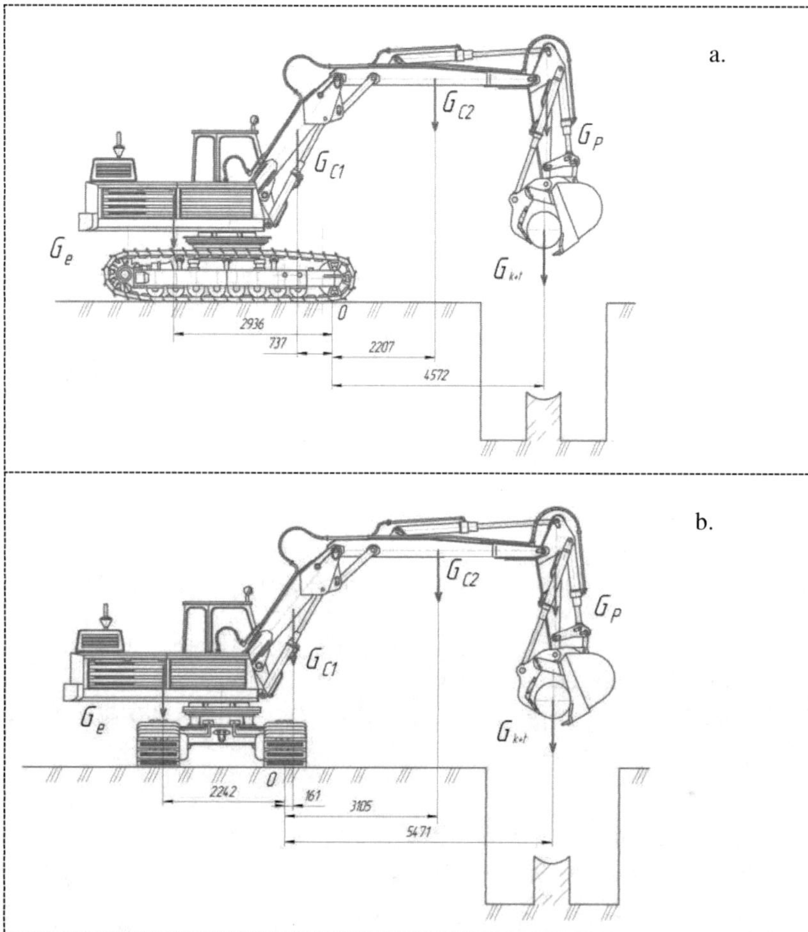

Figure 2. The design position of the excavator: a – location of the running gear perpendicular to the trench; b – location of the running gear parallel to the trench

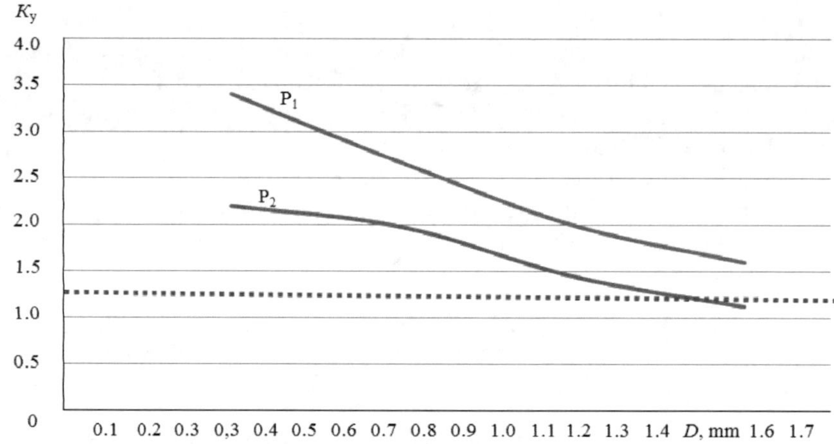

Figure 3. Dependence of the stability coefficient from pipe diameter at excavator positions: P_1 – the location of the running gear perpendicular to the trench; P_2 –arrangement of the running gear parallel to the trench

4. Conclusion

The presented calculated dependences make it possible to determine what engine power of the base machine is required for the productivity of the trencher.

Formulas for calculating the productivity of the machine, taking into account the speed of movement of the base machine and the remote capacity of the equipment, allow you to determine the pace of the repair cycle using the advanced design of the digging machine.

The design of the cutter drum for removing soil from under the main pipeline with a diameter of 720 mm has been developed. The cutter drum is installed on trencher with dual trencher implement, and is mounted perpendicular to a copying tool. This design allows you to exclude a digging machine from the standard fleet of vehicles, which reduces the cost of its operation and relocation.

And it also allows you to dig up the pipeline in one pass, which in turn saves time that was previously spent on installing the digging machine. The proposed technical solution allows for capital repairs of pipelines with the replacement of the outer insulation coating without lifting the pipeline from the trench and maintaining its position.

The design of the equipment of a trencher for opening the pipeline and cleaning the annular space from the ground can significantly increase the productivity of the construction equipment involved in the repair work and reduce the time spent carrying out this kind of work.

References

[1] Veselukha V M and Bogdanovich A V 2016 Methodology for assessing the ultimate state of pipes of the linear part of an oil pipeline with typical defects according to the criteria of crack resistance *Actual issues of engineering* (Minsk: United institute of mechanical engineering nan Belarus) **5** pp 214–217

[2] Letopolsky A B, Korchagin P A and Teterina I A 2020 Working equipment of the single-bucket excavator for the development of frozen ground *Mechanical Engineering and Materials Science (ICMTMTE 2019*) 044027 DOI: 10.1088/1757-899X/709/4/044027

[3] Klemba K V, Gritsenko A V, Averyanov Yu I and Plaksin A M 2020 Investigation of the indicators «operator-machine-environment» system in the technological process *Agribusiness of Russia.* (Chelyabinsk: South Ural State Agrarian University) **27** (1) pp 95–104

[4] Veselukha V M and Bogdanovich A V 2016 Methodology for assessing the ultimate state of pipes of the linear part of an oil pipeline with typical defects according to the criteria of crack resistance *Actual issues of engineering* (Minsk: United institute of mechanical engineering nan Belarus) **5** pp 214–217

[5] Prokhorova E V and Sevryugina N S 2015 Quick-detachable connecting device for the working equipment of the excavator *Technology of wheeled and tracked vehicles* (Moscow:Research Institute of Automobile Structures) **5** pp 26–32

[6] Korchagin P A, Letopolsky A B and Teterina I A 2018 Result of research of working capability of refined pipelayer equipment *Aviamechanical engineering and transport (AVENT 2018)* (Paris: Atlantis press) **158** pp 416–420

[7] Vashchuk I M, Utkin V I and Harkun B I 1989 *Pipelayers* (Moscow: Mechanical Engineering) p 180

[8[Gorelov V P 2011 *Fundamentals of engineering* (Novosibirsk: NGAVT)

[9] Pobegailo P A 2015 On the proportions of the working equipment of single-bucket hydraulic excavators *Mountain Information and Analytical Bulletin (scientific and technical journal*) (Ulan-Ude: Publishing house of VSGUTU) **2** pp 143–147

[10] Zenkov S A 2017 Method of defining rational parameters for excavator buckets vibrating devices in order to reduce soil adhesion *IOP Conf. Series: Earth and Environmental Science* **87** 082057 doi.org/10.1088/1755-1315/87/8/0820572017

[11] Tarasik V P and Savitsky V S 2019 Methodology for designing a mechanism for controlling the friction clutches of a hydromechanical transmission of a mining truck *Actual issues of engineering* (Minsk: Joint Institute of Mechanical Engineering of Belarus) **8** pp 45–52

[12] Teterina I A, Korchagin P A and Letopolsky A B 2019 Investigation of soil destruction by trench chain excavator cutting element process *Lecture Notes in Mechanical Engineering* **52** pp 2123–2132.
[13] Troshin D I and Nikolaev V A 2019 Increasing the efficiency of excavator buckets *Construction and road machines* (Moscow: sdm-press) **11** pp 22–25
[14] Lukashuk O N, Komissarov A P and Letnev K Yu 2018 Operation of single-bucket excavator transmission system *IOP Conf. Series: Materials Science and Engineering* chapter **323** 012016
[15] Unaspekov B A, Abdukalikova G M and Sivachenko L A 2017 Development of innovative equipment and ways of its use in engineering and technology of mechanical engineering *Modern problems of the theory of machines* (Novokuznetsk: Research Center "Engineering") **5** pp 101–105

Use of subminiature eddy current transducers for composite inspection

V Malikov[1], A Ishkov[2], A Katasonov[1], S Dmitriev[1], D Fadeev[1], A Sagalakov[1], A Grigorev[1]

[1] Physical-technical faculties, Altai State University, Barnaul, Russia
[2]Department of technology of construction materials and repair of machines, Altay State Agricultural University, Barnaul, Russia

E-mail: osys11@gmail.com

Abstract. The sensor for studying composite materials of the metal-polyethylene-metal type was developed on the basic of the designed ultra-miniature eddy-current transducer. The test of the developed sensor was carried out on standard defects of composite materials in the form of an inhomogeneity defect in the metal and (or) polyethylene layer. This sensor uses the original converter design, produced according to the differential circuit with the ability to localize the contact area of the sensor and the material up to 0.1-0.5 mm^2. The measurement procedure for detecting defects in composite materials is also given in the article. The article presents the correlations of the output signal of the developed sensor based on the aluminum-polyethylene-aluminum composite material, in which the defect was located. We also obtained standard images of model defects of a composite such as a layered discontinuity. The images obtained as a result of processing information from the developed sensor make it possible to draw conclusions about the topology of the field near violations of the internal structure, the type of this violation and its localization in layers of material.

1. Introduction

There are several basic methods of non-destructive testing in the industry, among which the most widespread are ultrasonic examinations, fluoroscopic scanning and eddy current method. The eddy current method will be considered in more detail in this study. This method is based on the analysis of the interaction of an external electromagnetic field with the electromagnetic field of eddy currents induced by a driving coil in an electrically conductive object of control by this field. In control systems that use the effects of eddy currents, the key role is assigned to special eddy current transducer sensors, since any change in the goals and scanning conditions requires changes in the design of such sensors. For example, one of the actual goals of scanning may be to determine the thickness of coatings, both capable of conducting an electric current and dielectric. Such coatings are usually placed on various substrates, both conductive and dielectric. In this case, eddy current monitoring converters are often implemented as clamp-on transformer sensors. As a rule, such circuits include exciting (transmitting) and measuring (receiving) coils. The first is designed to guide an alternating magnetic field in the scanned sample, which excites eddy currents in the object under study; the second – fixes the indicators of this field. Such schemes are also actively used to study the properties of non-ferromagnetic materials associated with electrical conductivity.

Eddy current non-destructive testing systems have their own specificity, which consists in the need for constant changes in the design of sensors in accordance with the conditions of a specific task. These circumstances are due to the fact that the key informative indicator of this approach is the voltage directed to the measuring winding of the converter. It correlates with the consolidated characteristic of the β0 sensor, which is a function of the nature of the substance, the type and parameters of internal disturbances in the structure, parameters and structure of the converter, as well as the frequency of the exciting alternating electromagnetic field. Also when calculating the electrical characteristics of the sensor, it is necessary to take into account the defect-sensitive material parameter, the location and parameters of possible violations of the internal structure of materials, and the locality of the eddy current converter.

Electromagnetic non-destructive testing (NDT) is widely used for the evaluation of conductive materials [1–3]. One common approach, eddy current testing (ECT), employs a system of coils that induces eddy currents in a conductive sample and subsequently monitors changes in the secondary magnetic field generated by the eddy currents [2, 4]. Defects in the sample distort the eddy currents, resulting in a change in the magnetic field detected by the coils. ECT has limitations, including shallow penetration depth, an inability to detect defects with interfaces parallel to the surface, and the requirement that the probe maintain minimal lift-off from the test material. Despite these limitations, ECT is widely used for NDT and inspection because of the advantages offered in terms of cost, portability and ease of interpretation of results. Currently, the practical utility of ECT is restricted to conductive materials. However, with the growing use of nonconductive materials in industrial applications that require non-destructive evaluation, there is a need for an approach for these materials with comparable advantages to ECT.

Measuring systems capable of scanning metal-polymer layered composites (MPLC), materials consisting of one or several metal layers alternating with dielectric layers, are of considerable interest. This is due to the widespread practical distribution of these materials in industry, everyday life and special industries. So MPLC are used in the production of sheet cladding materials, in heat and sound insulation of aviation, special equipment coatings that can neutralize the effect of radiation, as well as in materials for the production of printed circuit boards. In everyday life, metal-plastic pipes are often found that combine an aluminum layer and several layers of various thicknesses of cross-linked high or low pressure polyethylene (HPP, LPP) [2].

Typical defects of composites are disruption of the continuity of one or another layer, the thickness of these layers, violations of the surface of the metal layer, flaws in the interfaces between the metal and the polymer, the state of the bridges between the layers. At the same time, there are several parameters indicating the presence of these defects in materials. Among them is the magnitude of the local ability of a substance to conduct electric current and its distribution over the surface, associated with the topology of the electromagnetic field in the material interacting with the indicated defects [3].

Eddy current method (ECT) have been recently used for inspecting metals composites [5–9]. During ECT testing, the conductive or semiconductive material is excited by a coil with high frequency alternating currents. Defects including cracks, voids or delamination alter the eddy current distribution around the defect. ECT have been recently used for the inspection of composision materials [10-12]. For example, Liang Cheng et al. [13] compared ECPT with ultrasound and flash thermography for the detection of delamination in composites. However, to date, a quantitative comparison of the damage detection performance of eddy-current method for composite and metallic materials is still missing [14]

We have previously reported on the development of an effective system for scanning the ability of non-ferromagnetic materials to conduct current, which includes a Fourier analyzer of the incoming signal – MECNFMwFA-v5 (Meter of electrical conductivity of non-ferromagnetic materials with Fourier analyzer version 5) [4]. In this system, a subminiature ECT (SMECT) of a special structure was used, made according to the differential circuit for switching on the circuits of the transformer converter and allowing localizing the control boundaries down to 50 μm^2 and below. The system is capable of measuring the local electrical conductivity on the surface of the material with a parallel visual representation of the frequency response of the receiving circuit of the transducer in real time. This makes it possible for us to use this device in a defectoscopic mode, comparing the readings on the scanned sample with the image of the internal structural disturbance previously obtained on the model. The present study aims to study the prospects for detecting defects in metal homogeneity in a composite based on aluminum and LPP of the A1- (LPP-Me)п p-A1 type, as well as visual representation of the images of such defects using an MECNFMwFA-v5 device.

2. Materials and method

The study used an MECNFMwFA-v5 device equipped with a manual transducer [15, 16] with a locality of 0.1-0.5 mm^2. The connection between the sensor and the device was realized through the Bluetooth wireless communication protocol, which allows maintaining contact at a distance of up to 2.3 m. The software part of the scanning complex was installed on a computer for processing and storing the output signal. Converter excitation parameters: current 30-35 mA, voltage 3.5-4.5 V, operating frequency 0.3-300 kHz. To calibrate the device and the communication line with the sensor, copper with a certified electrical conductivity of 415 ± 2 MSm / m was chosen as a standard sample.

To test the performance of the scanning system, composites were made using copper or aluminum foil with a thickness of 20-100 μm. Foil layers were alternated with films of polyethylene or waxed paper, 20 and 50-100 μm thick, respectively. Further, within the framework of the technical process, the necessary work pieces with a size of 20x50 mm were cut out from the resulting materials. A package of the required structure was formed, a foil or dielectric layer with a defect was inserted, and it was pressed at a temperature of 110-120°C and a pressure of 5-10 MPa for 3.5 min.

Violations of the internal structure of the composite were reproduced by skipping or increasing the number of individual interlayers, and disruption of continuity and bridges – by creating a round or rectangular hole in the foil or dielectric. The thickness control of individual layers of the composite and the entire package was carried out using an MK 25 micrometer.

3. Results and discussion

Within the framework of the study, the main value of the composite, depending on the presence of violations of the internal structure, is the ability of the metal layer to conduct electric current. Earlier it was illustrated that it enters into the inhomogeneous Helmholtz equation for the vector potential of the eddy current field of a multilayer medium along with the magnetic permeability of the material. It should also be noted that this characteristic tends to the value of the electrical conductivity of a solid bulk material in accordance with an increase in the number of layers in the composite and as the thickness of non-conductive layers decreases. The influence that is felt on the measuring contour of the SMECT at high locality and the ratio of the radii of the exciting and measuring contours of at least 0.2-0.4 from each new layer is from 10 to 25%. This will be enough for registration by scanning systems with an absolute permissible error of up to 3.5%.

Close to these values of the influence on the level of the applied voltage of the converter shows an increase in the number of dielectric layers or thickness in the composite. At the same time, the conclusion about the admissibility of the scanned sample solely for this parameter requires the tuning of the scanning system from the size of the gap between the surface of the material and the transducer. This need is due to the fact that in the standard model its contribution to the measured parameter of the sensor is comparable to the values presented earlier. In this regard, we have fixed the dependence of the input data of the MECNFMwFA-v5 device on the distance between the surface of the composite and the SMECT.

At the same time, there are ways to ensure that the readings of the scanning system are detuned from the gap between the sensor and the sample surface in a wide range and to increase the sensitivity by placing a control sample on the surface of a more massive known and defect-free material.

Fig. 1. shows the correlation of incoming readings from MECNFMwFA-v5 when its sensor is removed from the surface of the A1-LPP-A1 composite, located on a copper base, 5 mm thick. In this case, the frequency during the measurement was 35 kHz, and the composite consisted of two layers of aluminum foil and one layer of low pressure polyethylene with a thickness of 20 μm each and alternating with each other. Copper was chosen as the base material.

Thus, the composite can be considered as a shielding element, which is located between the copper base and the sensor of the scanning system and gradually reduces the measured signal. At a distance of up to 250 μm from the surface of the composite, an area is recorded where these devices keep constant values. It can be assumed that at such a distance, the topology of the eddy current field of the scanned material is noted by the system sensor as a field from a homogeneous medium, and the presence of structural violations in the material will cause a deviation of its readings from the average value measured at a distance of 100-250 μm.

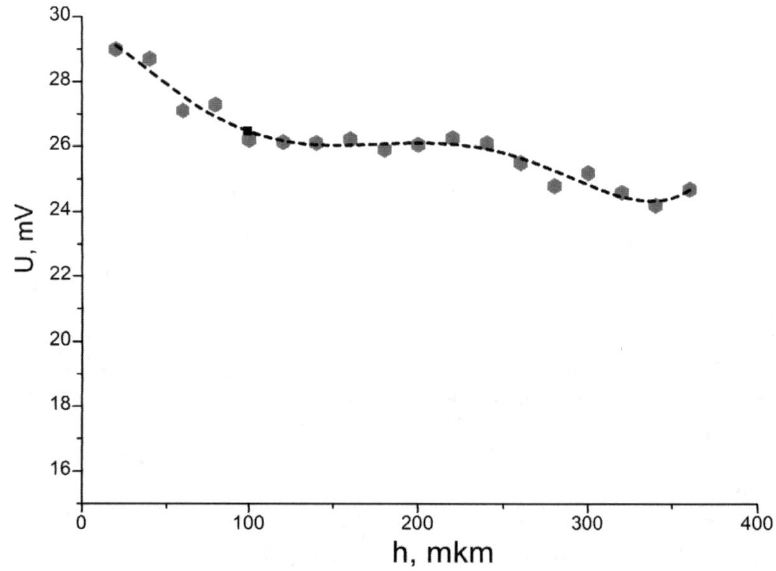

Figure 1. Correlation between the conditional electrical conductivity of a 3-layer composite L1-LPP-L1, placed on a copper base, and the gap to the SMECT.

In this case, scanning of defects is possible when debugging the readings of the device on an area known to be free of defects and using a hollow tip, which allows maintaining the distance between the surface of the scanned object and the transducer in the range of 100-250 μm.

Measurements carried out on defects caused by irregularities in the order of layers or their number for multilayer composites A1 (Cu) - LPP-A1 (Cu), showed the relationship between the value of the conditional electric current conductivity up to 3-7 mV in the defect zone, with an absolute permissible error devices 0.1-0.1-0.3 conventional units, or in other words 20-30 times excess of the defect signal over the noise. The presence of any of the listed violations in the scanned area leads to a significant change in the field topology and relative device data.

The amplitude-time correlation of the readings during the automated or manual examination of the material surface, which has already been introduced into MECNFMwFA-v5, can be considered as an image of a violation.

Determination of discontinuities in the composite, as well as interlayer closure and a number of others is available by direct contact measurement of the material surface with a sensor. In this case, the system converts the time into a coordinate of the sensor's position relative to the point of origin of movement, taking into account the speed of the sensor. In this case, the dependence of the change in the sensor signal on its position can serve as an image of a defect, and the main frequency of the signal and its amplitude are displayed on the screen, and the coordinate of the transducer is displayed along the X-axis.

Fig. 2. illustrates the standard images of model defects of a composite, such as a layered discontinuity, obtained by the device, when placing a sample of MPLC on a dielectric base.

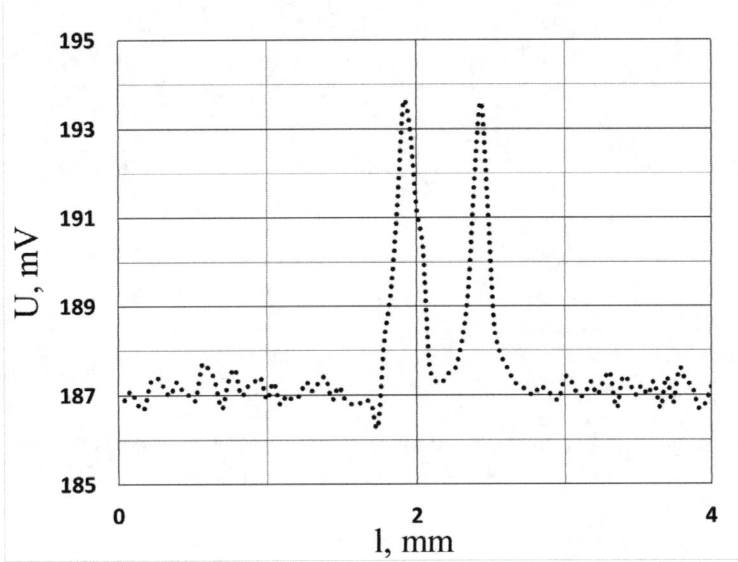

Figure 2.a. An image of a discontinuity in an insulated Al-layer.

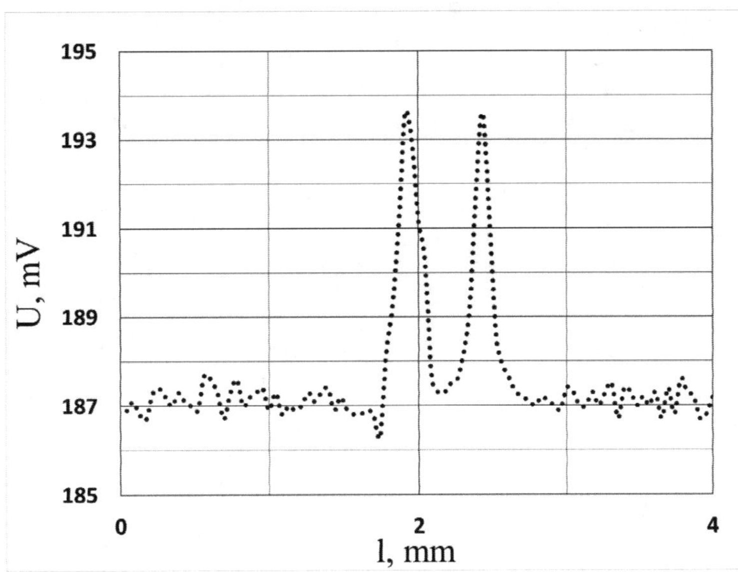

Figure 2.b. An image of a discontinuity in a 2-layer Al-LPP composite scanned from the dielectric side.

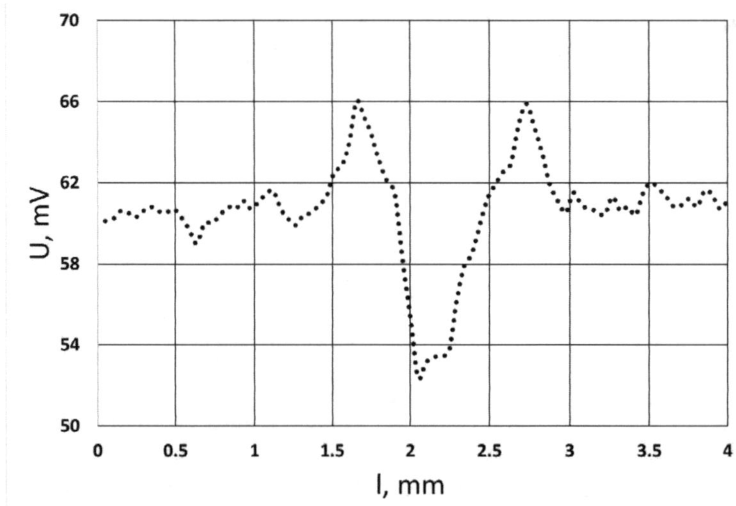

Figure 3.a. An image of a discontinuity in the bottom layer of a 3-layer composite Cu-LPP-Cu, scanned from the metal side.

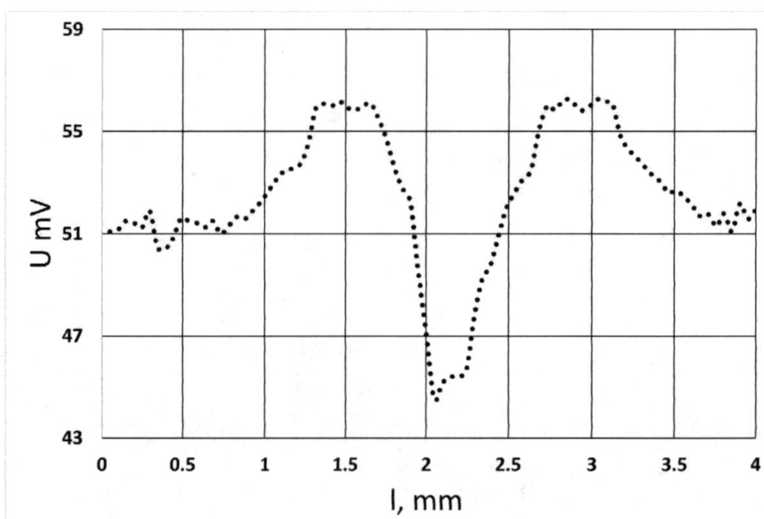

Figure 3.b. An image of a discontinuity in the bottom layer of a and a 5-layer composite $(Cu\text{-}LPP)_2\text{-}Cu$, scanned from the metal side.

Fig. 2, b clearly shows the fact that the measured signal amplitude from the defect of the interlayer discontinuity hidden in the 2-layer MPLC with the scanning mode set is identical to the change in the signal received from the visible defect (Fig. 2, a).

In the course of receiving a signal from such violations of the internal structure, MECNFMwFA-v5 shows a correlation between the signal amplitude and the position of the ECT. This makes it possible to perform the initial adjustment of the MECNFMwFA-v5 and train personnel using direct visualized models hidden between the layers of the composite. As the number of composite layers above the defect increases, the image display on the screen changes somewhat (Fig. 3.).

Fig. 3 shows the transmission of the incoming signal with constant measurement of the composite, while within the boundaries of its localization at the fundamental excitation frequency, an area of significant growth of the boundaries of the signal change appears, and symmetrically limited to the left and right by less intense signals.

The complication of the structure of the scanned sample and an increase in the distance to the defect leads to a gradual expansion of the region (Fig. 3b), which opens up prospects for increasing the accuracy of determining its localization both in the lower and in the inner layers. The connection of the signal frequency of the transducer with the level of its amplitude increases the convenience of reconstructing the topology of the fields of currents of an eddy nature near the defect and along the layers of the composite.

4. Conclusions

1. The MECNFMwFA-v5 device allows direct scanning of material defects according to indications due to the high locality of the sensor and the presence of a detected dependence of the electrical conductivity of the composite on model defects.

2. The MECNFMwFA-v5 device using a transducer with a locality of up to 50 μm^2 showed the independence of the measurements of the electrical conductivity of the composite type A1 (Cu) - (LPP-A1 (Cu))$_n$-A1 (Cu) from the gap in the range from 100 to 250 μm.

3. The image obtained by scanning with the MECNFMwFA-v5 device allows making conclusions about the topology of the field near violations of the internal structure, the type of this violation and its localization in the layers of material.

5. Acknowledgments

The reported study was funded by RFBR according to the research project № 20-38-90062.

References

[1] Sophian A and Tian G Y 2001 Insight **43** 302–306
[2] Auld B A and Moulder J C 1999 J. Nondestruct. Eval. **18** 3–36
[3] Garcia-Martin J, Gomez-Gil J and Vazquez-Sanchez E 2011 Sensors **11** 2525–2565
[4] Lee H, Jane E and Kevin M 2018 Journal of Nondestructive Evaluation **37** 58
[5] Xiaojuan Xu, Hongli Ji, Jinhao Qiu and Toshiyuki Takagi 2018 Int J Appl. Electromagn. Mech. **57(2)** 177–92
[6] Liu Zhiping, Lu Ge and Liu Xingle 2017 IEEE Instrum Meas Mag **20(4)** 34–44
[7] Schmitt G, Lenzmann C and Schneider N 2017 Mater. Corros **68** 1285–94
[8] Oswald-Tranta B 2018 Appl Sci **8** 257
[9] Ibarra-Castanedo C, Piau J, Guilbert Sand Avdelidis N P 2009 Res. Nondestruct. Eval. **20** 1–31
[10] Mizukami K, Mizutani Y, Kimura K and Sato A 2016 Composites Part A **90** 261–70
[11] Gao B, Lu P, Woo W L, Tian G Y, Zhu Y and Johnston M 2018 IEEE Trans. Ind. Electron. **65(10)** 8142–8152
[12] Abidin I Z, Tian G Y, Wilson J, Yang S and Almond D 2010 NDT E Int. **43(7)** 537–546
[13] Cheng L and Tian G Y 2002 J Sensors 408–437
[14] Zijun W, Junzhen Z, Gui Yun T and Francesco C 2019 NDT and E International **107** 102–155
[15] Dmitriev S F, Malikov V N, Sagalakov A M, Ishkov A V 2017 Mater. Sci. Forum 906 147-151
[16] Ishkov A, Dmitriev S, Malikov V and Sagalakov A 2016 AIP Conf. Proc. 1785 626-629.

IOP Conf. Series: Materials Science and Engineering 1061 (2021) 012026 doi:10.1088/1757-899X/1061/1/012026

Modeling of peat tractor semi-trailer motion

A V Mikhailov[1], **A I Zhigulskaya**[2], **YA Kasakov**[1]

[1] Mechanical Engineering Department, Saint Petersburg Mining University, 2, 21st Line V.O., Saint Petersburg, 199106, Russia

[2] Department of Technological Machines and Equipment, Tver State Technical University, 22, Nab. A. Nikitina, Tver, 170026, Russia

Corresponding author e-mail address: epc68@mail.ru

Abstract. Wheeled tractor vehicles are used to transport peat from open cast mines to storage stockpiles or processing facilities. This paper analyzes dynamic stability and control of an articulate vehicle. A mathematical model of linear dynamics is demonstrated for deformed supporting base and a two-tier transport unit consisting of an AWD tractor and a two-axle trailer. The functions obtained from analysis of maneuverability of the transport unit may be used in studies of operational and technical properties of the vehicles as calculation formulas to determine the trajectories of its components when transporting peat materials in peat mines. This analytic approach provided understanding of the propulsion system of the tractor vehicle and grounds for subsequent computer modeling of the process.

1. Introduction

Wheeled tractor vehicles are used to transport peat from open cast mines to storage stockpiles or processing facilities. The tractor transport unit is a combination of a power unit (tractor) and a transport unit (semi-truck) [1]. Transportation operations are related to process operation, usually they involve transfer of material without changes in its state.

Efficiency of movement of the tractor transport unit in peat extraction is facilitated by its cross-country performance and steerability. One of the most important operational properties of a tractor transport unit that influences its productivity is its steerability [2]. Characteristics of vehicle steerability on wet peat deposit significantly depend on its design, dimensions and weight of elements, the nature of their linkage, axle controlling laws, tire characteristics, modes and conditions of movement [3].

Energy property of a tractor transport unit is its ability to develop certain engine power and tractor pulling capacity. Tractor dynamics is related to mechanical modeling, as well as to mathematical description and analysis of the tractor systems [4, 5].

The general efficiency of a tractor transport unit operation depends on the efficiency of operation of its components. The process of substantiation of a quantitative composition of the tractor transport unit shall include the main characteristics of both the tractor and the semi-truck as separate test objects [3].

The subject of this paper is analysis of motion of the tractor transport unit used in peat extraction and consisting of an agricultural tractor and a semi-truck. The tractor dynamic models are important tools in research and development in the mining industry [6].

IOP Conf. Series: Materials Science and Engineering 1061 (2021) 012026 doi:10.1088/1757-899X/1061/1/012026

The objective of the work is to model the movement of the tractor transport unit in operational conditions in order to select the design parameters corresponding to the condition and modes of operation on weak soils.

2. Unit Description

The tractor transport unit is a mechanical linkage consisting of articulated transport elements. The principal kinematic and dynamic properties of the unit depend on physical phenomena taking place during its movement, which are in return determined by its structure. During consolidation of the tractor transport unit, its energy property does not have a fundamental importance in determining the quantitative composition of machinery in the unit, selecting their operational (in particular, speed-related) modes, etc. The main factor is flotation.

The flotation defines the ability of movement on weak soils without excessive immersion of wheels, when the possibility of movement depends on a ratio between the sliding surface resistance forces and the propulsive forces.

The main criteria in evaluation of unit's flotation is [7] the trailing weight (adhesion weight) of the tractor m_{tw} (G_{tw}) – a part of the tractor's weight that creates normal loads onto the tractor's driving wheels. The main factors influencing the traction between the tractor's wheels and the surface of peat deposit are:

- physical and mechanical properties of the soil (deposit type, shear strength, moisture content, etc.)
- weight per driving wheels – trailing weight G_{tw}, characterized by a normal reaction of soil surface R_s for the driving wheels of the tractor;
- design of the undercarriage, wheels type and size (diameter, rim width, tire pressure, etc.)

For known values of the wheel traction coefficient on weak soil μ, and the total movement resistance coefficient ψ of the peat deposit, the condition of passability for a tractor transport unit with a semi-truck, a total weight of M_{st} and the trailing weight of the tractor of M_{tw} is their conformance with the inequality [7]

$$\frac{M_{tw}}{M_{st}} \geq \frac{\psi}{\mu}.$$

For tractors on pneumatic tires, the traction coefficient between the tractor wheels and a weak soil is μ=0.35…0.55. For wheeled 4WD tractors, the operating weight is their trailing weight G_{tw}.

For agricultural general purpose wheeled tractor MTZ-1523 with wheels doubled and the body space of V_{tb}=8 m^3, the condition of passability is 0.67>0.40. The front wheels of the tractor are provided with 420/70R24 tires, the hind wheels – with 520/70R38 tires. Dimension of flotation tires of the semi-truck is 24.0/50-22.5.

Wheeled 4WD tractor is considered appropriate for peat extraction where the soil foundation is wet and soft, and tractor passability poses a serious problem.

3. System Description

When studying the movement modes of the unit, one may make a number of assumption that allow simplifying the movement equation and obtaining the formulas and criteria that determine the essential main parameters of maneuverability of the tractor transport unit with a constant hook load. Maneuverability analysis is performed at small speeds, without taking into account the dynamic effects, that is, from the point of view of kinematics of motion. The tractor transport unit is modeled as consisting of two articulated solid bodies that are moving in the horizontal plane at a constant travel speed [8].

At the first stage of solving the problem in the condition of small speeds and side accelerations, the influence of tire elasticity (lateral skid) on the trajectory of the tractor transport unit may be neglected. The movement of the tractor transport unit is assumed flat, parallel to a motionless supporting plane Q, without account for vertical movement of its parts. Motion of each segment of the unit is described by the motion of a figure that represents a projection of the segment onto the supporting plane, which has an OXY coordinate system rigidly fixed onto it. The tractor's and semi-truck's weight forces G_{tr} and G_{st} are applied to the centers of gravity of those segments.

Simplification of the actual dynamic systems shall be based upon three main principles: research objectives, operating modes and dynamic properties of the studied system [9].

It is possible to obtain a mathematical model of a wheeled machine by traditional methods using Lagrange equation of the second kind on the basis of the analytical model. In order to analyze dynamic parameters of the tractor under consideration, its design features, and to study the problem of interaction between the wheeled tractor and a semi-truck, a diagram is proposed (Figure 1), where the tractor and the semi-truck are idealized as two absolutely rigid bodies with their own centers of gravity.

The bodies in the unit are articulated with single-degree joint and are installed on elastic foundations, flexible in lateral, longitudinal and horizontal angular directions and characterized with stiffness coefficients of C_t, C_{st} [10, 11]. Its dynamics is described with a model that represents the movements of the semi-truck and the tractor deviating them from the course [8]. According to Figure 1, the local coordinate systems (x_1, y_1) and (x_2, y_2) are centered at the centers of gravity of the tractor and the semi-truck respectively [12].

When composing the motion equations for the system in question, the following assumptions may be made [10]:

- the wheeled tractor in the unit with the semi-truck is moving steadily, without descents or ascensions, there are no translational vibrations;
- the tractor is symmetric with respect of its longitudinal plane;
- the translational speed of the unit is constant;
- the wheels are involved in continuous motion without lateral skidding;
- tire stiffness of both tractor and its semi-truck are constant;
- the left and right doubled tires on the axles of the tractor and semi-truck may be merged into the equivalent tires;
- the angulation and the gliding angle of the tires are small, the motion equations are linear.
- no aerodynamic forces are influencing the tractor transport unit.

Assumption of a steady motion of the system without translational fluctuations is common in studying stability of motion and steerability of vehicles [13].

With these assumption, the position of the tractor transport unit with respect to the introduced coordinate system is determined by a course angle of the leading component and coordinates of the main trajectories of the points in the articulated component, which are functionally dependent on the position of the tractor's steerable wheels and the motion speed of the entire unit.

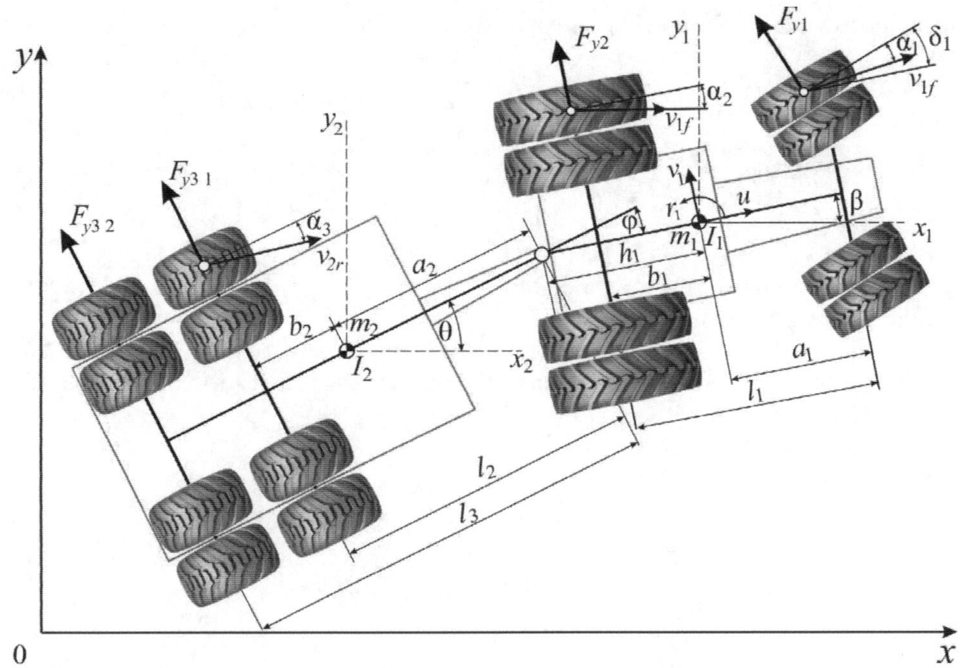

Figure 1. A coordinate system of the mathematical model for a curvilinear motion of the tractor transport unit.

Stability of motion and steerability of the tractor transport unit may be improved by determining its rational design and operational parameters. Analysis of this problem should employ mathematical modeling of main motions, allowing evaluating a degree of influence that various parameters have over stability and steerability. The mathematical model of the unit shall be as close to the actual unit as possible, taking into account the interactions between its components and those between the unit and the supporting surface.

In this model, lateral forces in the left and right wheels of both the tractor and the semi-truck were deemed equal and summed up together [14, 15], as the influence of the lateral load transmission onto the total turning force is small for small angles of steering wheel rotation at the unit's speeds (<1.5 m/s).

The mathematical description of the motion of the tractor transport unit motion may be represented with Lagrange's equation of the second kind [10, 11, 13]

$$\frac{d}{dt}\frac{\partial T}{\partial \dot{q}_i} - \frac{\partial T}{\partial q_i} + \frac{\partial U}{\partial q_i} = Q_i \qquad i = 1 \dots n,$$

where n is the number of degrees of freedom of the system; T is the system's kinetic energy; U is the system's potential energy; q_i is the i-th joint variable; Q is a generalized force along the i-ой joint variable.

The analysis uses a linear approach at small wheel slip angles. Let us apply the following coordinate relations reflected in Figure 1.

$$x = x_1 = x_2;$$

$$y_2 = y_1 - h_1\beta - a_2\Theta,$$

where h_1 is a distance between the tractor's center of gravity and the coupling point; β is the sliding angle of the semi-truck; a_2 is a distance between the semi-truck's center of gravity and the coupling point; Θ is the coupling point's sliding angle.

The kinetic energy of the system [11] T may be expressed as

$$T = 0.5\, m_1(\dot{x}^2 + \dot{y}_1^2) + 0.5m_2(\dot{x}^2 + \dot{y}_2^2) + 0.5I_1\dot{\beta}^2 + 0.5I_2\dot{\theta}^2 =$$
$$= 0.5(m_1 + m_2)(\dot{x} + \dot{y}_1^2) + 0.5m_2\left(h_1^2\dot{\beta}^2 - 2h_1\dot{y}_1\dot{\beta}\right)$$
$$+ 0.5m_2\left(a_2^2\dot{\theta}^2 - 2\dot{y}_1 a_2\dot{\theta} + 2h_1\beta\dot{a}_2\dot{\theta}\right) + 0.5I_1\dot{\beta}^2 + +0.5I_2\dot{\theta}^2,$$

where m_1 and m_2 are weights of the tractor and the semi-truck, respectively; I_1 and I_2 are the moments of inertia of the tractor and the semi-truck [12].

The mathematical description of the motion of the tractor transport unit has been made with an analytic method of engineering mechanics, namely, with Lagrange's equation of the second kind from the calculation diagram (Figure 1).

An expression for Lagrange's conditions

$$\frac{d}{dt}\frac{\partial T}{\partial \dot{x}} = (m_1 + m_2)\ddot{x};$$
$$\frac{d}{dt}\frac{\partial T}{\partial \dot{y}} = (m_1 + m_2)\ddot{y}_1 - m_2 h_1\ddot{\beta} - m_2 a_2\ddot{\theta};$$
$$\frac{d}{dt}\frac{\partial T}{\partial \dot{\beta}} = m_2 h_1\left(h_1\ddot{\beta} + \ddot{y}_1\right) + m_2 h_1 a_2\ddot{\theta} + I_1\ddot{\beta};$$
$$\frac{d}{dt}\frac{\partial T}{\partial \dot{\theta}} = m_2 a_2\left(a_{21}\ddot{\theta} + \ddot{y}_1 + h_1\ddot{\beta}\right) + I_2\ddot{\theta};$$
$$\frac{\partial T}{\partial q_i} = 0 \quad i = 1 \ldots n;$$
$$\frac{\partial U}{\partial q_i} = 0 \quad i = 1 \ldots n.$$

Using the virtual work, one may evaluate the generalized forces, related to the joint variables [16]. The virtual work with accounts for the generalized forces

$$\Delta W = \left(-F_{y1}\delta_1 - F_{y3}\varphi - F_{y4}\varphi\right)\Delta x + F_{y1}\Delta(y_1 + a_1\beta) +$$
$$+F_{y2}\Delta(y_1 - b_1\beta) + F_{y3}\Delta(y_1 - h_1\beta - l_2\theta) + F_{y4}\Delta(y_1 - h_1\beta - l_3\theta).$$

where δ_1 is the steerable wheels' turning angle; Δ is an angulation between the tractor and the semi-truck; F_{yki} is a lateral force created at the i-th tire on the axle k [12].

Generalized forces Q for each position

$$Q_x = -F_{y1}\delta_1 - F_{y31}\varphi - F_{y32}\varphi;$$
$$Q_{y1} = F_{y1} + F_{y2} + F_{y31} + F_{y32};$$
$$Q_\beta = a_1 F_{y1} - b_1 F_{y2} - h_1 F_{y31} - h_1 F_{y32};$$
$$Q_\theta = -l_2 F_{y31} - l_3 F_{y32}.$$

To express in the local coordinates with accounts for the speed of the tractor transport unit, let us apply the following correlations

$$\dot{x} = u - v_1\beta;$$
$$\dot{y} = v_1 + u\beta;$$
$$\dot{\beta} = r_1;$$
$$\theta = \beta + \varphi.$$

Thus, the forces on tires of the tractor and the semi-truck in local coordinates are [11, 6, 17] (Figure 1)

$$F_{Y1} = -\frac{1}{u}C_1(v_1 + a_1 r_r) + C_1\delta_1;$$

$$F_{y2} = -\frac{1}{u}C_2(v_1 - b_1 r_r);$$

$$F_{y31} = -\frac{1}{u}C_{31}(v_1 - h_1 r_r - l_2\dot{\theta}) + C_{31}\varphi;$$

$$F_{y32} = -\frac{1}{u}C_{32}(v_1 - h_1 r_r - l_3\dot{\theta}) + C_{32}\varphi.$$

where v_1 is a lateral speed in the tractor's center of gravity; r_1 is a yawing velocity of the tractor.

The motion equations may be expressed using only the local coordinates. System state vector

$$x = [v_1\ r_1\ \dot{\varphi}\ \varphi\ ]^T.$$

The first equation that describes the longitudinal dynamics is not taken into account, as the traveling speed u is assumed constant. The following shortcuts were used [16, 18]

$$C = C_1 + C_2;$$

$$C_t = C_{31} + C_{32};$$

$$C_{s_1} = a_1 C_1 - b_1 C_2;$$

$$C_{q_1^2} = a_1^2 C_1 + b_1^2 C_2.$$

where C is a total stiffness of the tractor in turns; $C_{1,2}$ is the stiffness on axles during turns; C_t is a total stiffness of the semi-truck during turns; s_1 is a distance along the longitudinal axis of the vehicle from its center of gravity to the non-slip point of the turn; q_1 is a length corresponding to the medium moment in turn.

Motion equations of a tractor transport unit [16, 11]

$$(m_1 + m_2)(\dot{v}_1 + ur_1) - \{m_2(h_1 + a_2)\}\dot{r}_1 - m_a a_2\ddot{\varphi} = -\frac{1}{u}[(C + C_t)v_1 + \{C_{s_1} - C_{31}(h_1 + l_2) -$$
$$-C_{32}(h_1 + l_3)\}r_1 - (C_{31}l_2 + +C_{32}l_3)\dot{\varphi} - C_t u\varphi] + C_1\delta_1 - hm_2(\dot{v}_1 + ur_1) + \{I_1 + +m_2 h_1(h_1 +$$
$$+a_2)\}\dot{r}_1 + m_2 h_1 a_2\ddot{\varphi} = -\frac{1}{u}[(C_{s_1} - C_t h_1)v_1 + \{C_{q_1^2} + C_{31}h_1(h_1 + l_2) + +C_{32}h_1(h_1 + l_3)\}r_1 +$$
$$+(C_{31}h_1 l_2 + C_{32}h_1 l_3)\dot{\varphi} + C_t h_1 u\varphi] + C_1 a_1\delta_1 - m_2 a_2(\dot{v}_1 + ur_1) + +\{I_2 + m_2 a_2(h_1 + a_2)\}\dot{r}_1 +$$
$$+(I_2 + m_2 a_2^2)\ddot{\varphi} = -\frac{1}{u}[-(C_{31}l_2 + C_{32}l_3)v_1 + \{C_{31}l_2(h_1 + l_2) + +C_{32}l_3(h_1 + l_3)\}r_1 + \{C_{31}l_2^2 +$$
$$+C_{32}l_3^2\dot{\varphi} + (C_{31}l_2 + C_{32}l_3)u\varphi],$$

where v_1 is a lateral speed in the tractor's center of gravity; r_1 is the yawing velocity of the tractor ($r_1 = \dot{\beta}$; φ is a relative angulation between the tractor and the semi-truck. The turning angle of the semi-truck is defined as θ when $\theta = \beta + \varphi$.

4. Conclusion

The mathematical model describing motion of a tractor with a semi-truck allows calculating current acceleration values from the values of forces and moments acting unto a peat transport unit. Passability conditions of the tractor transport unit were also studied in order to evaluate the parameters of the tractor's and semi-truck's tires. The functions obtained from analysis of maneuverability of the transport unit may be used in studies of operational and technical properties of the vehicles as calculation formulas to determine the trajectories of its components when transporting peat materials in peat mines. This analytic approach provided understanding of the propulsion system of the tractor vehicle and grounds for subsequent computer modeling of the process.

References

[1] Mikhailov A V and Kasakov Y A 2019 The fleet analysis at open-pit mining of peat. *Mining informational and analytical bulletin* **7** (special issue 20) p 16

[2] Zoz F M and Grisso R D 2003 Traction and tractor performance *ASAE Distinguished Lecture Series, Tractor Design* **27**

[3] Mikhailov A V et al 2017, Excavating and loading equipment for peat mining *IOP Conf. Ser.: Earth Environ. Sci.* **87** 022014

[4] Wong J Y 2001 *Theory of Ground Vehicles* 3-ed (John Wiley & Sons, New York, New York)

[5] Popp K 2010 *Ground vehicle dynamic Werner Schiehlen* (Springer Verlag Berlin Heidelberg) p 348

[6] Patil V 2017 Generic and complete vehicle dynamic models for open-source platforms *Master's thesis in Applied Mechanics. Department of Applied Mechanics* (Chalmers university of technology. Gothenburg, Sweden) p 112

[7] Kostyuchenkov N V and Plaksin A M 2010 Operational propertiess of mobile units [Ehkspluatacionnye svojstva mobilnykh agregatov] (Astana, KATU) p 204

[8] Taheri S 2014 *Steering Control Characteristics of Human Driver Coupled with an Articulated Commercial Vehicle* (Concordia University)

[9] Kochnev A M 2007 A theory of motion of wheeled skid machines [Teoriya dvizheniya kolesnykh trelevochnykh system] (St.Petersburg, Polytechnical University) p 612

[10] Cao Huy Giap, Kochnev A M and Nguyen Long Lam 2019 Mathematical model of the interaction of wheeled skidder with skid trail in the rotate mode *Izvestia SanktPeterburgskoj Lesotehniceskoj Akademii* **226** pp 90–98 (in Russian with English summary)

[11] Luijten M F J 2010 Lateral Dynamic Behaviour of Articulated Commercial Vehicles *Master's thesis* (Eindhoven University of Technology)

[12] Sharifzadeh M, Farnam A, Senatore A, Timpone F and Akbari A 2018 Delay-Dependent Criteria for Robust Dynamic Stability Control of Articulated Vehicles *Advances in Service and Industrial Robotics. RAAD 2017 Mechanisms and Machine Science* **49** (Springer, Cham)

[13] Gyachev L V 1981 Motion stability of agricultural vehicles and units [Ustojchivost dvizheniya selskokhozyajstvennykh mashin i agregatov] (Moskow, Mashinostroenie) p 206

[14] Feng L, He Y, Bao Y, and Fang H 2005 Development of trajectory model for a tractor implement system for automated navigation applications *Instrumentation and Measurement Technology Conference* (Ottawa, Canada)

[15] Karkee M, Steward B L 2010 Study of the open and closed loop characteristics of a tractor and a single axle towed implement system *Journal of Terramechanics* **47** pp 379–393

[16] Pacejka H B 2006 *Tyre and vehicle dynamics* (Butterworth Heimann) 2-nd edition

[17] Lee A, Max D, Michael H, Vassilios M and Craig S 1996 A Lateral Dynamic Model of a Tractor-Trailer: Experimental Validation *Retrieved from the University of Minnesota Digital Conservancy* http://hdl.handle.net/11299/155339

[18] Maksarov V, Efimov A and Golikov T 2020 Treatment of Titanium Alloys Based on Preliminary Plastic Impact // *Key Engineering Materials Submitted* **836** pp 63–70

Mathematical model of a micropolar lubricating stuff

M A Mukutadze, A M Mukutadze, A N Opatskikh, M V Novakovich and V I Poltinnikov

Rostov State Transport University, 2, Rostovskogo Strelkovogo Polka Narodnogo Opolcheniya sq., Rostov-on-Don, 344038, Russia

E-mail: opatskih@yandex.ru

Abstract. The article is devoted to the mathematical analysis model of the radial bearing of infinite length, lubricated with lubricating stuff and molten low-melting metal coating on the surface of the bearing bush, having the general rheological properties in the laminar condition of micro-polar lubrication properties, taking into account the dependence of viscosity rheological properties of micro-polar lubricating stuff, and the permeability of the porous coating from the pressure. The authors on the basis of the motion equations of a viscous incompressible fluid having micro-polar properties, and for "lamina", of the continuity equation, Darcy equation and determining from the expression for the dissipation rate of mechanical energy, the profile of molten contour of the bearing bush, while taking account of the dependence of the overall rheological properties of the lubricating stuff and melt the low-melting metal coating having the micro-polar properties and the permeability of the porous coating of the pressure found the asymptotic and exact self-similar solution of the differential equation system on the parameter, due to the melt and the dissipation rate of mechanical energy to zero (without melt) and first (including melt) approximation.

1. Purpose of study

Investigation of the field performance of a plain journal bearing based on the development of refined calculation models based on the classical Navier-Stokes, Darcy, continuity equations with additional consideration for the dissipation rate of mechanical energy, implementation of numerical analysis to create algorithms and software for solving practical tasks.

2. Scientific novelty

The proposed solution and refine design models is to assess the impact of the molten low-melting metal coating the surface of the bearing bush, a porous surface coating of the axle journal, as well as the dependence of viscosity of general rheological properties of the lubricating stuff and melt, with the micro-polar properties of the pressure.

3. Problem statement

Consider the motion of micro-polar lubricating stuff in the working gap of the radial bearing surface of the bearing bush is covered with fusible metal alloy, and the shaft neck – porous layer. The bearing bush made of a material with a low melting point is stationary, and the shaft covered with a porous layer rotates at a speed of Ω. All the heat during rotation is used to melt the material surface of the bearing bush, which is coated with a low-melting metal alloy. The rheological properties of the lubricating stuff used and the

IOP Conf. Series: Materials Science and Engineering 1061 (2021) 012027 doi:10.1088/1757-899X/1061/1/012027

permeability of the porous coating depend on the pressure and are given as:

$$\mu'=\mu_0 e^{\tilde{\alpha}p'},\ \kappa'=\kappa_0 e^{\tilde{\alpha}p'},\ \gamma'=\gamma_0 e^{\tilde{\alpha}p'},\ \tilde{k}'=\tilde{k}_0 e^{\alpha' p'}. \tag{1}$$

where μ' is the coefficient of dynamic viscosity of the lubricating stuff, κ', γ' is the viscosity coefficients of the micro-polar lubricating stuff; μ_0 - characteristic viscosity of Newtonian lubricating stuff; p' - hydrodynamic pressure in the lubricant layer, $\tilde{\alpha}$ - experimental constant, - permeability of the porous coating, - characteristic permeability.

In the polar coordinate system (Fig.1) $C_0: r' = r_0 - \tilde{H};\ C_1: r' = r_0;\ C_2: r' = r_1(1+H) + \lambda' f(\theta)$, (2)

where $H = \varepsilon\cos\theta - \frac{1}{2}\varepsilon^2\sin^2\theta + ...$, $\varepsilon = \frac{e}{r_0}$, r_0 - the axle radius, r_1 - the radius of the bearing bush, covered with fusible metal alloy, e – eccentricity, ε - the eccentricity ratio, $\tilde{H}$ – the thickness of the porous layer, $\lambda' f(\theta)$ - the function to be determined.

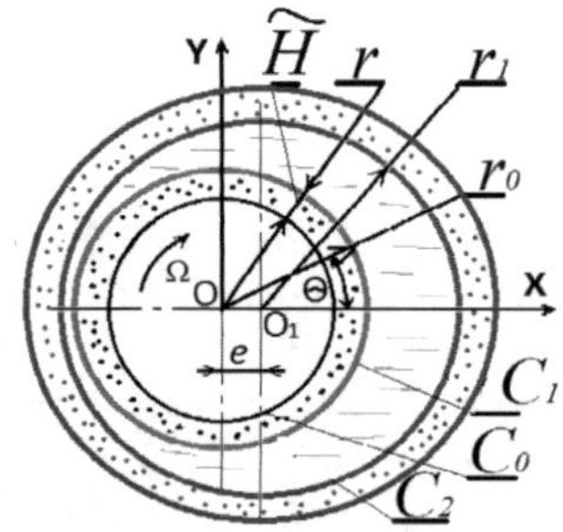

Fig.1. Working scheme

The initial equation is a system of dimensionless equations of motion of micro-polar lubricating stuff, Darcy's law, the continuity equation, and the equation that determines, taking into account the expression for the dissipation rate of mechanical energy, the profile of the molten contour of a bearing bush coated with a low – melting metal alloy with an accuracy of up to terms $O(\xi^2)$ - is written as:

$$\frac{\partial^2 u}{\partial r^2} + N^2\frac{\partial \upsilon}{\partial r} = \frac{dp}{d\theta},\quad \frac{\partial^2 \upsilon}{\partial r^2} = \frac{\upsilon}{N_1} + \frac{1}{N_1}\frac{\partial u}{\partial r},\quad \frac{\partial u}{\partial \theta} + \frac{\partial v}{\partial r} = 0,\quad \frac{\partial^2 P}{\partial r^{*2}};\quad \frac{\partial^2 v}{\partial r^2} = e^{-\alpha p}\frac{dp}{d\theta};\quad \frac{\partial u}{\partial r} + \frac{\partial v}{\partial \theta} = 0;$$

$$\frac{\partial^2 P}{\partial r^{*2}} + \frac{1}{r^*}\frac{\partial P}{\partial r^*} + \frac{1}{r^{*2}}\frac{\partial^2 P}{\partial \theta^2} = 0,\quad \frac{d\Phi(\theta)}{d\theta} = K\int_{r_0}^{1+\eta\cos\theta+\Phi(\theta)}\left(\frac{\partial v}{\partial r}\right)^2 dr; \tag{3}$$

where $K = \frac{2\mu\Omega r_0}{L'\delta}$, $\eta = \frac{e}{\delta}$, $\eta_1 = \frac{\lambda'}{\delta}$, $\Phi(\theta) = \eta_1 f(\theta)$.

For equation (3), the boundary conditions have the following form:

$u = 0,\ v = 0,\ \upsilon = 0$ at $r = 1 + \eta\cos\theta + \Phi(\theta)$,

$$\upsilon(0) = 0,\ u\big|_{r=0} = \tilde{M}\left.\frac{\partial P}{\partial r^*}\right|_{r^*=\frac{r_0}{\tilde{H}}},\ v(0) = 1,\ p = P \quad \text{при} \quad r^* = \frac{r_0}{\tilde{H}},\ \left.\frac{\partial P}{\partial r^*}\right|_{r^*=\frac{r_0}{\tilde{H}}-1} = 0,\ p(0) = p(2\pi) = \frac{p_g}{p^*} \tag{4}$$

where $\tilde{M} = -\frac{k^* r_0^2}{\tilde{H}\delta^3}$

the relationship between non-dimensional and dimensional values is given as:

$$r' = r_1 + \delta r,\ \delta = r_1 - r_0,\ v' = \Omega r_0 v,\ u' = \Omega\delta u,\ p' = p^* p,\ p^* = \frac{(2\mu_0 + \kappa_0)\Omega r_0^2}{2\delta^2},\ \upsilon' = \upsilon,\ \mu' = \mu_0\mu,\ \kappa' = \kappa_0\kappa,\ \gamma' = \gamma_0'\gamma',$$

$$N^2 = \frac{\kappa_0}{2\mu_0 + \kappa_0},\ N_1 = \frac{2\mu_0}{\delta^2 \kappa_0},\ l^2 = \frac{\gamma_0}{4\mu_0}. \tag{5}$$

Similarly, in a porous layer: $P' = p^* P,\ r' = \tilde{H} r^*,\ k' = k_0^* k^*$ (6)

Taking into account the smallness of the gap, as well as equality $\upsilon = 0$ on moving and stationary surfaces, we average the second equation of the system (3) over the thickness of the lubricant layer, we get:

(7)

The equation solution (7) will be found in the form:

$$\upsilon = A_1(\theta) r^2 + A_2(\theta) r + A_3(\theta) \tag{8}$$

Taking into account the boundary conditions (4), υ we obtain:

$$\upsilon = A_1(\theta)\left(r^2 - (h - \Phi) r - \Phi h\right) \tag{9}$$

Substituting (9) in (7) up to the terms $O\left(\frac{\Phi}{N_1}\right),\ O\left(\frac{\Phi}{N_1^2}\right)$ we get:

$$\upsilon = \frac{1}{2N_1 h}\left(r^2 - rh\right),\quad \frac{\partial \upsilon}{\partial r} = -\frac{1}{2N_1 h}(2r - h),\quad A_1 = \frac{1}{2N_1 h} \tag{10}$$

Taking into account (10), the equation system (3) in the approximation we have taken has the form:

$$\frac{\partial P}{\partial r} = 0,\ \frac{\partial^2 u}{\partial r^2} + \frac{N^2}{2N_1 h}(2r - h) = e^{-\alpha p}\frac{dp}{d\theta},\ \upsilon = \frac{1}{2N_1 h}\left(r^2 - rh\right)$$
$$\frac{\partial u}{\partial r} + \frac{\partial v}{\partial \theta} = 0,\ \frac{\partial^2 P}{\partial r^{*2}} + \frac{1}{r^*}\frac{\partial P}{\partial r^*} + \frac{1}{r^{*2}}\frac{\partial P}{\partial \theta^2} = 0,\ \frac{d\Phi(\theta)}{d\theta} = K\int_{-\Phi(\theta)}^{h(\theta)} \left(\frac{\partial u}{\partial r}\right)^2 dr \tag{11}$$

We will enter the designation $Z = e^{-\alpha p}$. Differentiating both parts of the equation, we get:

$$\frac{dZ}{d\theta} = -\alpha e^{-\alpha p}\frac{dp}{d\theta} \text{ or } e^{-\alpha p}\frac{dp}{d\theta} = -\frac{1}{\alpha}\frac{dZ}{d\theta}$$

Then the equation system (11) will be written as:

$$\frac{\partial Z}{\partial \theta} = 0,\ \frac{\partial^2 u}{\partial r^2} + \frac{N^2}{2N_1 h}(2r - h) = -\frac{1}{\alpha}\frac{dZ}{d\theta},\ \upsilon = \frac{1}{2N_1 h}\left(r^2 - rh\right)$$
$$\frac{\partial u}{\partial r} + \frac{\partial v}{\partial \theta} = 0,\ \frac{\partial^2 P}{\partial r^{*2}} + \frac{1}{r^*}\frac{\partial P}{\partial r^*} + \frac{1}{r^{*2}}\frac{\partial P}{\partial \theta^2} = 0,\ Z\frac{d\Phi(\theta)}{d\theta} = K\int_{r_0}^{1+\eta\cos\theta+\Phi(\theta)} \left(\frac{\partial v}{\partial r}\right)^2 dr \tag{12}$$

With appropriate boundary conditions:

$$u = 0,\ v = 0 \text{ at } r = 1 + \eta\cos\theta + \Phi(\theta),\ u\big|_{r=0} = \tilde{M}\left.\frac{\partial P}{\partial r^*}\right|_{r^* = \frac{r_0}{\tilde{H}}},\quad v(0) = 1 \tag{13}$$

$$p = P \text{ at } r^* = \frac{r_0}{\tilde{H}},\ \left.\frac{\partial P}{\partial r^*}\right|_{r^* = \frac{r_0}{\tilde{H}} - 1} = 0,\ Z(0) = Z(2\pi) = e^{-\alpha\frac{p_z}{p^*}}$$

The boundary conditions for non-dimensional velocity components u and v on the contour $r = -\Phi(\theta)$ can be written as:

$$v(1 + \eta\cos\theta + H(x)) = v(1 + \eta\cos\theta) - \left.\left(\frac{\partial v}{\partial y}\right)\right|_{r=1+\eta\cos\theta} \cdot H(\theta) - \left.\left(\frac{\partial^2 v}{\partial y^2}\right)\right|_{r=1+\eta\cos\theta} \cdot H^2(\theta) + \ldots = 0$$
$$u(1 + \eta\cos\theta H(x)) = u(1 + \eta\cos\theta) - \left.\left(\frac{\partial u}{\partial y}\right)\right|_{r=1+\eta\cos\theta} \cdot H(\theta) - \left.\left(\frac{\partial^2 u}{\partial y^2}\right)\right|_{r=1+\eta\cos\theta} \cdot H^2(\theta) + \ldots = 0 \tag{14}$$

The asymptotic solution of the differential equation system (12) taking into account the boundary conditions (13) and (14) is found in the form:

$$\begin{aligned} v(r,\theta) &= v_0(r,\theta) + Kv_1(r,\theta) + K^2 v_2(r,\theta) + \ldots \\ u(r,\theta) &= u_0(r,\theta) + Ku_1(r,\theta) + K^2 u_2(r,\theta) + \ldots \\ \Phi(\theta) &= -K\Phi_1(\theta) - K^2\Phi_2(\theta) - K^3\Phi_3(\theta) - \ldots \\ Z(\theta) &= Z_0(\theta) - KZ_1(\theta) - K^2 Z_2(\theta) - K^3 Z_3(\theta) - \ldots \end{aligned} \tag{15}$$

By substituting (15) into the differential equation system (12), taking into account the boundary conditions (13) and (14), we obtain the following equations:

For the zeroth-order approximation:

$$\frac{\partial^2 u_0}{\partial r^2} + \frac{N^2}{2N_1 h}(2r-h) = -\frac{1}{\alpha}\frac{dZ}{d\theta}, \quad \frac{\partial u_0}{\partial r} + \frac{\partial v_0}{\partial \theta} = 0, \quad \frac{\partial^2 P_0}{\partial r^{*2}} + \frac{1}{r^*}\frac{\partial P_0}{\partial r^*} + \frac{1}{r^{*2}}\frac{\partial P_0}{\partial \theta^2} = 0. \tag{16}$$

With boundary conditions $\upsilon_0 = 0,\ v_0 = 1$ at $r_0 = 0$, $u_0 = 0,\ \upsilon_0 = 0,\ v_0 = 0$ at $r_0 = 1 + \eta\cos\theta$,

$$u(0) = \tilde{M}\left.\frac{\partial P}{\partial r^*}\right|_{r^*=\frac{r_0}{\tilde{H}}}, \quad p_0 = P_0 \text{ at } r^* = \frac{r_0}{\tilde{H}}, \quad \left.\frac{\partial P_0}{\partial r^*}\right|_{r^*=\frac{r_0}{\tilde{H}}-1} = 0, \quad Z_0(0) = Z_0(2\pi) = e^{-\alpha\frac{p_e}{p^*}} \tag{17}$$

For the initial approximation:

$$\frac{\partial^2 u_1}{\partial r^2} = -\frac{1}{\alpha}\frac{dZ_1}{d\theta}, \quad \frac{\partial u_1}{\partial r} + \frac{\partial v_1}{\partial \theta} = 0, \quad \frac{\partial^2 P_1}{\partial r^{*2}} + \frac{1}{r^*}\frac{\partial P_1}{\partial r^*} + \frac{1}{r^{*2}}\frac{\partial P_1}{\partial \theta^2} = 0. \quad Z_0\frac{d\Phi_1(\theta)}{d\theta} = K\int_0^{1+\eta\cos\theta}\left(\frac{\partial u_0}{\partial r}\right)^2 dr \tag{18}$$

With boundary conditions:

$$\upsilon_1 = 0,\ v_1 = \left.\left(\frac{\partial u_0}{\partial r}\right)\right|_{r=0}\cdot\Phi_1(\theta);\ u_1 = \left.\left(\frac{\partial u_0}{\partial r}\right)\right|_{r=0}\cdot\Phi_1(\theta),\ \upsilon_1 = 0,\ v_1 = 0,\ u_1 = 0 \quad \text{at} \quad r = 1 + \eta\cos\theta,$$

$$u_1(0) = \tilde{M}\left.\frac{\partial P}{\partial r^*}\right|_{r^*=\frac{r_0}{\tilde{H}}}, \quad p_1 = P_1 \quad \text{at } r^* = \frac{r_0}{\tilde{H}}, \quad \left.\frac{\partial P_1}{\partial r^*}\right|_{r^*=\frac{r_0}{\tilde{H}}-1} = 0,\ Z_1(0) = Z_1(2\pi) = 0, \tag{19}$$

$K\Phi_1(0) = K\tilde{\alpha},\ \Phi(0) = \Phi(2\pi) = \tilde{\alpha}.$

For the zeroth-order approximation, we look for the exact solution in the form:

$$U_0 = \frac{\partial \psi_0}{\partial r} + V_0(r,\theta),\ V_0 = -\frac{\partial \psi_0}{\partial r} + U_0(r,\theta),\ \psi_0(r,\theta) = \tilde{\psi}_0(\xi),\ \xi = \frac{r}{h(\theta)},$$

$$V_0(r,\theta) = \tilde{v}(\xi),\ U_0(r,\theta) = -\tilde{u}_0(\xi)\cdot h'(\theta). \tag{20}$$

Substituting (20) in (16)-(17), we have:

$$\tilde{\psi}_0''' = \tilde{C}_2,\ \tilde{u}_0'' = \tilde{C}_1 - \frac{N^2}{2N_1}(2\xi - 1),\ \tilde{u}_0' + \xi\tilde{v}_0' = 0,\ \frac{dZ_0}{d\theta} = -\alpha\left(\frac{\tilde{C}_1}{h^2(\theta)} + \frac{\tilde{C}_2}{h^3(\theta)}\right). \tag{21}$$

And boundary conditions:

$$\upsilon(0) = 0,\ \tilde{\psi}_0'(0) = 0,\ \tilde{\psi}_0'(1) = 0,\ \left.\frac{\partial P}{\partial r^*}\right|_{r^*=\frac{r_0}{\tilde{H}}-1} = 0,\ p_0 = P_0 \text{ at } \quad r^* = \frac{r_0}{\tilde{H}},\ u_0\big|_{\xi=0} = \tilde{M}\left.\frac{\partial P}{\partial r^*}\right|_{r^*=\frac{r_0}{\tilde{H}}},$$

$$\upsilon(1) = 0,\ \tilde{\psi}_0'(0) = 0,\ \tilde{\psi}_0'(1) = 1,\ \int_0^1 \tilde{u}_0(\xi)d\xi = 0,\quad Z_0(0) = Z_0(2\pi) = e^{-\alpha\frac{p_e}{p^*}}, \tag{22}$$

By immediate integration we get:

$$\tilde{\psi}_0'(\xi) = \frac{\tilde{C}_2}{2}(\xi^2 - \xi),\ \tilde{u}_0(\xi) = \tilde{C}_1\frac{\xi^2}{2} - \frac{N^2}{2N_1}\left(\frac{\xi^3}{3} - \frac{\xi^2}{2}\right) - \left(\frac{N^2}{12N_1} - \frac{\tilde{C}_1}{2} + 1\right)\xi + 1,$$

$$Z_0 = -\alpha\left[\tilde{C}_1(\theta - 2\eta\sin\theta) + \tilde{C}_2(\theta - 3\eta\sin\theta)\right] + e^{-\alpha\frac{p_e}{p^*}}. \tag{23}$$

We get the following expression from the condition $Z_0(0)=Z_0(2\pi)=e^{-\alpha\frac{p_g}{p^*}}$:

$$\tilde{C}_2=-\tilde{C}_1 \tag{24}$$

Hydrodynamic pressure determination.
Substituting the value $\tilde{C}_2$ in equation (23) for Z_0, we get:

$$Z_0=-\alpha\tilde{C}_1(\eta\sin\theta)+e^{-\alpha\frac{p_g}{p^*}} \tag{25}$$

Using the asymptotic decomposition to determine *P*, we arrive at the following approximate equation:

$$1-\alpha p+\frac{\alpha^2p^2}{2}-\left(1-\alpha\frac{p_g}{p^*}+\frac{\alpha^2}{2}\left(\frac{p_g}{p^*}\right)^2\right)+\alpha\tilde{C}_1\eta\sin\theta=0 \tag{26}$$

To find the hydrodynamic pressure, we solve equations (26) up to $O(\eta^2)$, $O\left(\alpha^3\left(\frac{p_g}{p^*}\right)^2\right)$, we get

$$P=\frac{p_g}{p^*}+\left[1+\alpha\frac{p_g}{p^*}-\frac{\alpha^2}{2}\left(\frac{p_g}{p^*}\right)^2\right]\tilde{C}_1\eta\sin\theta \tag{27}$$

Given the expression (27), Darcy's equation takes the form:

$$P(r^*,\theta)=R(r^*)\tilde{C}_1\eta\sin\theta+\left[1+\alpha\frac{p_g}{p^*}-\frac{\alpha^2}{2}\left(\frac{p_g}{p^*}\right)^2\right]+\frac{p_g}{p^*} \tag{28}$$

Substituting (28) in Darcy's equation to determine the function $R(r^*)$, we proceed to the following differential equation:

$$R''(r^*)+\frac{R'}{r^*}-\frac{R'}{r^{*2}}=0 \tag{29}$$

and boundary conditions:

$$\left.\frac{dR}{dr^*}\right|_{r^*=\frac{r_0}{\tilde{H}}-1}=0,\ R\left(\frac{r_0}{\tilde{H}}\right)=1 \tag{30}$$

Immediate integration of (29) with (30) for the function $R(r^*)$ allows us to obtain the equation:

$$R(r^*)=\frac{r_0\tilde{H}r^*}{2r_0^2-2\tilde{H}r_0+\tilde{H}^2}+\frac{r_0\left(r_0^2-2\tilde{H}r_0+\tilde{H}^2\right)}{\tilde{H}\left(2r_0^2-2\tilde{H}r_0+\tilde{H}^2\right)r^*} \tag{31}$$

With the expression

$$\tilde{M}\left.\frac{\partial P}{\partial r^*}\right|_{r^*=\frac{r_0}{\tilde{H}}}=\int_0^1\tilde{\alpha}(\xi)d\xi \tag{32}$$

For $\tilde{C}_1$ we will have:

$$\tilde{C}_1=\frac{6r_0\left(2r_0^2-2\tilde{H}r_0+\tilde{H}^2\right)\left(1+\alpha\frac{p_g}{p^*}-\frac{\alpha^2}{2}\left(\frac{p_g}{p^*}\right)^2\right)}{12\tilde{H}^2\tilde{M}\left(r_0-\tilde{H}\right)\left(1+\alpha\frac{p_g}{p^*}-\frac{\alpha^2}{2}\left(\frac{p_g}{p^*}\right)^2\right)+r_0\left(2r_0^2-2\tilde{H}r_0+\tilde{H}^2\right)} \tag{33}$$

Then for p_0 we get:

$$p_0 = \frac{6r_0\eta\sin\theta\left(2r_0^2 - 2\tilde{H}r_0 + \tilde{H}^2\right)\left(1+\alpha\frac{p_g}{p^*} - \frac{\alpha^2}{2}\left(\frac{p_g}{p^*}\right)^2\right)}{12\tilde{H}^2\tilde{M}\left(r_0 - \tilde{H}\right)\left(1+\alpha\frac{p_g}{p^*} - \frac{\alpha^2}{2}\left(\frac{p_g}{p^*}\right)^2\right) + r_0\left(2r_0^2 - 2\tilde{H}r_0 + \tilde{H}^2\right)} + \frac{p_g}{p^*} \tag{34}$$

To determine $F_1(\theta)$, taking into account equation (23), we come to the following expression:

$$\frac{d\Phi_1(\theta)}{d\theta} = \frac{h(\theta)}{Z_0}\int_0^1\left(\frac{\tilde{\psi}_0''(\xi)}{h^2(\theta)} + \frac{\tilde{u}_0'(\xi)}{h(\theta)}\right)^2 d\xi. \tag{35}$$

Integrating (33) we get:

$$\frac{d\Phi_1(\theta)}{d\theta} = \frac{1}{Z_0}\left(\int_0^\theta \frac{\Delta_1 d\theta}{h^3(\theta)} + \int_0^\theta \frac{\Delta_2 d\theta}{h^2(\theta)} + \int_0^\theta \frac{\Delta_3 d\theta}{h(\theta)}\right). \tag{36}$$

where $\Delta_1 = \int_0^1 \left(\tilde{\psi}''(\xi)\right)^2 d\xi = \frac{\tilde{C}_2^2}{12}$; $\Delta_2 = \int_0^1 2\psi''(\xi)\cdot\tilde{u}'(\xi)d\xi = \frac{1}{6}\tilde{C}_1\tilde{C}_2$, $\Delta_3 = \int_0^1 \left(\tilde{u}'(\xi)\right)^2 d\xi = 4 + \frac{N^4}{720N_1^2}$

$$\sup_{[0\div 2\pi]} Z_0 = \sup_{[0\div 2\pi]}\left|\alpha C_1\eta\sin\theta + e^{-\alpha\frac{p_g}{p^*}}\right| = \frac{1}{68}; \tag{37}$$

Solving equations (36)-(37) taking into account $CF_1(\theta) = K\tilde{\alpha}$, we get:

$$\Phi_1(\theta) = \frac{1}{68}\left[(\theta - \eta\sin\theta)\left(-\tilde{C}_1^2 + 4 + \frac{N^4}{720N_1^2}\right) + \tilde{\alpha}\right] \tag{38}$$

Then for the initial approximation, we get:

$$u_1 = -\frac{\partial\psi_1}{\partial\theta} + U_1(r,\theta);\ v_1 = -\frac{\partial\psi_1}{\partial r} + V_1(r,\theta);\ \psi_1(r,\theta) = \tilde{\psi}_1(\xi);\ \xi = \frac{r}{h(\theta)};\ V_1(r,\theta) = \tilde{v}(\xi);\ U_1(r,\theta) = -\tilde{u}_1(\xi)\cdot h'(\theta);$$

$$\frac{dZ_1}{d\theta} = -\alpha\left(\frac{\tilde{\tilde{C}}_1}{h^2(\theta)} + \frac{\tilde{\tilde{C}}_2}{h^3(\theta)}\right),\quad h(\theta) = 1 + \eta\cos\theta. \tag{39}$$

Substituting (39) into the differential equation system (18), taking into account the boundary conditions (19), we obtain the following differential equation system:

$$\tilde{\psi}_1''' = \tilde{\tilde{C}}_2,\ \tilde{u}_1'' = \tilde{C}_1,\ \tilde{u}_1'(\xi) + \xi\tilde{v}_1'(\xi) = 0,\ \frac{dZ_1}{d\theta} = -\alpha\left(\frac{\tilde{C}_1}{h^2(\theta)} + \frac{\tilde{C}_2}{h^3(\theta)}\right). \tag{40}$$

And boundary conditions:

$$\tilde{\psi}_1'(0) = 0,\ \tilde{\psi}_1'(1) = 0,\ \tilde{u}_1(1) = 0,\ \tilde{v}_1(1) = 0,\ \upsilon_1(0) = 0,\ \upsilon_1(1) = 0,\ \tilde{u}_1(0) = M,\ \tilde{v}_1(0) = 0,\ \int_0^1 \tilde{u}_1(\xi)d\xi = 0,$$

$$\left.\frac{\partial P_1}{\partial r^*}\right|_{r^* = \frac{r_0}{\tilde{H}} - 1} = 0,\ p_1 = P_1 \text{ at } r^* = \frac{r_0}{\tilde{H}},\ u_1\big|_{\xi=0} = \tilde{M}\left.\frac{\partial P_1}{\partial r^*}\right|_{r^* = \frac{r_0}{\tilde{H}}},\ Z_1(0) = Z_1(2\pi) = 0, \tag{41}$$

By immediate integration we get:

$$\tilde{\psi}_1'(\xi) = \frac{\tilde{\tilde{C}}_2}{2}\left(\xi^2 - \xi\right),\ \tilde{u}_1(\xi) = \tilde{\tilde{C}}_1\frac{\xi^2}{2} - \left(-\frac{\tilde{C}_1}{2} + M\right)\xi + M,\quad \tilde{\tilde{C}}_1 = 6M \tag{42}$$

From the condition $Z_1(0) = Z_1(2\pi) = 0$ we get:

$$\tilde{\tilde{C}}_2 = -M\tilde{\tilde{C}}_1 \tag{43}$$

Where

$$M = \sup_{[0\div 2\pi]} \left.\frac{\partial u_0}{\partial r}\right|_{r=0} \cdot \Phi_1(\theta) = \sup_{[0\div 2\pi]} \left| \begin{matrix} \left[-\frac{1}{\eta\cos\theta} - \tilde{C}_1\eta\cos\theta + \frac{N^2}{6N_1} - (1-2\eta\cos\theta) - \frac{N^2}{4N_1}(1+\eta\cos\theta) \right] \times \\ \times \frac{1}{68}\left[(\theta - \eta\sin\theta)\left(\tilde{C}_1^2 + 4 + \frac{N^4}{720N_1^2} \right) + \tilde{\alpha} \right] \end{matrix} \right|$$

With (40) taken into account, Z_1 we get

$$Z_1 = -\alpha\tilde{\tilde{C}}_1 M\eta\sin\theta \tag{44}$$

The further solution is the same as for the zeroth-order approximation, we get:

$$\tilde{\tilde{C}}_1 = \frac{6Mr_0\left(2r_0^2 - 2\tilde{H}r_0 + \tilde{H}^2\right)\left(1 + \alpha\frac{p_g}{p^*} - \frac{\alpha^2}{2}\left(\frac{p_g}{p^*}\right)^2\right)}{12\tilde{H}^2\tilde{M}\left(r_0 - \tilde{H}\right)\left(1 + \alpha\frac{p_g}{p^*} - \frac{\alpha^2}{2}\left(\frac{p_g}{p^*}\right)^2\right) + r_0\left(2r_0^2 - 2\tilde{H}r_0 + \tilde{H}^2\right)} \tag{45}$$

$$p_1 = \frac{6Mr_0\eta\sin\theta\left(2r_0^2 - 2\tilde{H}r_0 + \tilde{H}^2\right)\left(1 + \alpha\frac{p_g}{p^*} - \frac{\alpha^2}{2}\left(\frac{p_g}{p^*}\right)^2\right)}{12\tilde{H}^2\tilde{M}\left(r_0 - \tilde{H}\right)\left(1 + \alpha\frac{p_g}{p^*} - \frac{\alpha^2}{2}\left(\frac{p_g}{p^*}\right)^2\right) + r_0\left(2r_0^2 - 2\tilde{H}r_0 + \tilde{H}^2\right)} \tag{46}$$

The results of the study:

$$R_x = p^* r \int_0^{2\pi} \left(p_0 - \frac{p_g}{p^*} + Kp_1 \right)\cos\theta d\theta = 0.$$

$$R_y = p^* r \int_0^{2\pi} \left(p_0 - \frac{p_g}{p^*} + Kp_1 \right)\sin\theta d\theta = \frac{\mu_0\Omega r_0^3}{\delta^2}\tilde{C}_1\eta\pi\left(1 + \alpha\frac{p_g}{p^*} - \frac{\alpha^2}{2}\left(\frac{p_g}{p^*}\right)^2 \right)(1+KM)$$

$$L_{\text{тр}} = \mu\int_0^{2\pi}\left[\left.\frac{\partial u_0}{\partial r}\right|_{r=0} + K\left.\frac{\partial u_1}{\partial r}\right|_{r=0} \right] d\theta = \mu_0\left(1 - \alpha p + \frac{\alpha^2 p^2}{2}\right)\left(1 - \frac{N^2}{6N_1} + \frac{N^2}{4N}\right)\left[-2\pi + \frac{K\pi^2}{34}\left(\tilde{C}_1^2 + 4 + \frac{N^4}{720N_1} \right) \right] \tag{47}$$

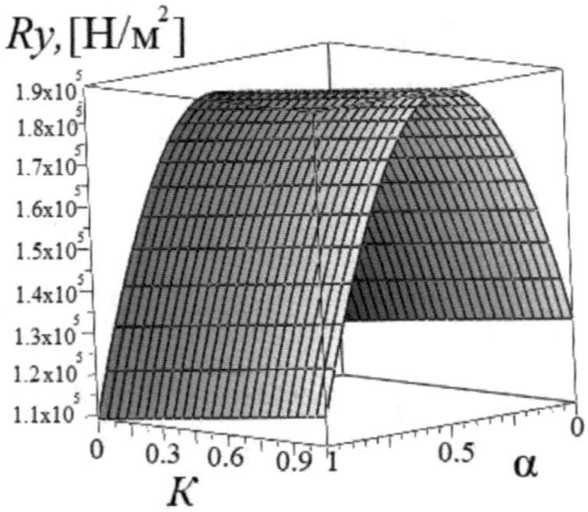

Figure 2. Dependency graph of vector component of the supporting force from the permeability parameters of the porous coating K the and α pressure-viscosity ratio.

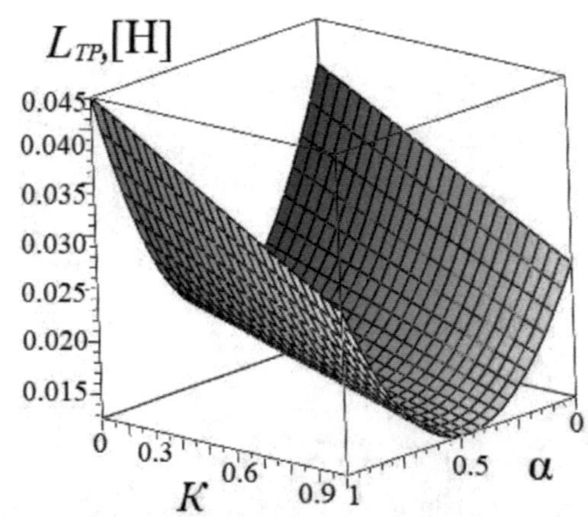

Figure 3. Dependency graph of the friction force on the permeability parameters of the porous coating K and α of the viscosity versus pressure.

Results

As a result, the velocity and pressure fields in the lubricant and porous layers were determined, taking into account the dependence of the rheological properties of the lubricating stuff and the melt of a low-melting metal coating. Having micro-polar properties, as well as the permeability of the porous coating from pressure. Also, the function $F_1(\theta)$ due to the melt of the bearing bush surface and the main operating characteristics: load-bearing capacity and friction force are determined. But the authors also determined the influence of the parameters κ due to the melt surface of the bearing bush, coated with low-melting metal alloy, $\tilde{H}$ the thickness of the porous coating, α characterizes the dependence of viscosity of the micro-polar lubricating stuff pressure, K characterize the permeability of the porous layer from the pressure of the structurally-viscous parameters, N_I and N^2 micro-polar lubricating stuff on the bearing capacity and friction force.

References

[1] Levanov I, Doykin A, Zadorozhnaya E and Novikov R 2017 Investigation of Antiwear Properties of Lubricants with the Geo-Modifiers of Friction *Tribology in Industry* **39** (3) pp 302–306

[2] Mukchortov I V, Pochkaylo K A and Zadorozhnaya E A 2016 The Influence of Anti-Wear Additives on the Bearings Hydro-Mechanical Characteristics *Procedia Engineering* **150** pp 607–611

[3] Zadorozhnaya E 2013 The Research of Non-Newtonian Properties and Rheology of Thin Lubricant Layers in Hydrodynamic Journal Bearings *Society of Tribologists and Lubrication Engineers Annual Meeting and Exhibition* pp 95–97

[4] Vasilenko V V, Lagunova E O, Mukutadze M A and Prikhodko V M 2017 International Journal of Applied Engineering Research ISSN 0973-4562 Vol 12 **19** pp 9138–9148

[5] Mukutadze M A, Lagunova E O, Garmonina A N and Vasilenko V V 2018 Radial Slip Bearing with a Pliable Supporting Surface *Russian Engineering Research* Vol 38 **3** pp 166–171

[6] Mukutadze M A 2016 Radial Bearing with Porous Elements *Procedia Engineering* **150** pp 559–570

[7] Mukutadze M A, Lagunova E O and Vasilenko V V 2018 Development of the Design Model of a Hydrodynamic Lubricating Material Formed during Melting of the Axial Bearing, in the Presence of Forced Lubrication *Journal of Machinery Manufacture and Reliability* Vol 47 pp 271–277

[8] Akhverdiev K S, Lagunova E O and Mukutadze M A Calculated Model of Wedge-Shaped Sliding Supports in Turbulent Friction Regime *Advances in Engineering Research (AER)* Vol 157 *International Conference "Actual Issues of Mechanical Engineering"* (AIME 2018) pp 346–353

[9] Akhverdiev K S, Lagunova E O and Mukutadze M A 2018 Calculated Model of Wedge-Shaped Sliding Supports Taking into Account Rheological Properties of Viscoelastic Lubricant *Advances in Engineering Research, Volume 158, International Conference on Aviamechanical Engineering and Transport (AviaENT 2018)* pp 246–253

[10] Lagunova E O and Mukutadze M A 2018 Radial Friction Bearings Conditioned by Melt *Proceedings of the 4th International Conference on Industrial Engineering, Lecture Notes in Mechanical Engineering (ICIE 2018)* pp 897–910

[11] Lagunova E O and Mukutadze M A Calculation of a Radial Slider Bearing with a Fusible Coating *Journal of Friction and Wear* 2019, Vol 40 **1** pp 88–94

[12] Mukutadze M A, Mukutadze A M and Vasilenko V V 2019 Simulation Model of Thrust Bearing with a Free-Melting and Porous Coating of Guide and Slide Surfaces *IOP Conf. Series: Materials Science and Engineering* **560** DOI:10.1088/1757-899X/560/1/012031

[13] Opatskikh A N, Mukutadze M A and Mukutadze A M 2019 V-Shaped Journal bearings Using Micropolar Lubricants Caused by a Melt Accounting for the Dependence of Lubricant Viscosity and Porous Layer Permeability on Pressure *Journal of Physics: Conference Series* **1353** 012025 DOI:10.1088/1742-6596/1353/1/012025

[14] Mukutadze M A, Vasilenko V V, Mukutadze A M and Opatskikh A N 2019 Mathematical Model of a Plain Bearer Lubricated with Molten Metal *IOP Conf. Series: Earth and Environmental Science* **378** 012021 DOI:10.1088/1755-1315/378/1/012021

[15] Akhverdiev K S, Alexandrova E V, Kruchinina E E and Mukutadze M A 2010 Stratified Course Of Two-Layer Lubrication In The Gap Of A Threshold Bearing, With An Increased Bearing Ability *Vestnik DGTU* Vol 10 **2** (45) pp 217–223

[16] Akhverdiev K S, Aleksandrova E E, Mukutadze M A and Kopotun B E Stratified Course Of Two-Layer Lubrication In The Radial Bearing Clearance With An Increased Bearing Capacity And Damping Properties *Vestnik RGURS* 2009 **4** (36) pp 133–139

[17] Akhverdiev K S, Aleksandrova E E and Mukutadze M A 2010 Stratified Two-Layer Lubrication Current In The Clearance Of A Complexly Loaded Radial Bearing Of A Final Length With An Increased Loading Capacity *Vestnik RGURS* 2010 **1** (37) pp 132–137

Investigation of the dynamics of a small spacecraft elastic element temperature change under a temperature shock considering a penumbral section

A V Sedelnikov[1], D I Orlov[1], Yu D Leskova[1]

[1] Samara National Research University, 34, Moskovskoye shosse, Samara, 443086, Russia

E-mail: axe_backdraft@inbox.ru

Abstract. In this paper, a study of the temperature shock of small technological spacecraft large elastic elements, considering the penumbral and shadow areas of its orbit, is carried out. A comparison of the dynamics changes in the temperature field of an elastic element with and without consideration of the penumbral section is presented. Conclusions are made about the significance of the penumbral section for modeling the dynamics of the small spacecraft under a temperature shock orbital motion in order to create favorable conditions for micro-accelerations for the successful implementation of gravitationally sensitive technological processes. The results of the work can be used in the small technological spacecraft design and the development of algorithms that reduce the negative impact of temperature shock on the micro-acceleration field of the small spacecraft inner environment.

1. Introduction

When designing small technological spacecraft, it is necessary to take into account the impact of the large elastic elements temperature shock on the micro-acceleration field of the inner environment. For successful implementation of gravity-sensitive processes, it is necessary to significantly limit the level of micro-accelerations in the zone of technological equipment placement [1-3]. In [5], the significance of the impact of temperature shock on the micro-acceleration field was studied without taking into account the penumbral section of the orbit. A simplified model of the Earth's shadow without the penumbral section is also studied in the works of other authors [6-9]. Considering the penumbral section of the orbit, the dynamics of large elastic elements temperature changes in solving the initial boundary value problem given in [5] will not change, if we assume that the accepted initial field of temperature distribution corresponds to the end of the penumbral section and the beginning of the light section of the small spacecraft orbit. However, in this case, it is more correct to change the initial conditions by assuming a uniform temperature field at the end of the shadow and the beginning of the penumbral sections.

This can be explained by the fact that the boundary conditions for all external surfaces of a large elastic element are the same inside the shadow section. Therefore, in the presence of a long shadow section, the temperature of large elastic elements can come to an equilibrium state, representing a uniform field. In the penumbral section, one of the surfaces will be illuminated by an increasing flow of solar radiation, so it is difficult to imagine a uniform temperature distribution.

Thus, the dynamics of the small spacecraft large elastic elements temperature change will differ from the evaluation made when solving the initial boundary value problem [5].

2. Models and methods of research

When evaluating the temperature dynamics of the small spacecraft large elastic elements, we will use a cone-shaped model of the Earth's shadow (figure 1).

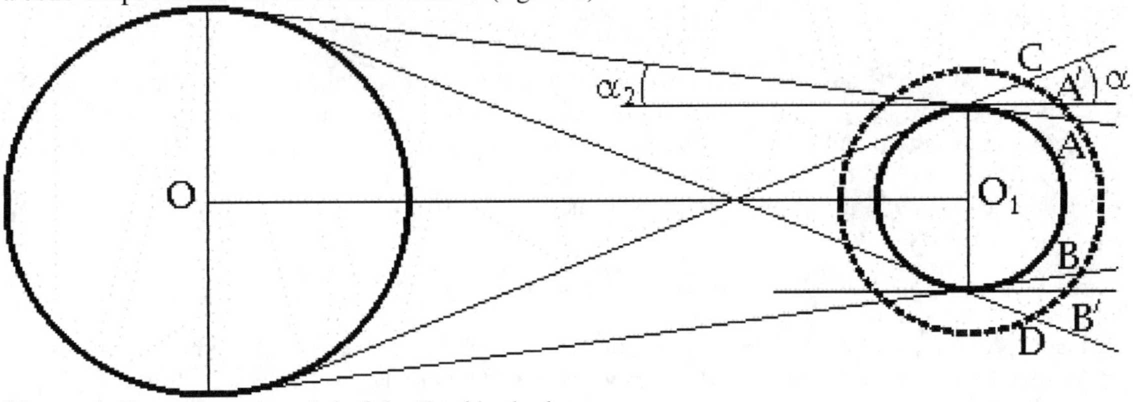

Figure 1. Cone-shaped model of the Earth's shadow

In figure 1, the dotted line indicates the orbit of a small spacecraft. The OO_1 line connects the centers of the Sun and Earth. AB and A'B' are the shadow section of the orbit for the cone-shaped and cylindrical models of the Earth's shadow, respectively. CA and DB are penumbral sections of the orbit for a cone - shaped model of the Earth's shadow.

Let's consider a solid central body with one elastic element rigidly attached to it as a model of a small spacecraft (figure 2).

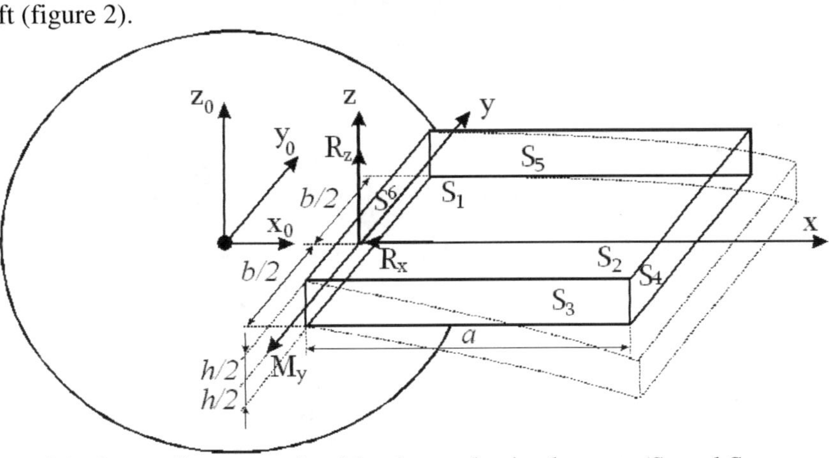

Figure 2. The model of a small spacecraft with a large elastic element: (S_1 and S_2 are respectively the upper and lower surfaces, and S_3–S_6 – the side surfaces of the elastic element; x_0 y_0 z_0 is the main associated coordinate system, xyz-the associated coordinate system, the center of which is located at the attachment point of the elastic element to the body of a small spacecraft, and the x axis lies in the plane of the elastic element in an undeformed position and passes through its center of mass)

Let's consider the temperature shock when the small spacecraft exits the Earth's shadow. This corresponds to the movement of the small spacecraft in its orbit along the BD section (figure 1). To do this, we set the initial boundary value problem of thermal conductivity. As in [5], we will use a one-dimensional model of thermal conductivity:

$$c\rho\frac{\partial T}{\partial t} = \lambda\frac{\partial^2 T}{\partial z^2}, (1)$$

where c – specific heat, ρ-density, λ – thermal conductivity.

We believe that as a result of the small spacecraft movement along the shadow section of the orbit AB, the temperature field of the elastic element became uniform with the temperature taken as the initial distribution of the temperature field in [5]:

$$T(x,\ y,\ z,\ 0) = 200\ K, (2)$$

where the moment of time $t_0=0$ corresponds to the arrival of the small spacecraft at point B on the border of the shadow and penumbral sections of its orbit (figure 1). In [5], a cylindrical shadow model was used, so this initial condition corresponded to the arrival of the small spacecraft at a point on the border of the shadow and light sections of its orbit (figure 1).

The boundary conditions for different parts of the orbit are the following:

a) shadow section AB:

$$\lambda\frac{\partial T}{\partial n} + \varepsilon\sigma\left(T^4 - T_s^4\right) = 0\ (x,\ y,\ z) \in \Omega, (3)$$

where ε – degree of blackness, σ – Boltzmann constant, T – temperature, T_s – ambient temperature, n – a normal to the surface of the solar panel power frame, Ω–boundary surface of the frame (S_1-S_6).

b) penumbra section BD:

$$\begin{cases} \lambda\frac{\partial T}{\partial n} + \varepsilon\sigma\left(T^4 - T_s^4\right) = Q_1(t)\ (x,\ y,\ z) \in \Omega_1 \\ \lambda\frac{\partial T}{\partial n} + \varepsilon\sigma\left(T^4 - T_s^4\right) = 0\ (x,\ y,\ z) \in \Omega_2 \end{cases}, (4)$$

Where $Q_1(t)$-variable solar radiation flux, varying from 0 to Q=1400 W/m^2, Ω_1 – boundary surface of the frame (S_1), Ω2 – boundary surface of the frame (S_2–S_6).

C) light section DC:

$$\begin{cases} \lambda\frac{\partial T}{\partial n} + \varepsilon\sigma\left(T^4 - T_s^4\right) = Q\ (x,\ y,\ z) \in \Omega_1 \\ \lambda\frac{\partial T}{\partial n} + \varepsilon\sigma\left(T^4 - T_s^4\right) = 0\ (x,\ y,\ z) \in \Omega_2 \end{cases}. (5)$$

Thus, the initial boundary value problem using a cone–shaped shadow model (1) - (5) is set for estimating the dynamics of small spacecraft large elastic element temperature change when it leaves the Earth's shadow.

3. Numerical modeling

To solve the initial boundary value problem, it is necessary to approximate the dependence $Q_1=Q_1(t)$ included in equation (4). Let's use the simplest linear approximation:

$$Q_1(t) = a\,t, (6)$$

where a is an unknown constant coefficient.

Since at point B $Q_1(t)=0$, the linear dependence (6) must not have a free term, we need to define the specific orbit of a small spacecraft and determine the duration of the penumbral section BD to estimate the coefficient a. In this work, we will consider the circular orbit of the small spacecraft with a height of 600 km. Such an orbit generally corresponds to the orbits of the implemented projects of technological and biomedical "Foton" [10, 11] and "Bion" medium-class spacecraft types [12, 13].

Simple calculations of angles α_1 and α_2 considering the known radiuses of the Earth and the Sun, as well as the distance between them (figure 1) show that $\alpha_1 \approx 0{,}268^0$ and $\alpha_2 \approx 0{,}263^0$. The length of the BD arc will be equal to (figure 3):

$$s_{BD} = (r_{\otimes} + h)(\gamma_2 - \gamma_1), \quad (7)$$

where $r_{\otimes}$ is the radius of the Earth, h is the height of the small spacecraft orbit.

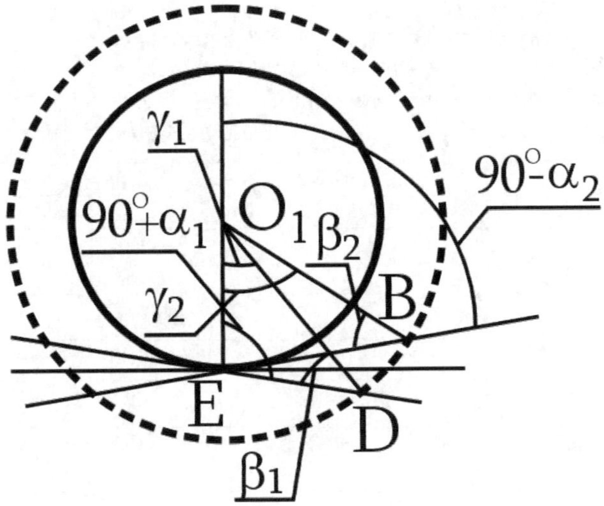

Figure 3. Calculating the length of the penumbra section of the small spacecraft orbit

By the law of sines from a triangle O_1DE:

$$\beta_1 = \arcsin\left[\frac{r_{\otimes}}{r_{\otimes} + h}\sin(90 + \alpha_1)\right] \approx 66{,}053^0. \quad (8)$$

Then:

$$\gamma_1 = 90 - \alpha_1 - \beta_1 = 23{,}679^0. \quad (9)$$

By the law of sines from a triangle O_1BE:

$$\beta_2 = \arcsin\left[\frac{r_{\otimes}}{r_{\otimes} + h}\sin(90 - \alpha_2)\right] \approx 66{,}053^0. \quad (10)$$

Then:

$$\gamma_2 = 90 + \alpha_2 - \beta_2 = 24{,}210^0. \quad (11)$$

Substituting the values obtained in (8)– (11) into expression (7):

$$s_{BD} \approx 64{,}605\ km. \quad (12)$$

The orbital speed of the small spacecraft is determined by the dependence:

$$v = \sqrt{\frac{GM_{\otimes}}{r_{\otimes} + h}} \approx 7{,}577\ km/s. \quad (13)$$

Assuming the orbital velocity (13) to be constant, let's estimate the transit time of the penumbral section of the orbit BD:

$$t_{BD} = \frac{s_{BD}}{v} \approx 8{,}526\ s. \quad (14)$$

Thus, the boundary conditions are distributed over time as follows:

– $t = 0$ initial condition (2) and boundary conditions (3);

– $0 < t \le 8{,}526\ s$ boundary conditions (4);

– $t > 8{,}526\ s$ boundary conditions (5).

Then the approximation of the solar radiation flux in the penumbra of the Earth (6), considering (14), will be the following:

$$Q_1(t) = \frac{t}{8{,}526} Q, \quad (15)$$

Numerical simulation was performed in the ANSYS package. The characteristics of the elastic element used in numerical modeling are shown in table 1 (quoted in [5]).

The parameter values used in the numerical simulation of solving the heat conduction problem

Table 1. characteristics of the elastic element used in numerical modeling

Parameter	Designation	Value	Dimension
Solar panel frame material	–	MA2	–
Coefficient of thermal conductivity	λ	96,3	$W/(m \cdot K)$
Stefan-Boltzmann constant	σ	$5{,}67 \cdot 10^{-8}$	$W/(m^2 \cdot K^4)$
External heat flux	Q	1400	W/m^2
Vacuum temperature	T_C	3	K
Initial temperature of the solar panel frame	$T(x, y, z, 0)$	200	K
Specific heat	c	1130,4	$J/(kg \cdot K)$
Density	ρ	1780	kg/m^3
Time step	Δt	0,04	s
Layer thickness	Δz	1,5	mm
Solar panel frame thickness	h	6	mm
Solar panel length	l	0,5	m

As a result of numerical simulation, the dynamics of the elastic element temperature field for a temperature shock when a small spacecraft exits the Earth's shadow is obtained. It is shown in figure 4. During the simulation, the entire plate was divided into four layers of thickness.

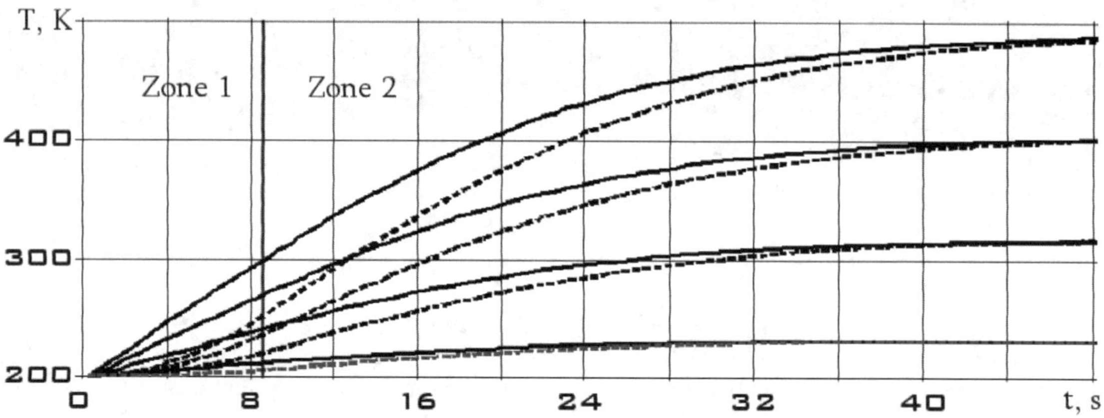

Figure 4. The temperature change dynamics of the inner layers (1–4) of the small spacecraft elastic element

In figure 4, a solid line shows the temperature dynamics in the case of using a cylindrical model of the Earth's shadow (quoted in [5]). The dotted line shows the temperature dynamics when using a cone-shaped model of the Earth's shadow. The penumbral section is indicated in figure 4 as Zone 1. Zone 2 is the light section of the small spacecraft orbit. The results obtained differ somewhat from the results presented in [5]. The heating of the elastic element, considering the penumbral section of the orbit, occurs more smoothly than in [5].

4. Conclusion and summary

The cone-shaped model of the Earth's shadow gives more adequate results of warming up the elastic element of the small spacecraft under a temperature shock than the cylindrical model of the shadow. In this case, the temperature shock itself does not occur at point B (figure 1) on the border of the shadow and penumbral sections of the orbit, but is shifted to the penumbral section. When considering the cone - shaped model of the Earth's shadow, we should talk about a section of the orbit for which the time derivative of temperature is sufficiently large, and not about a specific point at which a temperature shock occurs. By setting specific parameters of the small spacecraft, we can create a criterion for selecting this section based on restrictions, for example, on the level of micro-accelerations [3].

A number of conclusions can be drawn from the results of these studies.

– For the considered circular orbit with a height of 600 km, the penumbral section is significant. It substantially affects the dynamics of large elastic elements temperature changes under temperature shock. This orbit was chosen on the basis of actual projects of technological and biomedical spacecraft ("Bion-M" # 1 – 575 km, "Foton-M" # 4 – 525 km).

– Taking into account the penumbral section allows us to use the initial conditions in the form (2) more correctly than when using the cylindrical model of the Earth's shadow. On the one hand, the gradual solar radiation power growth in penumbral orbit section is taken into account, on the other hand, the initial condition of uniform temperature field at the moment, when the small spacecraft exits the shadow section, is retained.

– Micro-accelerations resulting from a temperature shock are somewhat reduced due to the «smearing» of the temperature shock time. If the temperature shock is a point in the cylindrical shadow model (the point in figure 1), then for the cone-shaped shadow model it is implemented at a certain section of the orbit with the time derivative of the temperature maximum values. In case of a point temperature shock, the maximum value of this derivative obviously corresponds to the small spacecraft exit point from the Earth's shadow (the point in figure 1). However, this disadvantage of the cylindrical Earth shadow model can be turned into an advantage. The overstatement of the micro-acceleration estimation gives it a certain margin of safety compared to the estimation of the cone-shaped shadow model. Therefore, from the point of view of estimating micro-accelerations, the use of different models of the Earth's shadow is under discussion - it depends on the specific task being solved.

References

[1] Bratsun D A, Krasnyakov I V and Zyuzgin A V 2018 Active control of thermal convection in a rectangular loop by changing its spatial orientation *Microgravity Science and Technology*. **30** 43–52

[2] Lyubimova T, Zubova N and Shevtsova V 2019 Effects of Non-Uniform Temperature of the Walls on the Soret Experiment *Microgravity Science and Technology*. **31(1)** 1–11

[3] Sedelnikov A V 2007 Fractal assessment of microaccelerations at weak damping of natural oscillation in spacecraft elastic elements. II *Russian Aeronautics*. **50(3)** 322–25

[4] Lyubimov V V 2009 Some features of disturbed rotational motion of a satellite with a strong onboard magnet *Russian Aeronautics* **52(2)** 172–75

[5] Gorozhankina A S, Orlov D I and Belousova D A 2020 Problems of development motion control algorithms for a small spacecraft for technological purpose taking into account temperature deformations of solar panels *Journal of Physics: Conference Series* 1546:012015

[6] Narasimha M, Appu Kuttan K K and Ravikiran K 2010 Thermally induced vibration of a simply supported beam using finite element method *International journal of engineering science and technology* **2(12)** 7874–879

[7] Zhang L, Chen Y 1999 The on-orbit thermal-structural analysis of the spacecraft component using MSC/NASTRAN *Aerospace Users' Conference Proceedings.* 1-8
[8] Johnston J D, Thornton E A 1998 Thermally induced attitude dynamics of a spacecraft with a flexible appendage *Journal of guidance, control and dynamics.* **4** 581-587
[9] Podvysotsky V V 2018 Superconductors applied in electrodynamic engine *International Review of Aerospace Engineering.* **11(3)** 120–126
[10] Shevtsova V, Melnikov D and Legros J C 2002 The Post Flight Study of Micro Accelerations On-Board of Russian Spacecraft FOTON-12 *European Space Agency Contract Report,* ESTEC, Noordwijk, Netherlands. p 135
[11] Abrashkin V I, Bogoyavlensky N L, Voronov K E et al. 2007 Uncontrolled motion of the Foton M-2 satellite and quasistatic microaccelerations on its board *Cosmic Research.* **45(5)** 424–44
[12] Kirilin A N, Akhmetov R N, Anshakov G P et al. 2013 Bion Project - TsSKB-Progress Contribution Into Space Biology And Medicine, Flight. **(11)** 3–16
[13] Abrashkin V I, Voronov K E, Piyakov I V, Puzin Yu Ya, Sazonov V V, Semkin N D and Chebukov S Yu 2015 Determining the rotational motion of the Bion M-1 satellite with the GRAVITON instrument *Cosmic Research.* **53(4)** 286–99

IOP Conf. Series: Materials Science and Engineering 1061 (2021) 012029 doi:10.1088/1757-899X/1061/1/012029

The upgrade of peen forming equipment for long-sized aircraft parts

A E Pashkov[1], A Y Malashchenko[1], A A Pashkov [1], A A Duk [1]

[1] Irkutsk National Research Technical University (INRTU), 84, Lermontova street, Irkutsk, 664074, Russia

E-mail: aa_pashkov@ex.istu.edu

Abstract. This article represents an overview of the main design and technological features of equipment for peen forming and grinding of long-sized parts of the aviation industry. Since the end of the 80s of the XX century, the Federal State Budgetary Educational Institution of Higher Education INRTU has been working on the creation of equipment for the manufacture of large-sized panels and aircraft skins for the leading enterprises of the aviation industry of the Russian Federation. Totally there were 4 industrial units created, which proved their efficiency and competitiveness in the production process of TU-204, TU-334 and BE-200 aircrafts. Based on experience, a universal installation has been developed, which can be used as part of the equipment complex of the industry competence center for the shaping of large-sized aircraft parts.

1. Introduction

An effective method of forming of large-sized panels and skins of aircraft is shot peen forming (SPF) [1-2]. The shot-peen forming process is based on the biaxial elongation of the surface layers by shot treatement. With this effect, the part takes on the shape of a double curvature surface [3-6]. To obtain required form of parts, in foreign practice, such technological methods are used as sequential processing of the outer surface of the part and its longitudinal edges on both sides, as well as the use of deflection, which allows increasing the curvature in the direction of preliminary elastic bending [7-9]. A large number of works are devoted to a comprehensive analysis of the stressed state of a part after shot peen forming using finite element methods [10-15], in which the main advantages of the process under consideration are noted: the formation of a uniform field of residual compressive stresses in the surface layer of the part, which has a positive effect on the product life, the absence of surface faceting and the effect of partial springback of the part after processing, which improves the accuracy of the product shape.

Irkutsk National Research Technical University (INRTU) has a rich experience of interaction with aviation enterprises in the creation of equipment for the implementation of the peen forming technology. Totally, 4 units of the UDF series (shot peening machine) were developed. Common features for all installations of the UDF series is using of the processing concept when the part is installed in the clamps of the fixing device, and the control system moves the working bodies relative to the workpiece and uses the working tools [16] that are in direct contact with the part using elastic seals with different widths:

- a shot-blasting machine for shaping operation. It uses stream of the shots to indent in the front surface of the part. As a result there is an elongation of the surface layer of the part and, as a consequence, the general bending deformation of the part;

- a grinding machine for ensuring the regulated roughness of the surface layer of the part after processing shot blasting.

These technological features competitively distinguish domestic equipment from foreign counterparts, which use shot-blasting in the closed chambers with an external shot circulation system [17-18]. These installations technologically very complex and expensive. The final shape of the parts is reached only upon completion of the entire machining cycle. In addition, the use of such installations solves the problem of mechanization only of preliminary shaping, and for the final formation of the contour, manual finishing is needed.

2. Stages of development of UDF series installations

The first domestic pilot installation represents a frame portal moving along rails in the longitudinal direction. Vertical feeding is carried out using a hand winch. For this purpose, guides and a movable carriage are provided, on which the transverse movement mechanism is mounted. The latter has cylindrical guides and a manual screw mechanism. The operating experience of the UDF-1 unit served as the basis for further improvement of such equipment.

Installation UDF-2

The development of the model range was the UDF-2 unit, designed for the formation of the first sets of wing panels for the TU-334 aircraft [16]. A general view of the processing unit of the UDF-2 installation, designed to fix and move the working tools of the installation: a shot-blasting machine and grinding machine, is shown in Fig. 1. Its main units are: a movable platform on which a supporting column with vertical and transverse displacement mechanisms are mounted, a control cabinet, two adjustable hydraulic pumps for the main rotation and longitudinal displacement drives. The adjustable parameters are the longitudinal feed rate and the main rotation frequency.

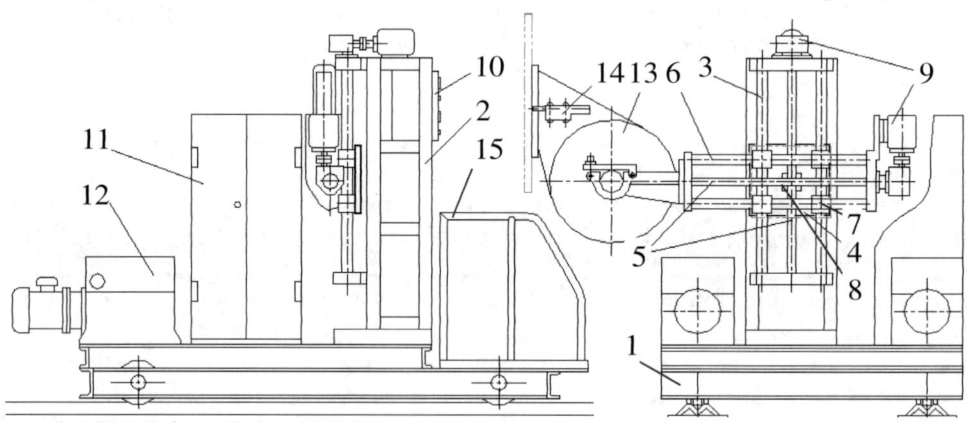

Figure: 1. UDF-2 installation: 1 - base; 2 - column; 3 - guides; 4 - carriage; 5 - lead screws; 6 - retractable rods of the transverse movement mechanism; 7 - supports; 8 - sliding nuts; 9 - drives for vertical and transverse movements; 10 - operator panel; 11 - control panel; 12 - hydraulic pump; 13 - working tool; 14 - pressure sensor; 15 - fence

Constant contact of the sealing device of the shot blasting machine with the treated surface is maintained using one contact pressure sensor responsible for the transverse movement of the seals of the working bodies.

The basic unit of the fixation system (Fig. 2) is a cross-beam containing clamping elements designed to fix the workpiece.

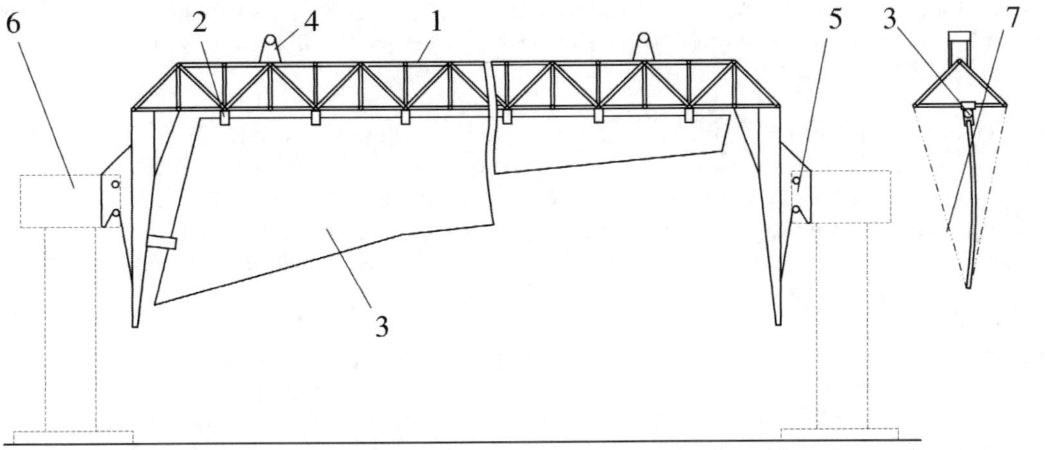

Figure: 2. Fixation system of the UDF-2: 1 - traverse; 2 - work piece; 3 - clamp; 4 - capture; 5 - locking device; 6 - stationary part; 7 - supporting rope

Exploitation of the UDF-2 unit in pilot manufacturing revealed the following disadvantages:

1. The arrangement of the processing unit with a fixed bearing column limits the technological capabilities of the installation - the magnitude of the transverse displacement of the working bodies and causes the occurrence of self-oscillations of the unit while peening a part.
2. The use of a hydraulic drive for the main rotation of the working bodies does not allow the use of a feedback device on the frequency of their rotation, which limits the accuracy of the shot speed control and does not allow controlling the metal removal during cleaning.
3. Fixing the part in the longitudinal contour is highly labor intensive and requires a special control equipment outline templates of longitudinal section.
4. Pressure control at one point limits the ability to process surfaces with a radius of curvature less than 4 m.

Installation UDF-3

The next stage of development was UDF-3, designed for serial technological process of forming panels and skins of the BE-200 aircraft (with dimensions up to 40 × 1200 × 12000 mm) and introduced at the Irkutsk Aviation plant. A general view of the processing unit of the UDF-3 installation is shown in Fig. 3.

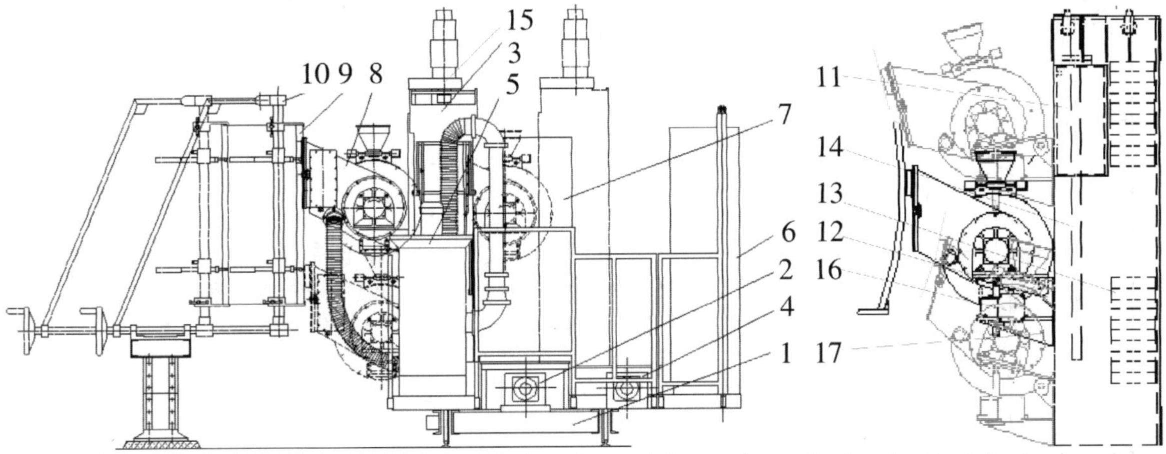

Figure: 3. General view of the UDF-3 installation: 1 - mobile platform; 2 - longitudinal feed drive; 3 - bearing column; 4 - cross feed drive; 5 - dust collecting unit; 6 - power electrical panel; 7 – control panel; 8 - shot blasting machine; 9 - work piece; 10 - fixing device; 11 - carriage; 12 - bracket with a platform for mounting working tools; 13 - counterweight; 14 - lead screw; 15 - vertical feed drive; 16 - tilt drive; 17 – bracket

The layout solution of the UDF-3 installation includes vertical feed drive of the working tools, that fixed on the movable bearing column. This makes possible to reduce the overturning moment, prevent the occurrence of self-oscillations and use the high-resource 2D400M shot-blasting machine as a working tool for the forming operation. The processing unit of the installation includes a movable platform on which a longitudinal movement drive mechanism is installed, a column of vertical feeds with a mechanism for its transverse movement; dust collection unit, power electrical cabinet and CNC system panel.

Electric motors were used as drives for longitudinal, transverse and vertical movement. A manual mechanism for tilting the working tools was added, which is a worm gear pair.

The pressure control is carried out in the same way by the contact method, but already at eight points.

Depending on the position of the device rods, control system generates the following operating modes: "Normal mode" "Movement in the transverse direction".

The panel fixation system on the UDF-3 unit consists of 12 lateral movement mechanisms that fix the part at the upper and lower points of the cross sections (Fig. 4). To reduce labor intensity and increase the accuracy of tuning to a given contour, a helical gear and a counting ruler are introduced into the design of these mechanisms.

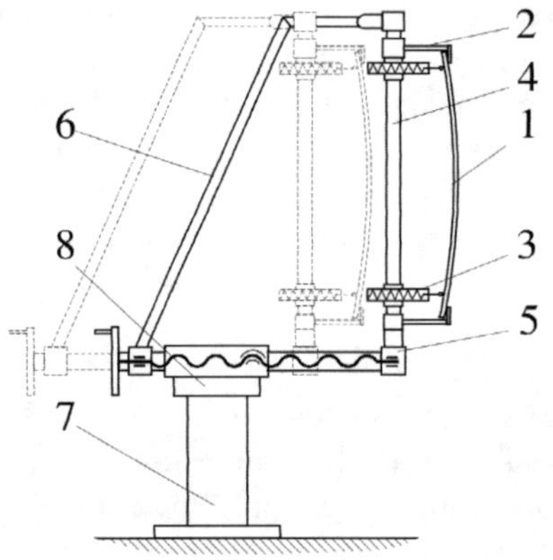

Figure: 4. Mechanism of lateral movement of the fixation system of the UDF-3 installation: 1 - workpiece to be processed; 2 - emphasis; 3 - clamp; 4 - rack; 5 - screw mechanism; 6 - brace; 7 - base; 8 – frame

Many years of research, as well as taking into account the accumulated experience, the increasing modern requirements for the type of aircraft parts being processed, general directions of development for the modernization of this equipment were formulated [19]:

1. Introduction of a modern control system with the possibility of a backup increase in the number of simultaneously performed functions, as well as an active feedback system during processing.

2. The design of the installation and its elements must ensure the prevention of the occurrence of self-oscillations in all operating modes of the working body (shot blasting machine and / or stripping head).

3. The design of the installation should be of sufficient rigidity to resist changing operational loads and to ensure consistency of accuracy and repeatability of the processing process.

Installation UDF-4

The UDF-4 installation was created to eliminate the disadvantages of the previous models (Fig. 5)

It is using the layout solution of UDF-3. The main differences are:

- using of high-precision linear and circular movement systems based on rolling guides, ball-screw drives, planetary gearboxes, couplings and other mechanical elements of increased rigidity and accuracy of movement produced by leading world manufacturers;
- using CNC system to control the four-axis positioning of the working body relative to the part;
- reinforcement of the structure to reduce the influence of bending loads, eliminate the counterweight system and increase the resistance to vibration effects;
- adding a drive for turning the working body;
- application of the adaptive clamping system [20].

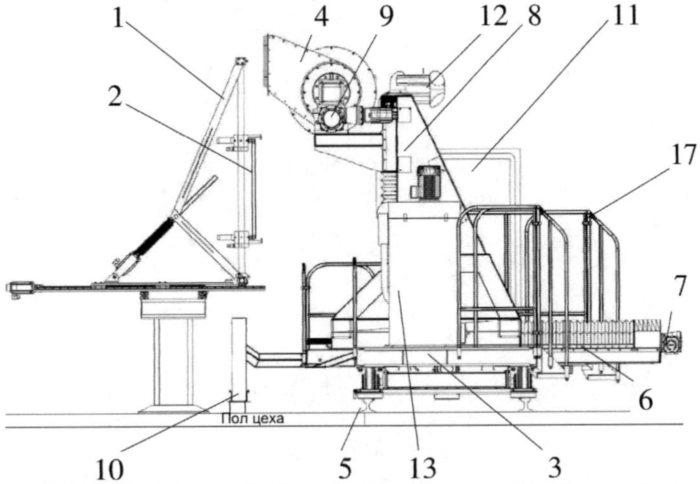

Figure: 5. General view of the UDF-4 installation: 1 - fixation system; 2 - work piece; 3 - trolley for moving the working body 4 along the rails 5 in the longitudinal direction (coordinate X); 6 - slide with rolling guides, ball screws and drive 7 for moving the working body in the transverse direction (Y coordinate); 8 - racks with rolling guides, ball screws and a drive for moving the working body in the vertical direction (coordinate Z); 9 - drive of rotation of the working body (coordinate A); 10 - flexible cable channel; 11 - electrical panel; 12 - control panel; 13 - dust collecting unit

A general view of the rack of the UDF-4 parts fixing system is shown in Fig. 6. The design includes the required number of racks, depending on the length and rigidity of the part, each of which fixes the workpiece 2 with the help of the stop 3 and the pneumatic clamp 1, providing a constant position of the cross-section during processing. With the help of a stepping motor 6 and guides 5, the rack moves laterally. The hinge assembly 4 in conjunction with the electric motor 7 is responsible for angular movements to ensure the fixation of parts having a twisted cross-section. The positioning of the racks in the longitudinal direction is carried out by moving them along the guides 8 with fixation with screw clamps.

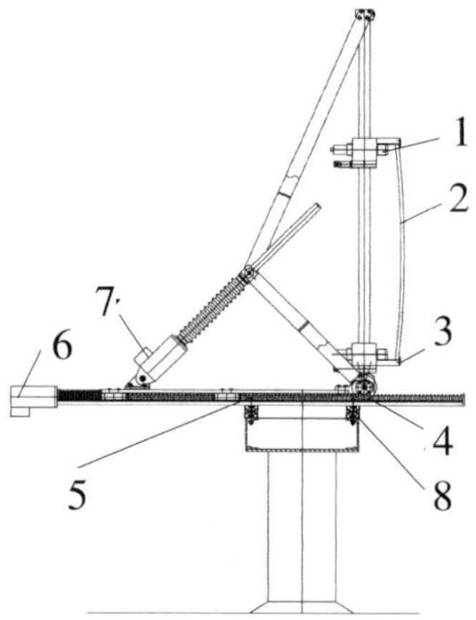

Figure: 6. The mechanism of transverse movement of the fixation system of the UDF-4 installation: 1 - pneumatic clamp; 2 - part; 3 - emphasis; 4 – hinge assembly; 5 - transverse guides; 6-7 - stepper motor; 8 - longitudinal guides

Installation UDF-5

Due to increasing in the complexity of the geometric shape of the long-sized aircraft parts, their dimensions (width over 2000 mm, length up to 30,000 mm) and shape accuracy, the concept of a new UDF-5 unit was proposed - the most advanced (among of the UDF series) (Fig. 7 -8). This installation should cover the needs of the entire domestic aviation industry in the field of shaping of the the long-sized aircraft parts. It is advisable to use it as part of the equipment complex of the industry center of competence for the shaping of large-sized parts. The main feature of the installation is a modular platform design and an additional column for attaching the grinding machine (Fig. 9), which has three translational and one rotational degrees of freedom and is installed on a platform, next to the column for shot-blasting machine, moving along fixing device for the part.

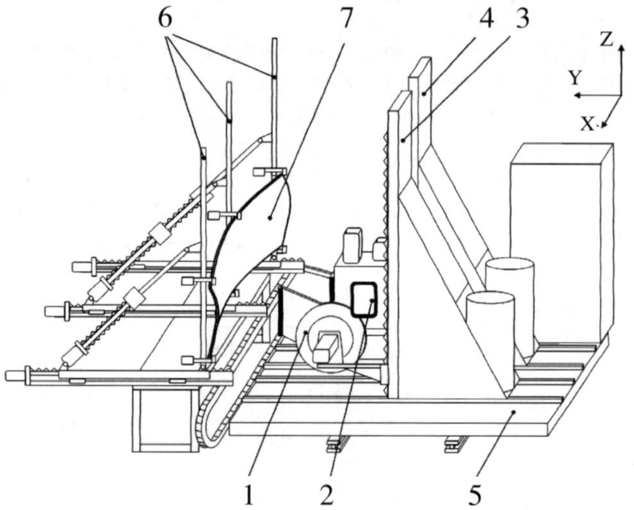

Figure: 7. General view of the UDF-5 installation: 1 - shot blasting machine; 2 – grinding machine; 3 - column for the shot blasting machine; 4 - additional column for the grinding machine; 5 - movable platform; 6 - fixing device; 7 – part

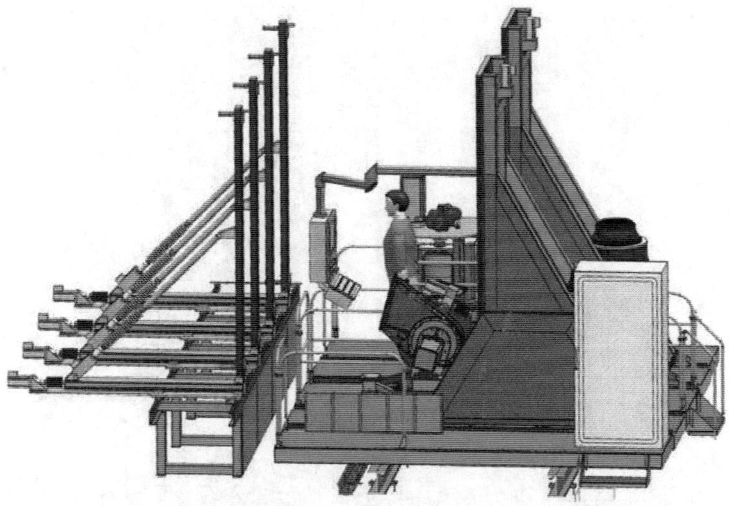

Figure: 8. Electronic model of the UDF-5 installation

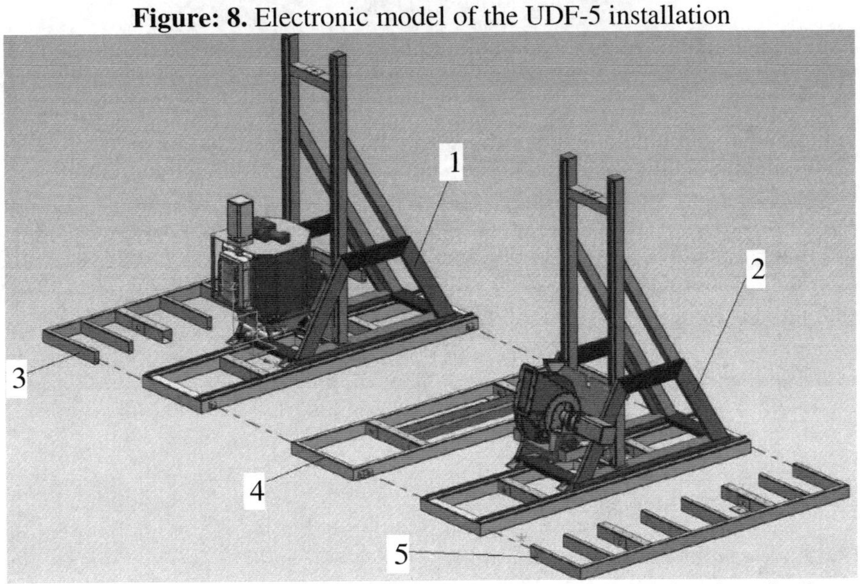

Figure: 9. Modular design of the UDF-5 platform: 1 - platform and column for the grinding machine; 2 - platform and column for attaching the shot-blasting machine; 3 - platform for servicing the grinding machine; 4 - operator platform; 5 - platform for servicing the shot blasting machine.

The shot-blasting machine module design is shown in Fig. 10. This solution provides the ability to move the working tool to the height required for forming a part with a width up to 2000 mm.

Figure: 10. The shot-blasting machine module design: 1 - shot-blasting machine; 2 - rotary base for attaching the shot-blasting machine; 3 - rack of vertical feeds; 4 platform; 5-6 - substrates for vertical and lateral movement guides; 7-8 - guides for vertical and lateral movement; 9-10 - carriages of vertical and lateral movement

Structural analysis using the finite element method has shown sufficient static strength of both individual units and whole installation (Fig. 11).

Figure: 11. Examples of distribution of residual stresses (von Mises) in platform and column

The automated fixation system of the UDF-5 installation is structurally similar to the previous model, but it provides the ability to fix parts up to 2000 mm wide, up to 20,000 mm long and includes:

- base columns;
- support frame;
- racks for fixing the part in the amount of 24 pcs., each of which includes a vertical guide with a tilt unit; pneumatic clamping devices (clamps) moving along the guide (2 pcs.); transverse movement mechanism with a controlled electric drive; tilt mechanism with controlled electric drive and feedback

device;

- a pneumatic system that supplies compressed air to the drive pneumatic cylinders of the struts clamps;
- a device that ensures the linear movement of the racks along the base during the assembly of the part, with a locking mechanism;
- control and power supply system, which includes a power cabinet with placed drive modules and control systems SIMOTION; operator console and industrial computer PCU.

Equipping the device for shot shaping and stripping with an additional stand for attaching the stripping head, equipped with mechanisms for moving the working body along three Cartesian and one angular coordinates, allows you to simultaneously perform shot processing and stripping or clean up after shot shaping without removing the part from the fixing device, which significantly reduces labor intensity and processing time.

The supply of the part fixation system with individual software-controlled mechanisms of orientation in space, allows you to automate the process of fixing the part and control the shaping process during processing.

3. Conclusion

Scientific team of the INRTU has unique competencies in the field of creating competitive equipment for forming long-sized aircraft parts. The experience began to form in the 80s of the last century, which made it possible to create an installation that has no analogues in the world. This installation - UDF-5 is a modern processing complex, which allows to significantly reduce labor intensity and processing time, as well as to increase the accuracy of the shape of the processed parts.

References

[1] Pashkov A E 2014 Technological complex for the formation of long panels and skins based on domestic equipment *Izvestiya of the Samara Scientific Center of the Russian Academy of Sciences* **1(5)** 1528–1535

[2] Andrey E Pashkov, Aleksander Y Malashchenko, Aleksander A Pashkov and Tatiyana V Zarak 2018 Parameters identification for combined forming of complex shaped sheet metal components *Advances in Engineering Research* **158** 323–329

[3] Baughman D L 1984 An overview of peen forming technology *Proceedings of the 2nd international conference on shot peening* 28–33

[4] Ebenau A, Macherauch E 1987 Influence of the shot peening angle on the condition of near–surface layers in materials *Proc. 3rd conf. shot peening (ICSP3)*253–260

[5] Gardiner T S and Platts M J 1999 Towards Peen Forming Process Optimization *Proc. 7th Conf. Shot Peening (ICSP7)*235–243

[6] Gariépy A, Miao H Y and Lévesque M 2014 Peen forming *Comprehensive Materials Processing* **3** 295–329

[7] Badrinarayan P Athreya, Narendra P Singh, Lingyun Pan and Wei Huang 2012 Computational approach for fatigue life prediction in shot peened welded specimens *Welding in the World* **57 5** 675–684

[8] Baragetti S 2001 Three–dimensional finite–element procedures for shot peening residual stress field prediction *International Journal of Computer Applications in Technology* **14 3** 51–63

[9] Barrett C, Todd R 1984 Investigation of the effects of elastic pre–stressing technique on magnitude of compressive residual stress induced by shot peening of thick Aluminum plates *Proceedings of the 2nd international conference on shot peening* 15–21

[10] Meguid S A, Shagal G, Stranart J C, Daly J 1999 Three–dimensional dynamic finite element analysis of shot peening induced residual stresses *Finite Elements in Analysis and Design* **31 3** 179–191

[11] Miao H Y, Larose S, Perron C, Evesque M 2009 On the potential applications of a 3D random finite element model for the simulation of shot peening *Adv. Eng. Softw* **40** 1023–1038

[12] Wang T A, Platts M J,Levers A 2006 Process model for shot peen forming *Journal of Materials Processing Technology* **172 2** 159–162
[13] Pashkov A E 2013 On increasing the efficiency of the contact type shot blasting machine *Proceedings of Irkutsk State Technical University* **10(81)** 46–52
[14] Pashkov Andrey E, Pashkov Aleksander A and Koltsov Vladimir P 2017 Complex method of peen forming and shot peening of aircraft structural components *Advances in Engineering Research* **133** 585–591
[15] Pashkov A E, Yu Malashchenko A, Pashkov A A, Bogdanov K V and Kryuchkin A V 2019 Development of digital manufacturing technologies for frame and casing parts *Materials Science and Engineering* **632 012104** 1–16
[16] Pashkov A E 2005 Technological connections in the process of manufacturing long sheet metal parts *Mater. Sci.* 5–22
[17] Horowitz Z I 1979 Das «Shot Peening» - verfahren, seine Grundlagen, Begriffe und die MeS-Methoden zugehorigen *Fachber. Huttenprax. Metallweiterverarb* **3** 174–181
[18] Vaccary J A 1985 Peen forming enter the computer age *American mechanist* June, **6** 91–94
[19] Gavrilov A S 2011 On the creation of an automated contact-type installation for shot-impact shaping and stripping of long panels of aircraft *Materials of the All-Russian scientific and practical seminar with international participation* 111–117
[20] Likhachev A A, Pashkov A A and Gerasimov V V 2015 Implementation of the control system for the process of shot-impact shaping on contact-type installations *Proceedings of Irkutsk State Technical University* **2** 42–47

Formation of spatial responses of the optical field in the region of location of the optical antenna array

Yu G Pasternak[1], K S Safonov[1], F S Safonov[1], S M Fedorov[1], A V Volodko[1], E A Ischenko[1]

[1] Voronezh State Technical University, 14, Moscow ave., Voronezh, 394026, Russia

E-mail: fedorov_sm@mail.ru

Abstract. This article describes a model of an optical antenna array operating in the mode of receiving electromagnetic waves in the frequency range 30-300 THz (in this case, the wavelength varies from 10 □m to 1 □m). In the course of numerical experiments (to solve the system of Maxwell equations in the integral formulation, the Weiland method [31] was used) the accuracy of approximation of the spatial structure of the optical field at the points of the "virtual" antenna array elements using the auxiliary field sources method was studied. A study was also carried out on the possibility of using a virtual antenna array formed on the plane of arrangement of elements of a real array to increase the resolution of a multichannel optical system. At the upper frequency of the analyzed range of 300 THz, the wavelength is 1 □m. The lattice period was chosen equal to half the minimum wavelength of d=0.5 □m, so that it was possible to approximate the spatial distribution of the field in the space between the elements of the antenna array, in accordance with the requirement of the Kotelnikov-Shannon theorem.

1. Introduction

During the simulation, the following parameters of the optical antenna array (figure 1) were considered: w=0.25 □m crystal width; crystal height h=0.25 □m [1, 2, 3]; the lattice period in coordinates x, y is d=0.5 □m; the wave with vertical polarization falls in the plane $y0z$, the angle of the point of incidence of the wave is measured from the z axis. The material of the crystals is silicon dioxide. The substrate is the perfect metal. It was believed that the intensity E_y of the field components was measured at points corresponding to the midpoints of the upper edges of the optical lattice crystals (z=0.25 μm).

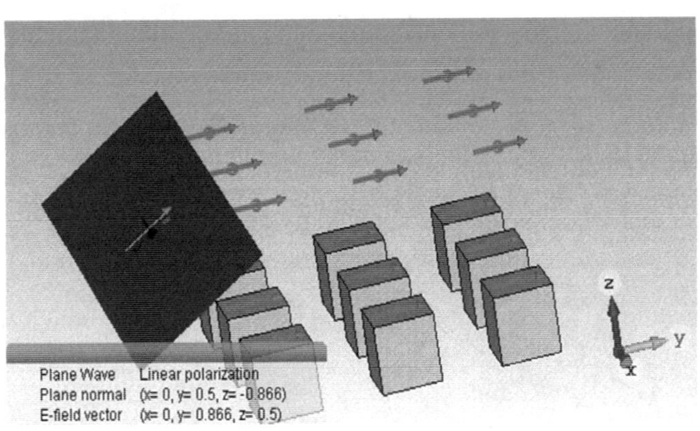

Figure 1. Geometry of the optical antenna array, plane incident wave and field probes (shown by arrows)

In the course of numerical experiments, we study an accuracy of approximating the spatial structure of the optical field at locations of the "virtual" antenna array elements using the method of auxiliary field sources. Also, we study a method to increase the resolution of a multi-channel optical system by use of virtual antenna array, formed at the same plane as real array elements were located.

At the upper frequency of the analyzed range (300 THz), the wavelength is 1 □m. The array period was equal to half the minimum wavelength (0.5 □m), so we can approximate the spatial distribution of the field in space between the antenna array elements in accordance with the requirement of the Nyquist–Shannon theorem.

The upper arrows in figure 1 correspond to field probes at the points of formation of the “virtual” antenna array (in fig. 1, the “virtual” antenna array is elevated along the z coordinate by 0.5 □m from the upper edges of the optical grating (z=0.75 □m)). The coordinates x, y the «real» antenna array and the “virtual” antenna array, as well as at the auxiliary point sources of the field (not shown in fig. 1) in this case coincide (later we will also consider other cases of the arrangement of the “virtual” antenna array and auxiliary field sources) [4].

Auxiliary sources of the field (figure 1) are located above the “virtual” antenna array: $z_{auxill.array.} > z_{VAA}$ In addition, when calculating the approximated field, the model takes into account the specular reflection of auxiliary sources in the ideal metal substrate.

It was believed that the amplitudes of specular reflections of auxiliary sources are equal to the amplitudes of auxiliary sources, and the phases of specular reflections of auxiliary sources are opposite to the phases of auxiliary sources [5].

To find the complex amplitudes of the auxiliary sources of the field, find the solution of the following system of linear algebraic equations:

$$\sum_{n=1}^{N}\left(U_n^{auxill}\cdot\frac{\exp\left(-ik_0\sqrt{\left(x_n^{auxill}-x_k^{RAA}\right)^2+\left(y_n^{auxill}-y_k^{RAA}\right)^2+\left(z_n^{auxill}-z_k^{RAA}\right)^2}\right)}{\sqrt{\left(x_n^{auxill}-x_k^{RAA}\right)^2+\left(y_n^{auxill}-y_k^{RAA}\right)^2+\left(z_n^{auxill}-z_k^{RAA}\right)^2}}-\right.$$
$$\left.-U_n^{auxill}\cdot\frac{\exp\left(-ik_0\sqrt{\left(x_n^{auxill}-x_k^{RAA}\right)^2+\left(y_n^{auxill}-y_k^{RAA}\right)^2+\left(-z_n^{auxill}-z_k^{RAA}\right)^2}\right)}{\sqrt{\left(x_n^{auxill}-x_k^{RAA}\right)^2+\left(y_n^{auxill}-y_k^{RAA}\right)^2+\left(-z_n^{auxill}-z_k^{RAA}\right)^2}}\right)=U_k^{RAA}, \quad (1)$$
$$k=1,\dots,N,$$

Where N- the number of analyzed elements of the "real" antenna array (RAA); U_n^{auxill} - complex amplitude of n-th auxiliary source located at the point with coordinates $(x_n^{auxill}; y_n^{auxill}; z_n^{auxill})$; k_0 - free space wave number; U_k^{RAA} - complex amplitude of the signal at the output of k-th element of the RAA, located at a point with coordinates $(x_k^{RAA}; y_k^{RAA}; z_k^{RAA})$.

To approximate the field at the point $(x_k^{VAA}; y_k^{VAA}; z_k^{VAA})$ of the location of the “element” of the “virtual” antenna array (VAA) (spatial reference of the field), the following expression is used:

$$\sum_{n=1}^{N}\left(U_n^{auxill}\frac{\exp\left(-ik_0\sqrt{\left(x_n^{auxill\_VAA}-x_k\right)^2+\left(y_n^{auxill\_VAA}-y_k\right)^2+\left(z_n^{auxill\_VAA}-z_k\right)^2}\right)}{\sqrt{\left(x_n^{auxill\_VAA}-x_k\right)^2+\left(y_n^{auxill\_VAA}-y_k\right)^2+\left(z_n^{auxill\_VAA}-z_k\right)^2}}\right.$$

$$\left.-U_n^{auxill}\frac{\exp\left(-ik_0\sqrt{\left(x_n^{auxill\_VAA}-x_k\right)^2+\left(y_n^{auxill\_VAA}-y_k\right)^2+\left(-z_n^{auxill\_VAA}-z_k\right)^2}\right)}{\sqrt{\left(x_n^{auxill\_VAA}-x_k\right)^2+\left(y_n^{auxill\_VAA}-y_k\right)^2+\left(-z_n^{auxill\_VAA}-z_k\right)^2}}\right)=U_k^{VAA}, \quad (2)$$

$k = 1,...,K,$

where K - the number of formed "elements" of VAA (it can be any, including - substantially larger than the number of elements of RAA N).

Figure 2 shows the frequency relationships of the real and imaginary parts of the relative dielectric constant of the material of optical lattice (silicon dioxide) crystals.

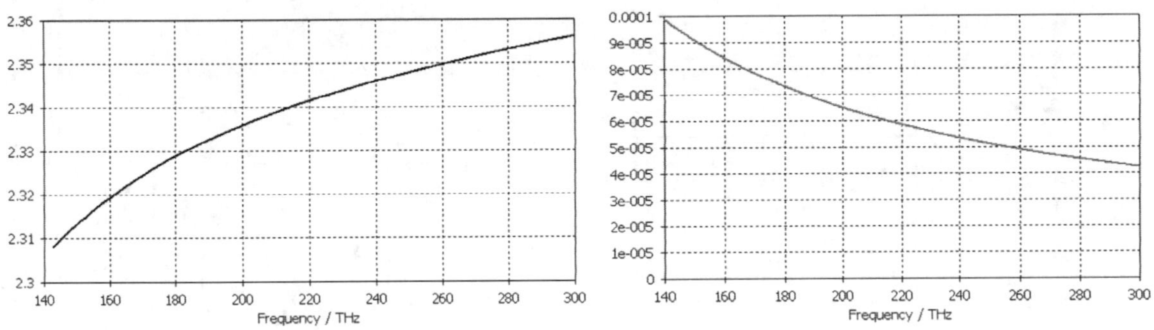

b)

a)

Below are the relationships that characterize the accuracy of the approximation of the optical field in the central element of the «virtual» antenna array in the frequency range from 30 to 300 THz ($\lambda_0 \in [10;1]$ □m).

Solid lines indicate the frequency relationships of the amplitudes and phases of the true value E_y — the field components at the analyzed point, corresponding to a rigorous numerical analysis of the electrodynamic structure using the Weyland method [6]; the dashed lines show the approximated relationships obtained by forming a "virtual" antenna array using the auxiliary field sources method.

2. The study of the accuracy of the approximation of the optical field in relation to the number of elements of the antenna array

In this subsection, the relationship of the accuracy of the approximation of the phases and field amplitudes, carried out using the auxiliary field sources method, in relation to the number of elements of the antenna array (in accordance with figure 2, for an array with $N \times N$ elements, $2 \times N \times N$ auxiliary sources is formed, including specular reflections fields of auxiliary sources) [7].

Figure 3 shows the frequency relationships of the phases and field amplitudes at the point with the coordinates (0;0;0.75) □m (this point is located above the central element of the antenna array and is raised to a height of 0.5 □m relative to its upper edge). Grids with dimensions 3x3, 5x5 and 7x7 elements were considered.

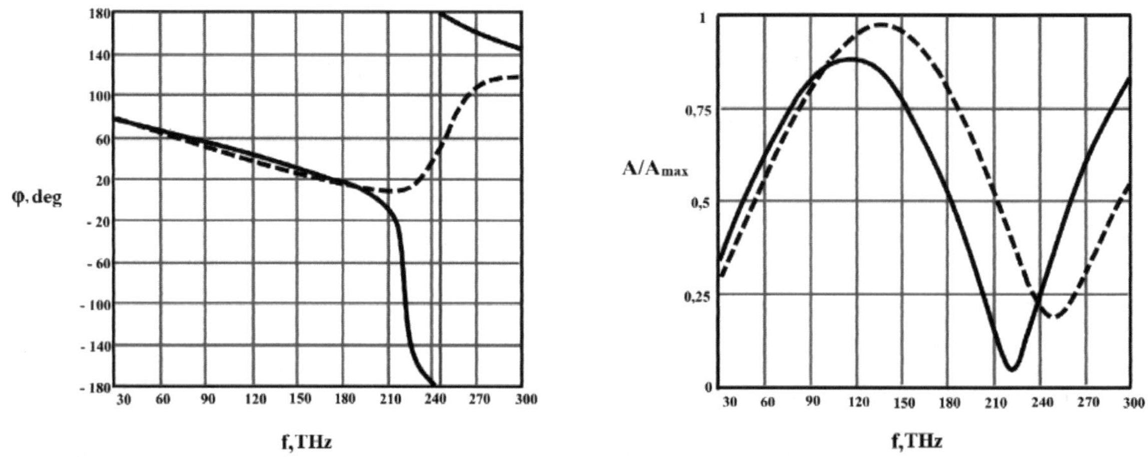

a)

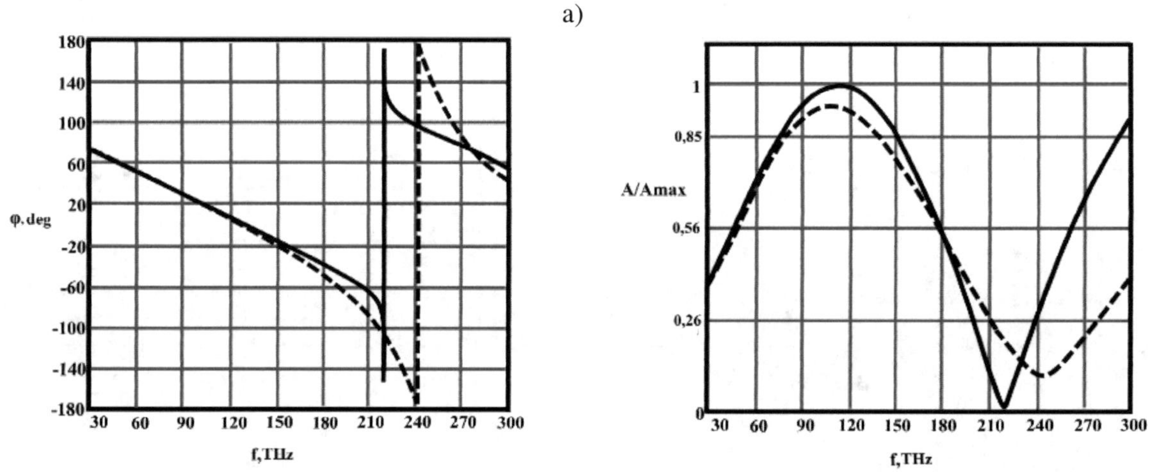

b)

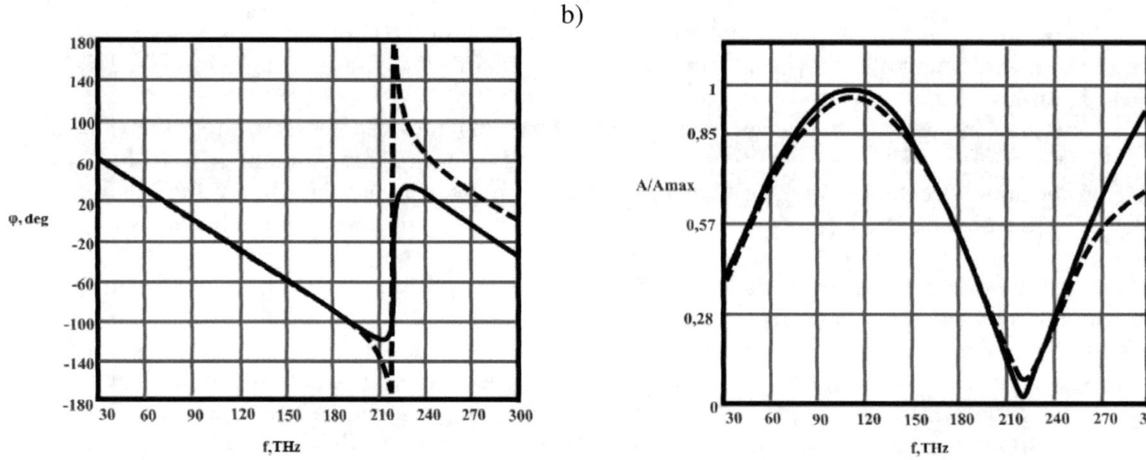

c)

Figure 3. The frequency relationships of the phases and amplitudes of the field at the point with (0;0;0.75) □m coordinates: a) a 3x3-element grid; b) a grid of 5x5 elements; c) a grid of 7x7 elements

In figure 4, 5, it can be seen that, with an increase in the number of elements of the antenna array (and a proportional increase in the number of auxiliary sources of the field), the approximation accuracy rapidly increases.

IOP Conf. Series: Materials Science and Engineering 1061 (2021) 012030 doi:10.1088/1757-899X/1061/1/012030

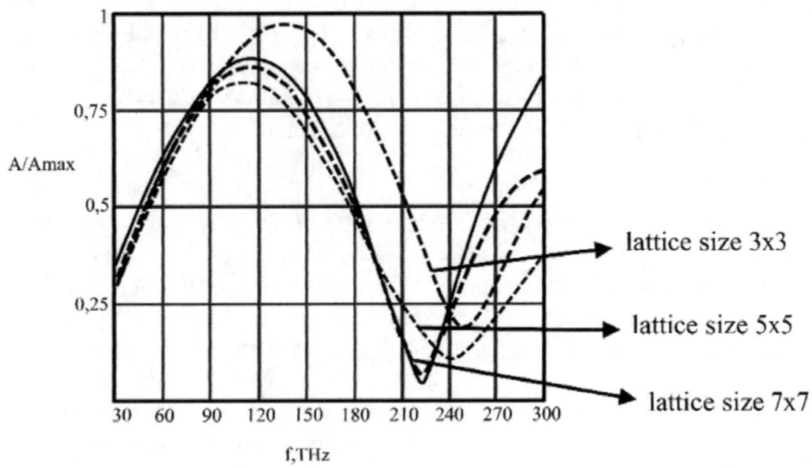

Figure 4. The family of frequency relationships of the normalized field amplitudes at the analyzed point with (0;0;0.75) □m coordinates for various lattice sizes (the full line is the true value)

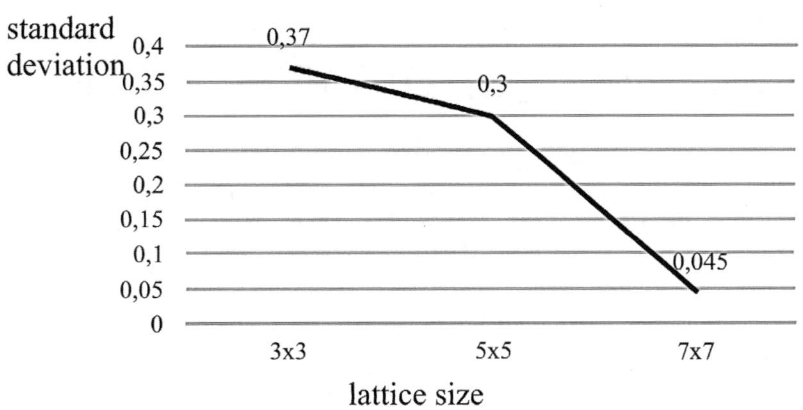

Figure 5. The standard deviation of the approximated values of the normalized field amplitude from its true value for a different number of lattice elements

The studies allow us to establish a very obvious fact - with an increase in the deviation of the incidence angle from the normal, the approximation accuracy decreases, because the phase shift increases from element to element of the array in the plane of wave incidence. The frequency dependences of the field amplitude in case of array with minimum number of elements (3×3) can be approximated with an acceptable error only at small values of the elevation angle (no more than 30°).

The frequency dependences of the phase can be predicted with acceptable accuracy in the long-wave region of the studied frequency band, up to an almost grazing incidence of the wave along the array - at elevation angles up to 70°. Note that in optical location and in measurement of the near-field structure of an optical field, the phase dependences are, as a rule, more informative. Therefore, significant errors in measuring the amplitudes during approximation of the observed field can be acceptable in some practical applications.

Study has shown that with an increase in the distance between the "real" and "virtual" arrays from 0 to 0.4 microns, the approximation accuracy monotonically increases, and with further distance increase between the generated spatial samples and the antenna array, accuracy starts to decrease.

This tendency is easy to explain: at short distance between the "real" and "virtual" arrays, the maximum contribution to the approximated field is made by the field of the nearest auxiliary source. As the distance increases further, two opposing factors act: on the one hand, the contribution of all auxiliary field sources becomes more uniform; on the other hand, with increasing distance from the point at which the value of the function is known, the approximation error is also increase.

3. Conclusion

Thus, in the present work, on the basis of the obtained data of electrodynamic modeling carried out by numerical solution of the system of Maxwell equations with given boundary conditions, it can be concluded that the use of methods and approaches for the formation of VAA developed for the radio frequency range can be used in the optical wavelength range features of the dispersion properties of materials. To convert a holographic image obtained in one layer into a volume hologram (analogue of Denisyuk hologram) we can approximate an optical field measured by the optical antenna array at points lying above the plane of the optical array elements. Study has shown the possibility of increasing the resolution of optical systems with additional "virtual" channels for receiving signals. We studied an accuracy of the field approximation on the plane of the optical antenna array elements, as well as above the plane of the optical antenna array.

4. Acknowledgments

This research was funded by Russian Science Foundation, grant number 19-79-10109.

References

[1] Hale U K 1980 Frequency assignment. *Theory and applications.* **68** 112
[2] John R B 1990 Electronic System Design: Interference and Noise Control Techniques (Moscow: Mir)
[3] Slivkin V G 2004 *Electromagnetic compatibility of electrical equipment of information technologies under the influence of pulsed electromagnetic fields* (Samara, SGTU)
[4] Tryapitsyn A V *Development of the methodology for the design of wireless local area networks with EMC* (Moscow: Mir)
[5] Williams T E 2004 *EMC for Product Designer*s (Moscow: Publishing house "Technologies")
[6] Habiger E H 1995 *Electromagnetic Compatibility. Fundamentals of its provision in technology* (Moscow: Energoatomizdat)
[7] Kechiev L N 2007 *Proektirovanie pechatnykh plat dlya tsifrovoy bystrodeystvuyushchey apparatury* (Moscow: Gruppa IDT)

IOP Conf. Series: Materials Science and Engineering 1061 (2021) 012031 doi:10.1088/1757-899X/1061/1/012031

Use of "virtual" antenna array methods for reduction of DOA estimation systematic error caused by the scattering of waves on mobile carrier body

Yu G Pasternak[1], A V Ashikhmin[2], Yu A Rembovsky[2], S M Fedorov[1], K A Razinkin[1], S A Antipov[1]

[1]Voronezh State Technical University, 14, Moscow ave., Voronezh, 394026, Russia
[2]IRCOS JSC, 21, Zvezdny blvd., Moscow, 129626, Russia

E-mail: fedorov_sm@mail.ru

Abstract. Article shows that "virtual" antenna arrays can be used to effectively improve the DOA estimation accuracy of mobile direction finders. "Virtual" antenna arrays are the samples of the spatial field distribution calculated by approximation of the field near a three-dimensional scattering body that supports physical ("real") antenna system. Signals from the outputs of the "real" array are used as input data for the approximation process. Information on the geometry and material properties of the medium is not required. This procedure can be based on the main provisions of electrodynamics (Kirchhoff-Huygens principle, the Lorenz lemma, method of equivalent currents and fields) or on the use of quasi-static approximation of the field based on the apparatus of the theory of analytic functions of a complex variable (the Cauchy integral and Laurent series). "Virtual" arrays can reduce the error caused by the carrier body without costly and labour-intensive calibration procedures.

1. Introduction

The calibration of receiving antenna in order to compensate the field diffraction distortions introduced by the carrier body is a complex and expensive procedure. In addition, during equipment operation, calibration effect will be lost very quickly, due to vibration, mechanical damage of the carrier body or oxidation of its parts. Also, there is possibility of uncontrolled change in the scatterer geometry due to opening of doors, hatches, etc. Therefore, development of method for calculation the field near antenna carrier body, which does not require information about its geometry, is an actual problem [1-9].

In this article, we researched an effective way for forming the "virtual" antenna array (VAA) based on analytic continuation of the field measured by the antenna using methods of theory of analytic functions of complex variable $z = x + iy$, in particular, the Cauchy integral or the Laurent series. This approach is applicable only in case of sufficiently small electrical size of the VAA location region.

2. Method description

On the contour of an equidistant circular antenna array with radius R, the E_z – component can be described, based on the measured field values $U_1, U_2, U_3, \ldots, U_n$, using following polynomial:

$$U_{AA}(z = R \cdot \exp[i\phi]) = \sum_{n=1}^{N+1} B_{n-1} \cdot \exp[i(n-1)\phi/(N+1)], \quad (1)$$

which coefficients are found by the least squares method.

Using the Poisson integral (a special case of the Cauchy integral for a contour in form of circle), we can deduce an integral equation of the first kind, which connects fields on the contours of "real" and "virtual" arrays:

$$U_{AA}(z=R\cdot\exp[i\phi])=\frac{1}{2\pi}\int_0^{2\cdot\pi}U_{VAR}(r\cdot\exp[i\psi]])\frac{r^2-R^2}{r^2-2rR\cdot\cos(\psi-\phi)+R^2}d\psi=$$
$$=\frac{1}{2\pi}\int_0^{2\cdot\pi}\left\{\sum_{k=1}^{K+1}X_{k-1}\cdot\exp\left[\frac{i(k-1)\psi}{K+1}\right]\right\}\frac{r^2-R^2}{r^2-2rR\cdot\cos(\psi-\phi)+R^2}d\psi \quad (2)$$

which can be solved by the collocation method, after the Tikhonov regularization procedure performed (or – after use of the limited damage method).

Another way for field approximation is based on the expansion of function $U_{AA}(z)$ in circle $R-\delta\leq r\leq R+\delta$ (where $\delta\ll R$) in the Laurent series:

$$U_{AA}(z)=\sum_{k=-\infty}^{\infty}c_k z^k\text{ , where }c_k=\frac{1}{2\pi i}\int_\Gamma\frac{U(\zeta)d\zeta}{\zeta^{k+1}} \quad (3)$$

Next, with use of (3) the derivatives of functions $U_{AA}(z)$ in the radial directions $\partial U(\varphi,r)/\partial r$,$\partial^2 U(\varphi,r)/\partial r^2$, etc. are evaluated, and the values of function $U_{VAA}(\phi,R+L)$ are estimated by use of segment of the Taylor series:

$$U(\varphi,R+L)\approx U(\varphi,R)+L\cdot\partial U(\varphi,R)/\partial r+0.5L^2\cdot\partial^2 U(\varphi,r)/\partial r^2+\cdots \quad (4)$$

The possibility of VAA formation follows from the Lorentz lemma, which allows, according to the known tangential field components on a closed surface S, calculate all field components in a volume V bounded by it; in the derivation of the Lorentz lemma, data on properties of the medium are not used. The scalar analogue of the Lorentz lemma is the Kirchhoff integral (the 3d integral Green's formula), which makes it possible to calculate field at any point of the volume V bounded by closed surface S on which values of the field and its normal derivative are known. We can move from the original three-dimensional problem to equivalent two-dimensional problem of estimation field values on the contour ζ from its known values $u(P_n)$ at several points N of contour L, Fig. 1

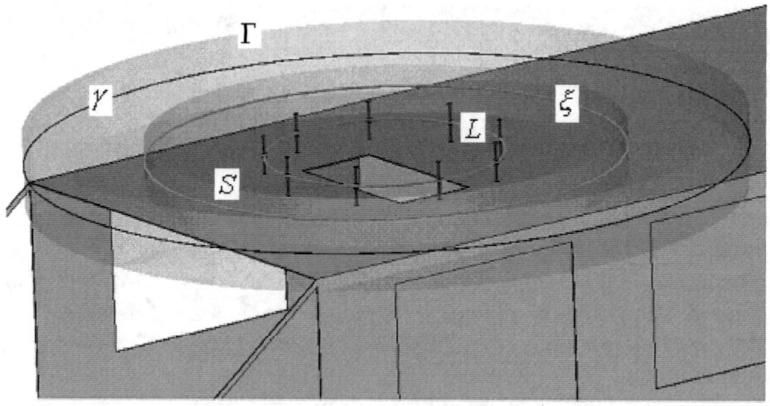

Figure 1. Antenna array located on the roof of a car.

On outer contour γ there are auxiliary field sources $\varphi_m(Q_i,q_m)=-(\pi\cdot i/2)\cdot H_0^1(k_0 r_{Q_i,q_m})$, which linear combination (with unknown complex amplitudes U_{q_m}) simulates the field inside contour γ.

The VAA forming procedure is reduced to solving the system of linear algebraic equations with respect to U_{q_m}:

$$\sum_{m=1}^{N} U_{q_m} \cdot \sum_{i=1}^{I} \left\{ G(P_n, Q_i) \cdot \frac{\partial \varphi_m(Q_i, q_m)}{\partial \vec{n}} - \varphi_m(Q_i, q_m) \cdot \frac{\partial G(P_n, Q_i)}{\partial \vec{n}} \right\} \Delta\zeta_i =$$
$$= u(P_n),\ \mathrm{n} = 1,2,\ldots,\mathrm{N}, \tag{5}$$

where: $G(P_n, Q_i) = -(\pi \cdot i/2) \cdot H_0^1(k_0 r_{P_n,Q_i})$; $Q_i \in \zeta$; $P_n \in L$; $q_m \in \gamma$; I – number of contour γ partitions; $\Delta\zeta_i$ – length of element of circle ζ.

Since the φ_m functions are solutions of the Helmholtz equation, expression (5) is equivalent to the following system of linear algebraic equations:

$$\|A\| \cdot \overrightarrow{u_q} = \overrightarrow{u_p}\ ;\ (\|A\|)_{n,m} = H_0^1(k_0 r_{P_n,q_m});\ (\overrightarrow{u_q})_m = U_{q_m}\ ;$$
$$(\overrightarrow{u_p})_n = U_{P_n}\ ;\ m, n\ \in 1, \ldots, N. \tag{6}$$

Fields on planar contours L, ζ and γ (u(P), $U_Q(Q)$ and $U_q(q)$ respectively, defined by Fourier series with coefficients a_n, b_n and c_n) are also related by the following formulas:

$$u(P) = \sum_{n=-\infty}^{\infty} a_n \exp(in\varphi_L) = \oint_\gamma \sum_{n=-\infty}^{\infty} c_n \exp(in\varphi_\gamma)\, H_0^1(k_0 r_{P,q}) d\gamma_q =$$
$$= \textstyle\sum_{n=-\infty}^{\infty} c_n \oint_\gamma \exp(in\varphi_\gamma) H_0^1(k_0 r_{P,q}) d\gamma_q\ ,$$

$$U_Q(Q) = \sum_{n=-\infty}^{\infty} b_n \exp(in\varphi_\zeta) = \oint_\gamma \sum_{n=-\infty}^{\infty} c_n \exp(in\varphi_\gamma)\, H_0^1(k_0 r_{Q,q}) d\gamma_q =$$
$$= \textstyle\sum_{n=-\infty}^{\infty} c_n \oint_\gamma \exp(in\varphi_\gamma) H_0^1(k_0 r_{Q,q}) d\gamma_q\ ,$$

Coefficients a_n, b_n and c_n are describe functions that satisfy the Helmholtz equation, i.e. at least twice continuously differentiable. Taking into account the Riemann-Lebesgue theorem, the following relations hold:

$$|a_n| < \frac{C_1}{n^2},$$
$$|b_n| < \frac{C_2}{n^2},$$
$$|c_n| < \frac{C_3}{n^2}, \tag{7}$$

where C_1, C_2 and C_3 – arbitrary constants.

It was found that, when fields on the contours L, ζ and γ are calculated with help of a finite number of Fourier harmonics, the following restrictions are valid:

$$a_n \exp(\mathrm{i}n\varphi_L) < \frac{C_1}{n^2},$$
$$b_n \exp(\mathrm{i}n\varphi_\zeta) < \frac{C_2}{n^2},$$
$$c_n < \frac{C_3}{n^2}, \quad (8)$$
$$\oint_\gamma \exp(in\varphi_\gamma)\, H_0^1(k_0 r_{P,q})d\gamma_q < \alpha_1 \exp(-\beta_1 n),$$
$$\oint_\gamma \exp(in\varphi_\gamma)\, H_0^1(k_0 r_{Q,q})d\gamma_q < \alpha_2 \exp(-\beta_2 n),$$

Expressions for the errors of estimation of the field on contours L, ζ and γ:

$$\delta u_a < 2C_1\left\{\frac{\pi^2}{6} - \sum_{n=1}^{N}\left[\frac{1}{n^2}\right]\right\},$$
$$\delta U_c < 2C_3\alpha_2\{Li_2(\exp(-\beta_2)) - \sum_{n=1}^{N}[\frac{\exp(-\beta_2 n)}{n^2}]\}$$
$$\delta u_c < 2C_3\alpha_1\{Li_2(\exp(-\beta_1)) - \sum_{n=1}^{N}[\frac{\exp(-\beta_1 n)}{n^2}]\} \quad (9)$$

At $\frac{r_\zeta}{r_L} = 3$ $(r_L = 0.5\text{ m}, r_\zeta = 1.5\text{ m})$ and $\frac{r_\gamma}{r_L} = 10$ $(r_\gamma = 5\text{ m})$ to reduce the error of field extrapolation on the contour of VAA to the error level of estimation of the field on the antenna array contour, it is necessary to increase 1.8 times the number of used auxiliary sources.

Fig. 2 illustrates increase in DOA estimation accuracy achieved by use of developed method of the VAA.

To calculate bearings, we use the field strength values $\vec{U} = [U_1, U_2, U_3, \dots, U_N]^T$, known value of wavelength λ_0, and known values of coordinates of phase centers of array elements x_n and y_n; bearing value corresponded to maximum modulus of correlation function between measured amplitude-phase distribution $\vec{U} = [U_1, U_2, U_3, \dots, U_N]^T$ and calculated under assumption of absence of incident wave diffraction distortions:

$$D(\varphi) = \left|\sum_{n=1}^{N} U_n \cdot \exp[-\mathrm{i}\cdot\frac{2\cdot\pi}{\lambda_0}\cdot(x_n\cdot\cos(\varphi) + y_n\cdot\sin(\varphi))]\right|$$

where angle φ varies from 0 to $2\cdot\pi$ radians.

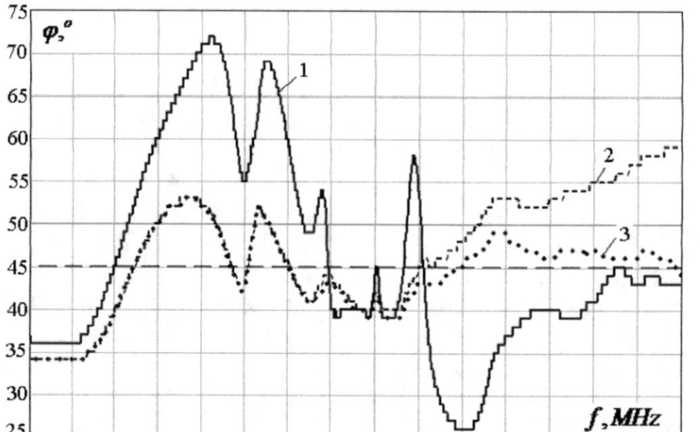

IOP Conf. Series: Materials Science and Engineering 1061 (2021) 012031 doi:10.1088/1757-899X/1061/1/012031

Figure 2. Dependence of bearings from frequency at $\phi_{Source} = 45^{\circ}$: 1) "real" antenna array, $N = 18$, $r_L = 0.5$ m; 2) VAA, $N = 36$, $r_\zeta = 1.5$ m, formed by 9 field samples measured by "real" antenna array (1, 3, ..., 17 antenna elements); 3) VAA, $N = 36$, $r_\zeta = 0.5$ m, formed by 18 field samples measured by "real" antenna array.

3. Experiment validation of the proposed method

Studies of the proposed method of DOA estimation using field experiment data obtained by the mobile radio direction finder based on chassis of the "Gazel" minibus equipped with the seven-element antenna array with diameter of 1.08 m (Fig. 3) has shown a significant reduction (up to 2-4 times) in maximum and RMS components of DOA estimation systematic error caused by the electromagnetic waves scattering on the antenna array carrier body. Amplitude-phase field structure at the antenna array plane was approximated at points on a circle with diameter of 2 m.

Figure 3. The ARK-MK1M radio monitoring complex with additional ARK-MK4 antenna system on the telescopic mast (JSC IRCOS (Moscow)).

Zero azimuth was directed along the longitudinal axis of the car; frequency dependencies of measured bearings and absolute errors of direction finding for the incidence azimuthal angles equal to 30° and 60° are shown at Fig. 4. Bold line show direction-finding characteristics of "real" seven-element antenna array with diameter of 1.08 m (the correlation phase method of DOA estimation was used); lines with circles correspond to the VAA with diameter of 2 m (same correlation phase method of DOA estimation was used); lines with diamonds show the results of DOA estimation using the correlation amplitude-phase method of direction finding and the VAA with diameter of 2 m. Maximum errors of direction finding at $\varphi_{Source} = 30^{\circ}$ obtained during experiment are: $\delta\phi^{max}_{real.AA} = 16^{\circ}$; $\delta\phi^{max}_{VAA(ampl.-phase)} = 7^{\circ}$; $\delta\phi^{max}_{VAA(phase)} = 9^{\circ}$. Mathematical expectation of the bearings: $M\phi_{real.AA} = 32.6^{\circ}$, $M\phi_{VAA(ampl.-phase)} = 28.6^{\circ}$, $M\phi_{VAA(phase)} = 28.5^{\circ}$. Standard deviation of the bearings: $\sigma\phi_{real.AA} = 6.2^{\circ}$, $\sigma\phi_{VAA(ampl.-phase)} = 3.2^{\circ}$, $\sigma\phi_{VAA(phase)} = 3.5^{\circ}$

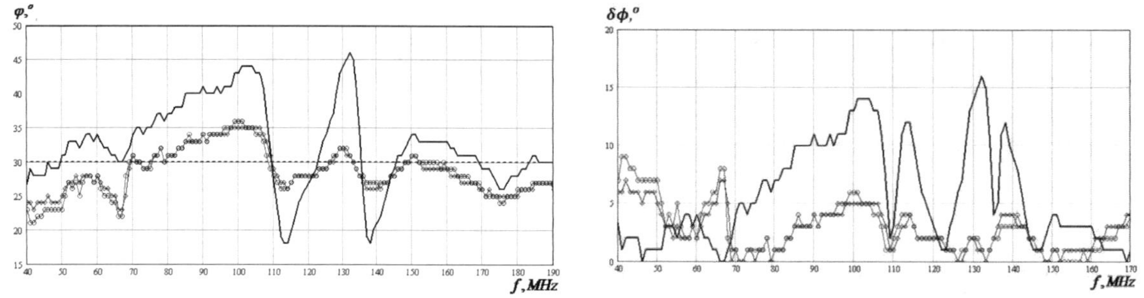

a) b)

Figure 4. Dependence of bearings (a) and bearing errors (b) from frequency of "real" array ($R =$ 0.54 m; $N = 7$) and the VAA ($r = 1$ m; $N = 56$) at $\phi_{Source} = 30^{\circ}$. Rejection of fourth harmonic was used.

4. Conclusion

It was found that use of increased number of the VAA elements (compared to the "real" array), as well as the procedure of rejecting the higher harmonics of the FFT spectrum of measured amplitude-phase distribution, makes it possible to get mathematical expectation of the reconstructed bearing closer to its true value.

5. Acknowledgments

This research was funded by Russian Science Foundation, grant number 19-79-10109.

References

[1] Friedlander B 1990 Direction finding with an interpolated array *Proc. IEEE Int. Conf. Acoust., Speech, Signal Processing*

[2] Pesavento M, Gershman A B and Luo Z-Q 2002 Robust array interpolation using second-order cone programming *IEEE Signal Processing Letters* **9(1)** 8-11

[3] Buhren M, Pesavento M and Bohme J F 2003 A new approach to array interpolation by generation of artificial shift invariances: interpolated ESPRIT *Proceedings IEEE Int. Conf. Acoustics, Speech, and Signal Proces. (ICASSP)* **5** 205-208

[4] Hyberg P 2005 Antenna Array Mapping for DOA Estimation in Radio Signal Reconnaissance. PhD thesis *Royal Institute of Technology (KTH)*

[5] Belloni F, Richter A and Koivunen V 2005 Reducing Excess Variance in Beamspace Methods for Uniform Circular Array *Proceedings of the IEEE Workshop on Statistical Signal Processing (SSP)* 17-20

[6] Belloni F and Koivunen V 2006 Beamspace Transform for UCA: Error Analysis and Bias Reduction *IEEE Transactions on Signal Processing* **54(8)** 3078-3089

[7] Belloni F, Richter A and Koivunen V 2007 DOA Estimation via Manifold Separation For Arbitrary Array Structures *IEEE Trans. Signal Processing* **55(10)** 4800-4810.

[8] Tuncer E and Friedlander B 2009 Classical and Modern Direction-of-Arrival Estimation. *USA: AP*. 429 p.

[9] Antenna Array Including Virtual Antenna. P. van Rooyen, P. Roux. Patent US 7,605,755B2, 20.10.2009 (prior publication data US 2008/0303719 A1, 11.12.2008).

Approximation of EMF spatial distribution near mobile phone in order to form a "virtual" antenna array

Yu G Pasternak[1], V V Negrobov[1], A V Ostankov[1], S M Fedorov[1], E A Ischenko[1], K A Razinkin[1], S A Antipov[1]

[1]Voronezh State Technical University, 14, Moscow ave., Voronezh, 394026, Russia

E-mail: fedorov_sm@mail.ru

Abstract. Article describes the methods of approximation of electromagnetic field spatial distribution, which spatial samples are measured using an annular antenna array located near scatterer with a priori unknown geometry and material properties. As example, case of mobile phone located at user's head were studied. The model of the human head and hand was simulated as a layered structure, the outer layer imitated the skin, and water was used as the inner filling. The study results show good performance of modified method of forming a "virtual" antenna array, which can significantly reduce the field approximation error, thereby making it possible to successfully predict the electromagnetic field phase structure at distances exceeding the radius of the used antenna array by three times. Proposed method can significantly increase the efficiency of subspace methods of direction finding and methods of adaptive signal processing in multichannel systems (in particular, the MIMO) due to the multiple increase in the number of virtual channels which are functionally equivalent to additional real channels, and also - increasing the electrical dimensions of the antenna system ("virtual" antenna array).

1. Introduction

The user's body has a significant impact on the functioning of receiving antenna system of a mobile phone, caused by scattering and absorbing electromagnetic waves as a result of which spatial structure of the measured electromagnetic field is distorted. It is obvious that due to the a priori uncertainty of the geometric parameters and material properties of the user's body the calibration of mobile phone antenna system, considering the influence of the scatterer, is impossible. We also note the complexity of scatterer morphology which is the user's body.

Another important aspect is that size of antenna array is limited by the size of a mobile phone's body. That is why problem of developing a method that can reduce or even completely eliminate negative impact of nearby scatterers on radio monitoring and radio direction finding systems, without using any information about geometry of scatterers or their electrodynamic properties, is very actual task. The result of using these methods should be a significant increase in resolution of the complexes, reducing errors in DOA estimation, and eliminating blunders.

Practical application of the methods is to use the data on spatial distribution of total (incident and scattered waves) field, measured by a physically existing radio direction-finding antenna system, to form a “virtual” antenna array (VAA) which is a system of spatial samples of complex amplitudes of the recorded field. Using this array, it is possible, by calculating the corresponding direction-finding characteristic, to significantly increase accuracy of DOA estimation and angular coordinates resolution of radio monitoring complexes of mobile, airborne, and stationary bases.

The use of the virtual antenna arrays apparatus for radar-location is of particular interest, such an application is considered in a number of studies [1-4]. In works [1-3] antenna arrays are considered, in which an additional virtual antenna array is formed, which allows to improve the characteristics of the considered antenna system; in [4] mathematical algorithms that allow improving the antenna system characteristics are considered.

It should be noted that scope of these methods' applicability is not limited to radio direction finding tasks. In this paper, the possibility of using the methods of approximating structure of measured electromagnetic field in cellular communications equipment in case of a mobile phone receiving system is considered and studied.

2. Methods of "virtual" antenna array formation

These methods are based on hypotheses about the possibility of representing a field on an antenna array contour using the theory of methods of analytic functions of a complex variable $z = x + i \cdot y$, in particular, the Cauchy integral or the Laurent series [5]. In this case, the amplitude-phase distribution of the field on the elements of N -element antenna array can be considered as samples $U_n = U(\phi_n)$ ($\phi_n = 2\pi(n-1)/N, n = 1,2,\dots,N$) of periodic function $U(\phi)$ with period of 2π describing the continuous distribution of the field around a circle of radius R (R - is radius of the antenna array) taken through the intervals $1/(2F)$ (where $F = 1/(4\pi/N)$). In turn, function $U(\phi)$ can be represented as a Kotelnikov series: $U(\phi) = \sum_{n=-\infty}^{\infty} U_n \frac{sin(2\pi F(\phi-\phi_n))}{2\pi F(\phi-\phi_n)}$. Then use the decomposition of function $U_{\text{AP}}(z) = U(\phi)$ in ring $R - \delta \le r \le R + \delta$ (where $\delta << R$) in the Laurent series: $U_{AA}(z) = \sum_{k=-\infty}^{\infty} c_k\, z^k$, where $c_k = \frac{1}{2\pi i}\int_\Gamma \frac{U(\zeta)d\zeta}{\zeta^{k+1}}$. After that, the derivatives of function $U_{\text{AP}}(z)$ are found in the radial directions $\partial U(\phi,r)/\partial r, \partial^2 U(\phi,r)/\partial r^2$ etc., and values of function $U_{VAA}(\phi, R+L)$are estimated using a segment of the Taylor series: $U(\phi, R+L) \approx U(\phi,R) + L \cdot \partial U(\phi,r)/\partial r + 0.5L^2 \cdot \partial^2 U(\phi,r)/\partial r^2 + \dots$

However, the approach described above are applicable only in case of sufficiently small electrical dimensions of region where VAA is formed.

Other methods are following directly from the principles of electrodynamics. In particular, from the Lorentz Lemma, according to which it is possible to calculate all field components in volume V bounded by closed surface S if the tangent field components are known on a given surface [6]. The scalar analogue of the Lorentz lemma is the expression for the Kirchhoff integral, which allows to calculate the field at any point of volume V if values of the field and its normal derivative are known on closed surface S bounding given volume. In this case, it is possible to move from a volume problem to a flat one.

In addition, it is possible to represent the amplitude-phase distribution on elements of real antenna array (RAA) using expression:

$$E_{z_n} = E_z(x_n, y_n) =$$

$$= \sum_{m=1}^{M} B_m \cdot \left(exp\left\{ -ik_0 \sqrt{\left(x_n - x_m^q\right)^2 + \left(y_n - y_m^q\right)^2} \right\} \Big/ \sqrt{\left(x_n - x_m^q\right)^2 + \left(y_n - y_m^q\right)^2} \right),$$

where $x_n, y_n; n = 1,2,\dots,N$ - N-element AA coordinates; x_m^q, y_m^q ; $m = 1,2,\dots,M$ - coordinates of M points auxiliary sources; $\vec{B} = [B_1, B_2, \dots, B_M]^T$ - complex amplitude vector of auxiliary sources; $k_0 = 2\pi/\lambda$ - free space wave number; λ - free space wavelength; $M_{n,m} = exp\left(-ik_0 \sqrt{\left(x_n - x_m^q\right)^2 + \left(y_n - y_m^q\right)^2} \right) / \sqrt{\left(x_n - x_m^q\right)^2 + \left(y_n - y_m^q\right)^2}$ - matrix $N \times M$, elements of which depend on frequency and coordinate values of elements of the antenna array and auxiliary point sources. In this case $N = M$, therefore, the matrix is square.

However, authors of this article found out that the most promising method of VAA generation is the method using the pseudo-solution procedure, in which $N$is not equal to M. In this case, vector of the complex amplitudes of the auxiliary sources calculated as $\vec{B} = (M^H \cdot M)^{-1} \cdot M^H \cdot \vec{U}$, where $\vec{U}$- is vector of the complex amplitudes of elements of the physically existing antenna array.

3. Research and results

The authors researched the possibility of using the VAA formation methods based on the pseudo-solution search procedure for approximating the phase structure of the electromagnetic field near handset antenna system and user's head. The electromagnetic field of the incident wave was measured at 18 points, located equidistantly along a circle with radius of $R_{AA} = 15$ mm near the surface of the phone. These field readings were taken as samples measured by the telephone antenna system. In addition, on circles with radius of $R_1 = 22.5$ mm, $R_2 = 30$ mm, $R_3 = 45$mm, also at 18 equidistant points, complex amplitudes of the E component of electromagnetic wave were measured. The human head model was presented as a layered structure imitating skin, bone and human brain [7]. The phone model was represented as an ideal metal parallelepiped with dimensions of 100×40×18 mm, Fig.1.

Front of electromagnetic wave was directed from the face of the user, and vector E was parallel to the vibrators.

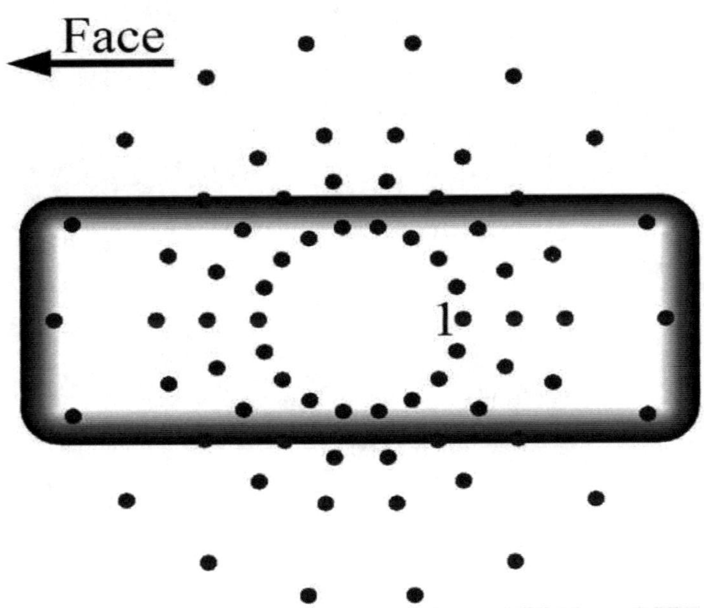

Figure 1. Location of spatial samples in which EMF was recorded relative to the mobile phone casing.

Using the complex amplitudes measured at radius R_1, and the pseudo-solution procedure described above, the phase values at the corresponding points on circles R_2, R_3 were predicted. Fig. 2 and Fig. 3 show graphical dependencies of the predicted phases. The circles indicate the phases in each of K points measured experimentally, diamonds – indicate the phases obtained during the approximation. In addition, the table 1 presents numerical values of errors (maximum and average) within the studied frequency bands.

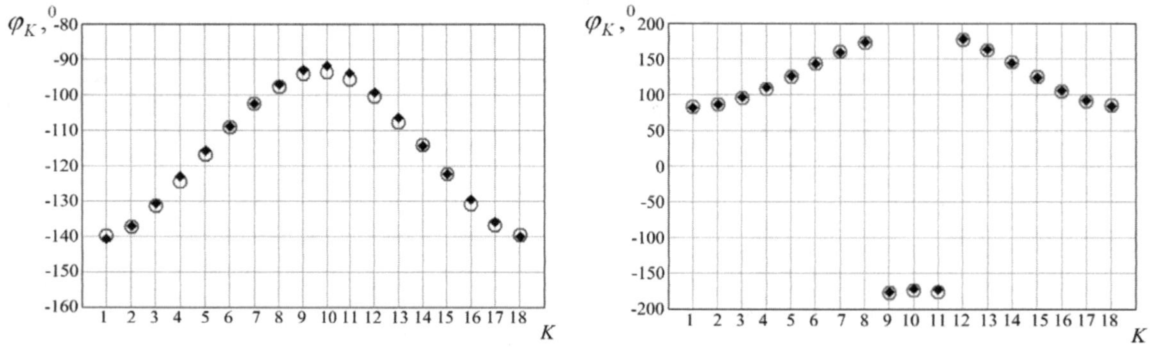

a) b)

Figure 2. Phase values at points within radius of $R_1 = 22.5$ mm at a) $f = 900$ MHz, and b) $f = 1800$ MHz.

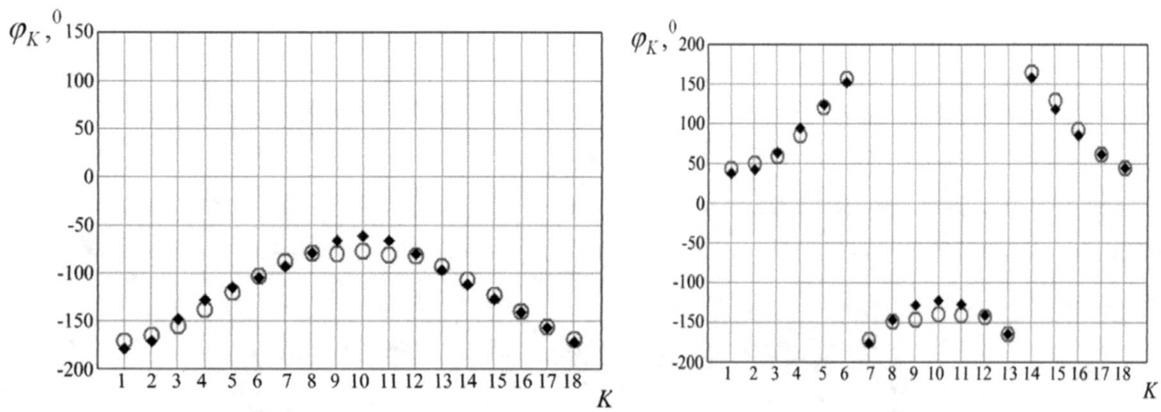

a) b)

Figure 3. Phase values at points within radius of $R_3 = 45$ mm at a) $f = 900$ MHz, b) $f = 1800$ MHz

Table 1. Approximation of error values

Radius, mm	Error type	Frequency range, MHz			
		890-915	935-960	1710-1785	1805-1880
22,5	Max.	4,6°	4°	4,8°	5,7°
	Avg.	2,1°	2,3°	1,8°	2,2°
30	Max.	6,4°	6°	7,9°	8,9°
	Avg.	2,2°	2°	2,9°	3,4°
45	Max.	9,3°	8,6°	16,8°	17,3°
	Avg.	3,8°	3,7°	6,3°	6,5°

4. Conclusion

As can be seen from the data presented above, the VAA formation method can quite successfully approximate phase structure of electromagnetic field at distances up to three times the radius of antenna array which was receiving electromagnetic wave. The approximated complex amplitudes of the field can be interpreted as elements of a hypothetical "virtual" antenna array and used for adaptive processing of received information: estimating the angular coordinates of radio emission sources, suppressing interference, increasing the signal-to-noise ratio at receiver input, increasing angular coordinate resolution. A key feature of the method is absence of the need in any information about scatterer (user`s body) and high efficiency of its implementation in a portable computer system of real-time processing.

It should be noted that use of proposed method can significantly increase efficiency of applying the methods of super resolution of radio emission sources and methods of adaptive signal processing in multichannel systems (in particular, MIMO systems) due to the possibility of a significant increase in the number of virtual channels functionally equivalent to additional real channels, and also - an increase in the electrical size of the antenna system ("virtual" antenna array).

Acknowledgments
This research was funded by Russian Science Foundation, grant number 19-79-10109.

References

[1] Yahia A A and Elkamchouchi H M 2019 Design of virtual antenna array for direction of arrival estimation using real antenna array system *34th International Technical Conference on Circuits/Systems, Computers and Communications 2019* pp 1-3
[2] Yaqoob M A, Tufvesson F, Mannesson A and Bernhardsson 2013 B Direction of arrival estimation with arbitrary virtual antenna arrays using low cost inertial measurement units *IEEE International Conference on Communications Workshops (ICC) 2013* pp 79-83
[3] Mbugua A W, Fan W, Ji Y and Pedersen G F 2018 Millimeter wave multi-user performance evaluation based on measured channels with virtual antenna array channel sounder *IEEE Access 2018* pp 12318-12326
[4] Li W and Zhao Y 2016 A null broadening beamforming method of virtual antenna array *IEEE/ACES International Conference on Wireless Information Technology and Systems (ICWITS) and Applied Computational Electromagnetics (ACES) 2016* pp 1-2
[5] Korn G A and Korn T M 2000 *Mathematical handbook for scientists and engineers: definitions, theorems, and formulas for reference and review* (Dover Publications)
[6] Jackson J D 1998 *Classical electrodynamics* (Wiley)
[7] Khalatbari S, Sardari D, Mirzaee A A and Sadafi H A 2006 Calculating SAR in two models of the human head exposed to mobile phones radiations at 900 and 1800 MHz *Progress in electromagnetics research symposium 2006* pp 104-109

IOP Conf. Series: Materials Science and Engineering 1061 (2021) 012033 doi:10.1088/1757-899X/1061/1/012033

Analysis of the functioning of the orientation device to the Earth of a telecommunications space satellite

M N Petrov[1]

[1] Siberian State University of Science and Technology named after M.F. Reshetneva, 31, prospect. newspapers Krasnoyarsk Rabochy, Krasnoyarsk, 660037, Russia

E-mail: mnp_kafaes@mail.ru

Abstract. This article presents the results of statistical observations of an operating communications spacecraft. The functioning of the orientation system to the Earth is considered. For ensure the reliability of communication between the satellite and ground stations, it is important to have a constant direction to the coverage area. For this, the spacecraft has an orientation system to the Earth. The orientation aims to create a reliable connection on the territory of the country of the spacecraft owner. The article examines the analysis of the operation of the satellite guidance system to the Earth using the example of a telecommunications communication satellite of the Republic of Kazakhstan. At the same time, the satellite orientation system to the Sun works in parallel to organize a stable and stable connection. The statistics of the operation of the orientation system to the Earth are presented. For this, the satellite has devices responsible for positioning – the orientation devices to the Earth. There are two given devices, namely, POZ-1 and POZ-2. The analysis of real statistics is given. The forecast of the system operation for the next nine years has been made.

1. Introduction

For ensure the reliability of communication between the satellite and ground stations, it is important to have a constant direction to the communication coverage area [1-6]. For this, the spacecraft has the orientation system to the Earth.

2. Analysis of the statistics of the operation of the orientation device to the Earth.

In [1], the analysis of the operation of the satellite guidance system in the Sun is considered on the example of a telecommunication communication satellite of the Republic of Kazakhstan. At the same time, the satellite orientation system to the Earth works in parallel to organize stable and stable communication. Consider the statistics of the orientation system to the Earth. For this, the satellite has devices responsible for positioning: the orientation devices to the Earth. These devices are two: POZ-1 and POZ-2.

Figure 1 shows the moments in time at which the Sun falls into the field of sight of the orientation devices to the Earth. This information is calculated in software, orientation and stabilization systems, taking into account the data coming from the initial orientation device unit. This graph shows that the exchange between the initial orientation device unit and the orientation and stabilization system software is carried out normally and without comments.

Figure 2 shows the moments at which the Earth falls in the field of view of the device for orientation to the Earth. This testifies to the normal functioning of the device, even in those moments when the Sun fell into the field of view.

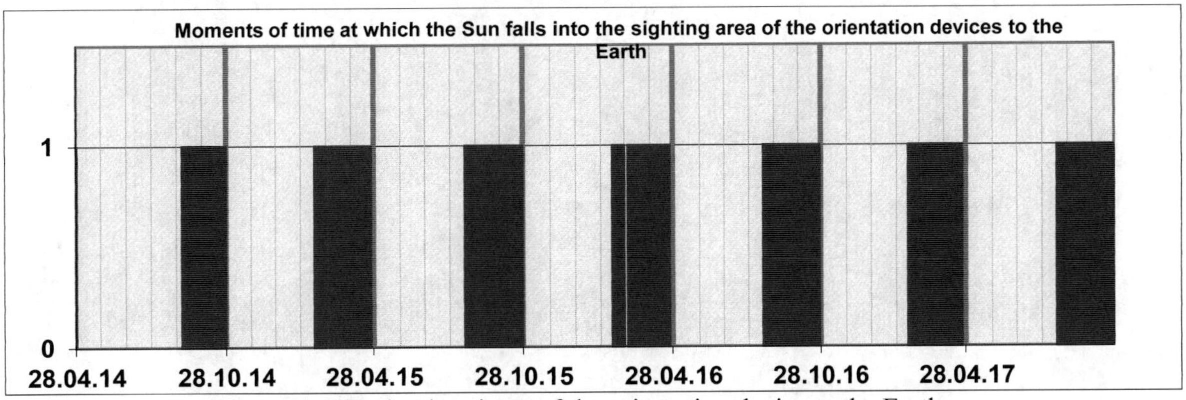

Figure 1. Illumination times of the orientation device to the Earth

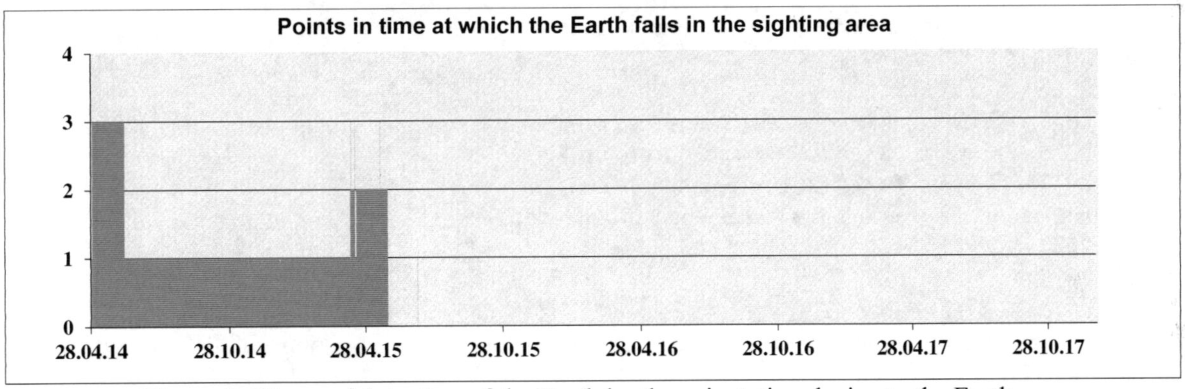

Figure 2. Times of detection of the Earth by the orientation device to the Earth

where

0 –orientation device to the Earth POZ1 and orientation device to the Earth POZ2 do not observe the Earth;

1 – the device for orientation to the Earth POZ1 observes the Earth, the device of orientation to the Earth POZ2 does not observe the Earth;

2 – the device for orientation to the Earth POZ2 observes the Earth, the orientation device to the Earth, POZ1 does not observe the Earth;

3 –orientation device to the Earth, POZ1 and orientation device to the Earth POZ2 observe the Earth.

It is crucial to monitor the temperature of the Earth orientation device. Figure 3 shows the statistics data.

IOP Conf. Series: Materials Science and Engineering 1061 (2021) 012033 doi:10.1088/1757-899X/1061/1/012033

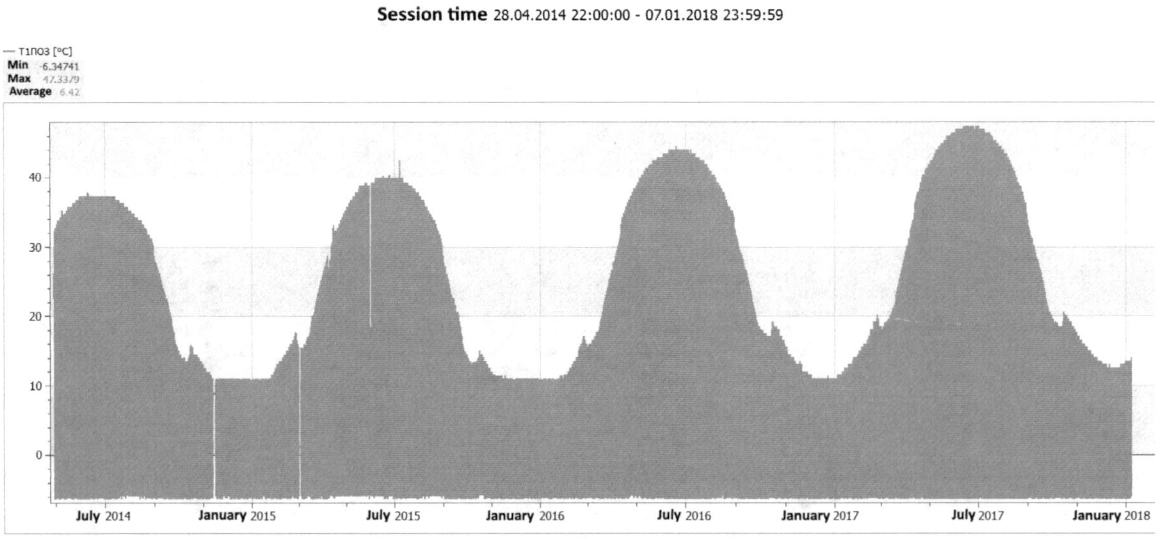

Figure 3. Change in the temperature of the radiator of the orientation device to the Earth, POZ1 (T1POZ)

The maximum temperature at the summer solstice was:

- 2014 - 37.72 ° C;
- 2015 - 40.25 ° C;
- 2015 - 42.27 ° C (when the device is turned on);
- 2016 - 44.30 ° C (when the device is turned on);
- 2017 - 47.34 ° C (when the device is turned on).

The degree of increase in the thermal background from year to year is:

- 2014 → 2015 - 2.53 ° C;
- 2015 → 2016 - 2.03 ° C;
- 2016 → 2017 - 3.04 ° C.

In 2015, the upper limit of control of the temperature parameter -TM-parameter T1CO3 + 40 ° C was recorded (Fig. 3).

According to flight data on a spacecraft with similar thermal circuits of the Earth orientation device and with temperature sensors installed at the POS seats, the difference in temperature between the device design and the seat is from 6 to 6.8 ° C. According to the analysis, the difference between the switched on and off device is 7.1 ° C – 8 ° C for the devices of orientation to the Earth POZ2 and POZ1, respectively, it is necessary to compare it with the TM-parameters TPOZ1 / 2. The data indirectly indicate that the temperature of the POZ2 instrument seat on the spacecraft reached ~ 40 ° C during the specified period.

The reason for the parameter T2POZ2 exceeding control tolerances (over 40 ° C) is as follows. The optical window of the Earth orientation devices absorbs additional thermal radiation from the surface of the EVTI shell and lock. At the same time, the heat flux emitted by the optical windows of the device for orientation to the Earth into space is reduced. The reason is the negative influence of the overhanging design of the shell and the lock on the visibility factor of outer space by the devices of the Earth orientation device. Due to the closer location of the orientation device to the Earth POZ2 to the shell structure, the temperature of the orientation devices to the Earth POZ2 is higher than the temperature of the orientation device to the Earth POZ1. That is, when designing and calculating the heat balance, the developers did not take into account that the transition system has a conical shape, and not cylindrical; as a result, the real heat flux falling on the device for orientation to the Earth POZ2 is greater than the calculated one. This circumstance caused the temperature of the radiator of the device of the Earth orientation device POZ2 to go beyond the upper limit of the TM parameter control.

Table 1. Predicted temperatures of the Earth orientation device during the summer solstice during hardware stabilization system and the maximum per day duration of the stay of the Earth orientation device seats at temperatures above 40 ° C.

	Predicted temperatures, °C						Duration of stay at temperatures above 40 °C, hh: mm	
	TPOZ1	Seat POZ1	TPOZ2	Seat POZ2	Seat POZ1 Average integrated per day	Seat POZ2 Average integrated per day	Seat POZ1	Seat POZ2
June 2014	41.4	35.4	45.4	39.4	15.6	14.0	0:00	0:00
June 2015	44.0	38.0	48.0	42.0	16.1	14.5	0:00	2:11
June 2016	46.0	40.0	50.0	44.0	16.6	15.0	0:00	3:11
June 2017	47.0	41.0	51.0	45.0	16.8	15.2	1:48	3:11
June 2018	47.6	41.6	51.6	45.6	16.9	15.3	1:48	3:25
June 2019	48.0	42.0	52.0	46.0	17.0	15.4	1:48	3:41
June 2020	48.4	42.4	52.4	46.4	17.1	15.4	2:11	3:41
June 2021	48.7	42.7	52.7	46.7	17.2	15.5	2:44	3:41
June 2022	49.0	43.0	53.0	47.0	17.2	15.6	2:44	3:41
June 2023	49.3	43.3	53.3	47.3	17.3	15.6	2:44	3:41
June 2024	49.6	43.6	53.6	47.6	17.4	15.7	2:44	3:54
June 2025	49.9	43.9	53.9	47.9	17.4	15.8	2:44	3:54
June 2026	50.1	44.1	54.1	48.1	17.5	15.8	3:01	3:54
June 2027	50.4	44.4	54.4	48.4	17.5	15.9	3:01	3:54
June 2028	50.6	44.6	54.6	48.6	17.6	15.9	3:24	3:54
June 2029	50.8	44.8	54.8	48.8	17.6	16.0	3:24	4:11

Taking into account the actual operating temperatures of the POZ for 2014 – 2017 and the degree of degradation of the optical coating of the spacecraft radiator and the POZ radiator, the maximum predicted temperature at the end of the hardware stabilization system during the summer solstice will be (the analysis data are given in curly brackets):

- body of the device for orientation to the Earth POZ1 – 50.8 ° C (48.8 ° C);
- the seat of the device for orientation to the Earth POZ1 – 44.8 ° C (51.2 ° C);
- body of the device for orientation to the Earth POZ2 – 54.8 ° C (52.3 ° C);
- the seat of the device of orientation to the Earth POZ2 – 48.8 ° C (54.3 ° C).

3. Conclusion

During normal operation, attitude control is carried out according to the information of the star device. The orientation device to the Earth is not obligatory in the basic version of the triaxial stabilization mode, and its absence does not affect the characteristics of the spacecraft during normal operation.

When analyzing the reliability of the orientation and stabilization system in the triaxial stabilization mode of the Earth orientation device is not included in the structural reliability diagram. Therefore, the absence of the Earth orientation device does not affect the probability of failure-free operation of the orientation and stabilization system during normal operation.

Since the devices for orientation to the Earth are back-up during the implementation of the normal mode of the triaxial stabilization mode, then for periods when the temperature of their seats is predicted to go beyond the upper limit, they can be turned off. These periods during the hardware stabilization system are the periods from May to July of each year of an operation lasting up to 3.2 hours per day in 2016 and up to 4.2 hours per day at the end of the hardware stabilization system.

The operating time of the device of orientation to the Earth POZ1 is as of 01.01.2018 – 8335 hours, which is about 6% of the total working resource of the device. The operating time of the device of orientation to the Earth POZ2 is 2149 hours as of 01.01.2018, which is about 1.5% of the total working resource of the device.

The spacecraft developer made a change to the attitude control system to change the temperature control boundaries of the radiators of the orientation devices to the Earth POZ1 and the orientation devices to the Earth POZ2 from + 40 ° C to + 55 ° C.

According to the analysis of the spacecraft at the end of the hardware stabilization system, the maximum temperature of the instruments will be 48.8 ° C and 52.3 ° C in terms of TPOZ1 / 2 parameters with a tolerance of 50 ° C. It is necessary to take measures to prevent an increase in temperatures above permissible or to increase the border of this temperature above 52.5 ° C.

References

[1] Petrov M N and Anarov M Zh 2018 *Analysis of the functioning of the attitude control and stabilization system of a geostationary communication satellite* (Krasnoyarsk, Polycom. St. Vavilov)

[2] Ignatov A I and Sazonov V V 2018 Stabilization of the solar orientation mode of an artificial Earth satellite by an electromagnetic control system *Cosmic research* Vol 56 **5** pp 375–383

[3] Ovchinnikov M Yu, Penkov V I, Roldugin D S, Varatarao R and Ryabikov V S 2017 Motion of a satellite equipped with a pitch flywheel and magnetic coils in a gravitational field // *Cosmic research* Vol 55 **3** pp 218–225

[4] Zavoli A, Giulietti F, Avanzini G and de Matteis G. 2016 Spacecraft dynamics under the action of Y-dot magnetic control law *Acta Astronautica* Vol 122 pp 146–158

[5] Lyubimov V V and Podkletnova S V 2016 Optimal control laws for decreasing the angular momentum and damping the angular velocity of nanosatellites and microsatellites with magnetic coils on board *Bulletin of Samara State Aerospace University. Acad. S. P. Korolev* Vol 15 **2** pp 57–67

[6] Ovchinnikov M Yu and Roldugin D S 2019 A survey on active magnetic attitude control algorithms for small satellites *Progress in AerospacPSciences* DOI: 10.1016 / j.paerosci.2019.05.006

IOP Conf. Series: Materials Science and Engineering 1061 (2021) 012034 doi:10.1088/1757-899X/1061/1/012034

Effect of electrode pressing force on the amount of energy supplied to interelectrode space in resistance projection welding with constant current welding machine

A Yu Paliakou

Belarusian-Russian University, 43, Prospect Mira, Mogilev, 212000, Republic of Belarus

E-mail: mortis2008@mail.ru

Abstract. The paper describes the procedure for theoretical and experimental determination of the graphical regularity of the effect of the electrode pressing force on the amount of energy supplied to the interelectrode space when producing overlap joints of plates with equal thicknesses by using resistance projection welding. The regularity reflects a sufficiently strong connection between the parameters of the welding mode with the heat balance equation of the interelectrode space and the Joule-Lenz law.

1. Introduction

Resistance projection welding (RPW) is used to create strong permanent joints of sheet metal with a thickness of up to 5 + 5 mm in the production of bodies for cars and trucks, railroad cars, aircraft, heating radiators [1].

The RPW process used for joining plates involves a number of operations performed in the following sequence: 1) the plates to be welded are placed between the electrodes of the resistance welding machine; 2) the plates are held together with a clamping force exerted by electrodes. 3. When the plates are pressed together between the electrodes, an alternating or direct current is passed through the interelectrode space in order to ensure an electric thermal deformation cycle of a heating operation applied to the metals and the formation of a permanent area of their interaction (the metals of the plates are joined). 4. When the current is switched off, the parts are kept under pressure to reduce residual stresses in weld and heat-affected zones [2, 3].

Unlike in the case of resistance spot welding (RSW), with RPW, due to a high concentration of current flow lines with a density of more than 400 A/mm^2, it is possible to effectively weld three or more parts in one cycle, i.e., to obtain multi-layer (packet) joints [4 , 5].

According to recommendations from the literature sources, the RPW process used for producing overlap joints of plates does not present any difficulties from the technological point of view. For a specified thickness of the sheets being welded, the recommended sizes of dies and punches for creating embossed round projections of a given diameter d_P and a height h_P as well as values of the pressing force of electrodes F_W, the welding current I_W and the length of time the current is supplied τ_W are selected. In the front section, the projection can have a spherical cap or a truncated conical shape (depending on the shape of the punch and die) (Figure 1).

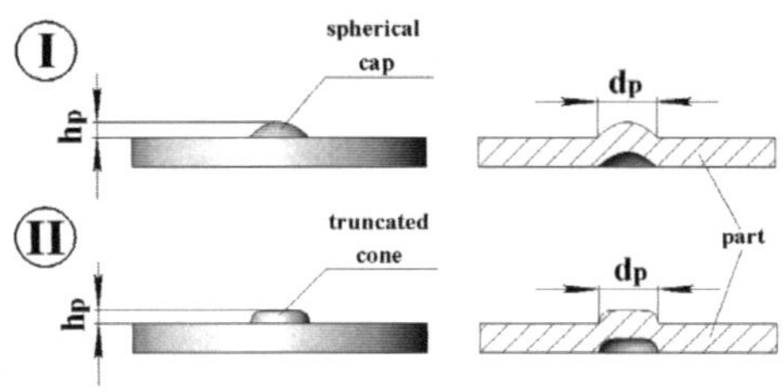

Figure 1. View and front section of the part with embossed round projection: I – spherical cap; II – truncated cone.

Currently, most of the leading industrial enterprises, including assembly and welding shops, mainly use labor- and energy-intensive arc welding techniques for large batch and mass production of various elements from thin sheet metal with a thickness of 1 to 5 mm. In the Republic of Belarus, these are body parts for cars and trucks manufactured by SZAO BelGee and OAO Minsk Automobile Plant; elements and fasteners for agricultural machinery by JSC Gomselmash, JSC Minsk Tractor Works; wall panels and brackets for elevators by JSC Mogilevliftmash; parts of railway passenger cars by CJSC Gomel Carriage Works, consumer goods by OJSC Mogilev Strommashina Plant.

This is due to the absence in the technical literature of scientifically substantiated parameters of the RPW mode for similar joints closely connected with the analysis of the heat capacity of the metal of the interelectrode space according to the calculation of the heat balance equation (HBE).

In some cases, enterprises try to use RPW instead of arc welding techniques. However, the main parameters of this mode are chosen according to the recommendations developed many years ago, by taking values from the section "Resistance Spot Welding" for similar thicknesses of sheets being welded or on the basis of the best RPW practices in relation to joints of the same type accumulated by large enterprises of different countries or manufacturers of welding equipment [6]. Previously, the mode parameters were determined experimentally in the context of ensuring the strength of joints and the absence of defects such as "incomplete penetration", but without assessing the energy intensity of the welding process.

In the second half of the 20th century, because of a lack of welding machines with sufficient power, most enterprises tried to ensure compliance of the weld nugget diameter with GOST 15878-79 by choosing lower I_W and higher τ_W values as the main parameters of the resistance welding method ("soft welding modes"). With these modes, it is not possible to achieve the required diameter of weld nuggets and ensure high strength of joints, especially with the lowest possible energy intensity of the welding process. Despite this, if strength characteristics of projection joints obtained by RPW in such modes met the requirements of enterprises, they were recognized as standards and referred to as recommended values in the literature on pressure welding [2, 6].

Consequently, when performing the above RPW modes the weak relationship between the calculated heat capacity of the metal of the interelectrode space and the observance of the Joule – Lenz law was recognized by manufacturers and scientists as being acceptable.

Yet, currently operators of welding machines erroneously and unreasonably set the τ_W parameter to a higher value than the recommended one, which does not affect the growth of the weld nugget but only causes an increase in the heat transfer from the interelectrode space and the power consumption of the welding machine.

The situation is complicated by the lack of graphs representing dependencies of the parameters of heat input into the interelectrode space (power P_{EE} and energy Q_{EE}) on the parameters τ_W and F_W, which makes it impossible to determine the optimal value of I_W according to the Joule-Lenz law taking into account the dynamics of variations in the resistance of the interelectrode space R_{EE}.

In addition, there are no separate recommendations for the main parameters of the RPW mode in the books on pressure welding for two fundamentally different cases: 1. Welding technique involving mutual melting and mixing of the metals of parts being welded. 2. Solid-state welding (the metals of the parts being welded do not reach their melting temperature).

At present, in the context of the widespread policy to reduce material and the energy intensity of manufacturing, the above approach is unacceptable. Saving even 5% of electricity when producing one projection weld nugget (without compromising its strength) can have a significant economic impact, especially as part of large batch and mass production of different elements.

A decrease in heat input into the interelectrode space with respect to the value calculated using the HBE can help ensure the strength of projection welded joints by obtaining common permanent solid-phase zones of metals alternating with mutual metal melting zones.

For these reasons, it is necessary to keep the calculated heat capacity of the metal of the interelectrode space (based on thermophysical properties of materials of parts being welded and electrodes) strongly connected to the main parameters of the RPW mode when energy is supplied to this space according to the Joule-Lenz law (taking into account the dynamics of variation in R_{EE}).

It should be noted that for RPW processes the value R_{EE} is a purely conditional concept before the current is switched off, since every single welding operation (even for joints of the same type) is fundamentally different. This is due to improperly prepared surfaces of the parts being welded, shapes of projections (caused by slightest distortions of dies and punches generated during their stamping) and positions of the parts with respect to each other in pieces of equipment [7, 8].

In the literature on pressure welding there is a generally accepted HBE used for approximate calculation of the amount of energy Q_{EE} to be supplied to the interelectrode zone for performing RSW of overlap joints consisting of two parts with the same thickness.

It is commonly known that, in accordance with the generally accepted HBE, the value Q_{EE} includes the useful energy used to heat the metals of the parts being welded, the energy of heat transfer to the metal and the energy of heat transfer to the metal of electrodes. The values of the Q_{EE} components are determined on the basis of current carrying areas of the contacts of the parts and the contacts of the electrode and the part, a geometric estimate of volumes of metals of the parts and electrodes involved in effective heating and heat transfer, and their thermophysical properties [2].

Over the past 50 years scientists involved in research on resistance welding processes have continued to argue that this equation is applicable to the RPW process of overlap joints with embossed or coined projections.

Additionally, there are no specific algorithms and mechanisms for such calculations in the literature. The geometric shape of the weld nugget for both welding methods is not specified, and the calculation is based on the assumption that the useful Q_{EE} component in both cases is used for heating and melting a hypothetical metal cylinder part bounded in the horizontal plane by front surfaces of the parts being welded and in the vertical plane by the diameter of the spot electrode (for RSW) or the projection (for RPW). Initially, the coefficient of the nonuniform heat transfer to the metals of the parts being welded and the coefficient of the electrode contact surface shape were experimentally determined exclusively for the RSW process.

After determining the value Q_{EE} according to HBE, the subsequent calculation of the value I_W based on the Joule-Lenz law, which is recommended in the literature, is rather conditional, (since the parameter F_W is taken into account indirectly) and is performed using the approximate value R_{EE} at the moment when the welding current is switched off (the curve of variation in R_{EE} in each welding operation is different).

However, in practice, the variation of the parameter F_W when performing RPW affects the ratio of the parameters $I_W - R_{EE}$ even with the same current set on the weld timer. Accordingly, for a specific value τ_W used in HBE, theoretically, there can be quite suitable values of the parameter F_W which makes it possible to obtain the calculated heat input with a calculated value I_W (using the HBE), an increased value I_W and a reduced value I_W (with respect to the calculated one) in the interelectrode space.

Thus, until recently, it was impossible to definitely recommend the values of I_W and F_W parameters closely connected to the parameter τ_W and the calculation of the heat capacity of the metal of the interelectrode space for RPW of a particular overlap joint, especially due to the need for the obligatory presence of a common zone of mutual metal melting to obtain joints with required strength characteristics. This did not allow scientists involved in the study of resistance welding to develop RPW techniques without mutual metal melting zones with the lowest possible energy intensity of the process.

IOP Conf. Series: Materials Science and Engineering 1061 (2021) 012034 doi:10.1088/1757-899X/1061/1/012034

2. Materials and methods

The following experiment was carried out. RPW of overlap joints consisting of two plates 3 + 3 mm thick from cold-rolled low-carbon steel was performed with the MT-3201 alternating current resistance welding machine. One of the plates had an embossed round projection (6 mm in diameter and 1.5 mm in height) with a truncated conical shape (Figure 1, II).

Two main signals of the process (the voltage proportional to the welding current and the voltage of the interelectrode space U_{EE}) were recorded using a DTPH-32000 current sensor (according to the Hall effect) and shielded cables soldered to the electrodes. The current sensor was calibrated according to the indications of the Rogowski coil by measuring the highest short circuit current of the welding machine.

Signal processing with subsequent real-time calculation of the value Q_{EE} was performed by the ENERGY System device consisting of the National Instruments NI USB 6251 analogue and digital data acquisition device and the LabView 2017 graphical programming environment [9].

The time of the preliminary electrode pressing force τ_C, the time the parts remained pressed by electrodes after the current was switched off τ_F and the parameter τ_W were set on the RKS-801 weld timer.

Figure 2 shows a general view of the experimental setup (a), ENERGY System device (b) and RKS-801 weld timer (c).

a)

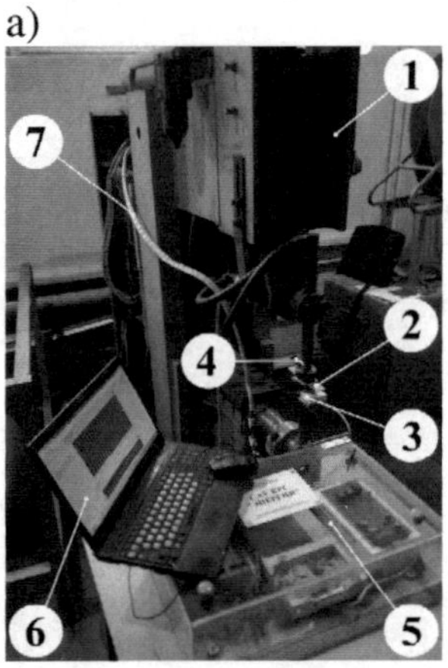

b)

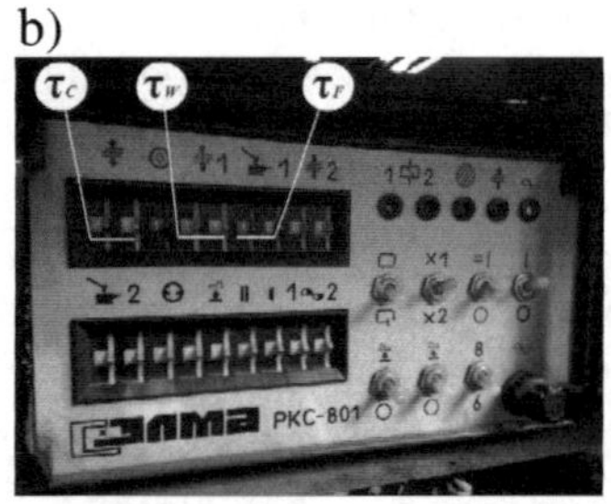

c)

1 – MT-3201 welding machine; 2 – parts being welded; 3 - current sensor; 4 - cable for recording the voltages in the interelectrode space; 5 - ENERGY System; 6 - personal computer; 7 - cable for external control of the welding machine; I - National Instruments NI USB 6251 device; II – unit for coupling and amplifying signals; III - power supply for the current sensor (± 12 V)

Figure 2. Experimental equipment.

During the welding process, the following mode parameters were set as constant: $\tau_C = 0.8$ s; $\tau_W = 0.36$ s (according to [10]); $\tau_F = 0.6$ s.

In accordance with the generally accepted HBE ($Q_{EE} = 8395$ J) and the experimentally recorded value $R_{EE} = 130$ μOhm (with an approximate current value of 13 kA), the parameter I_W was set by a DC voltage signal (3.4 V) using ENERGY System. Thus, throughout the experiment, when specimens of the same type were welded, the parameter I_W varied over a range of values due to deviations in the values R_{EE} when the same current was set.

During the experiment, RPW of specimens of the same type was performed with different values of the parameter F_W (kN): 7.4; 6.8 (according to the recommendations [6]); 6.4; 5.8; 5.2; 4.8; 4.2; 3.6; 3.1.

For the MT-3201 welding machine, these values corresponded to the following compressed air pressure readings (atm) on the manometer: 6; 5.5; 5; 4.5; 4; 3.5; 3; 2.5; 2.

3. Results

When recording the signals of the RPW process, curves of the temporal variations of effective values of the parameters I_W, U_{EE}, R_{EE}, P_{EE} и Q_{EE} were created.

Further, the values corresponding to the last peak value of the welding current before it was switched off were marked on the curves of variations in the parameters $I_W = f(\tau_W)$, $R_{EE} = f(\tau_W)$ и $Q_{EE} = f(\tau_W)$.

The marked values of the parameters Q_{EE}, I_W and R_{EE} were superimposed for all samples of the same type when performing RPW with different values F_W. As a result, characteristic curves $Q_{EE} = f(F_W)$, $I_W = f(F_W)$, $R_{EE} = f(F_W)$ were constructed (Figure 3).

The characteristic curve $Q_{EE} = f(F_W)$ was divided into two sections based on Q_{EE} calculated using the HBE: I - the range of values of the parameter F_W (up to 5.5 kN), providing an increased Q_{EE} when performing RPW; II - the range of values of the parameter F_W (greater than 5.5 kN), providing the values Q_{EE} not higher than the calculated value when performing RPW. In section II, two values of the parameter F_W (5.5 and 6.8 kN), at which the calculated Q_{EE} is supplied to the interelectrode space, were marked (Figure 3, bottom left).

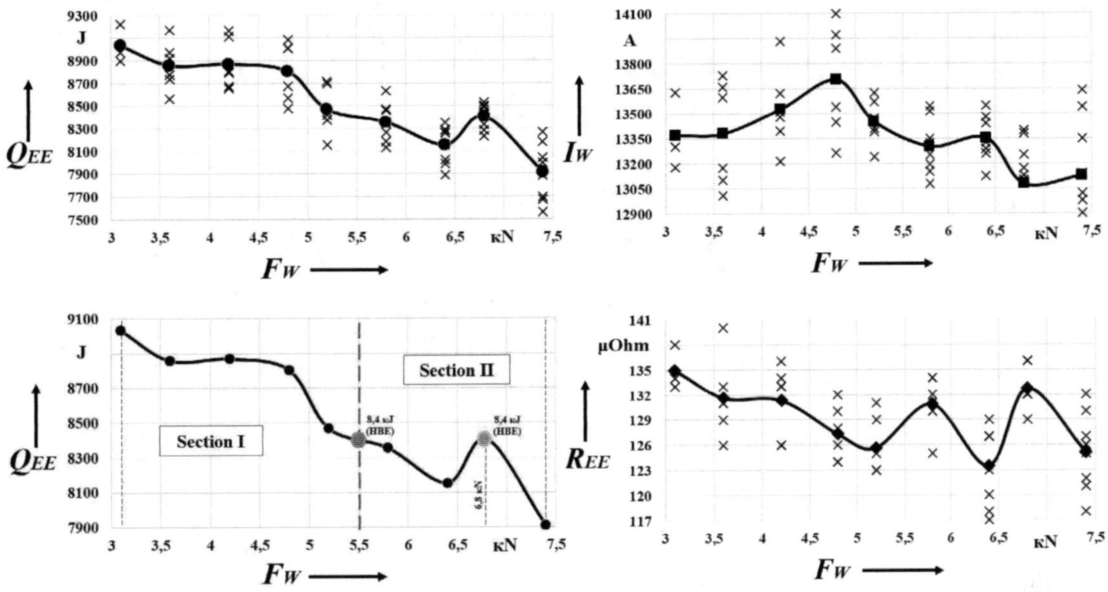

Figure 3. Characteristic curves $Q_{EE} = f(F_W)$, $I_W = f(F_W)$ and $R_{EE} = f(F_W)$ and two sections in the curve $Q_{EE} = f(F_W)$.

Comparison and analysis of the characteristic curves $Q_{EE} = f(F_W)$, $I_W = f(F_W)$, $I_W = f(F_W)$ showed that the fluctuation of the parameter I_W in the last peak before switching off the current is no more than 5% (< 0.7 kA) over the entire range of variation in the value of the parameter F_W (with a constant current set on the weld timer). In this regard, for both marked values of the parameter F_W (5.5 and 6.8 kN), the deviation from the welding current calculated using the HBE and the Joule-Lenz law is no more than 0.4 ... 1.2% (Figure 4, top).

In addition, the fluctuation of the parameter R_{EE} in the last peak before switching off the current is no more than 11 ... 12% (< 15 μOhm) over the entire range of variation in the value of the parameter F_{CB} (with a constant current set on the weld timer). When performing the calculation and carrying out the experiment, the deviation in the value R_{EE} is no more than 0.8 ... 2.3% for both marked values of the parameter F_W (Figure 4, bottom).

Thus, a graphical regularity of the effect of the parameter F_W on the amount of heat input into the interelectrode space with a constant current set on the welding machine was determined. This regularity rather accurately reflects the relationship between the parameters τ_W, I_W and R_{EE} with reference to the Joule-Lenz law.

This facilitates determining the optimal value of the parameter I_W for a specific value of the parameter τ_W depending on the required (calculated) value of the heat capacity in the interelectrode space taking into account the dynamics of variation in its resistance.

This, in turn, makes it possible to study the structure and strength of the projection welded joints in the range of values of the parameter F_W (in this case, 5.5 ... 6.8 kN), at which the amount of energy that does not exceed the value calculated using the HBE is supplied to the interelectrode space.

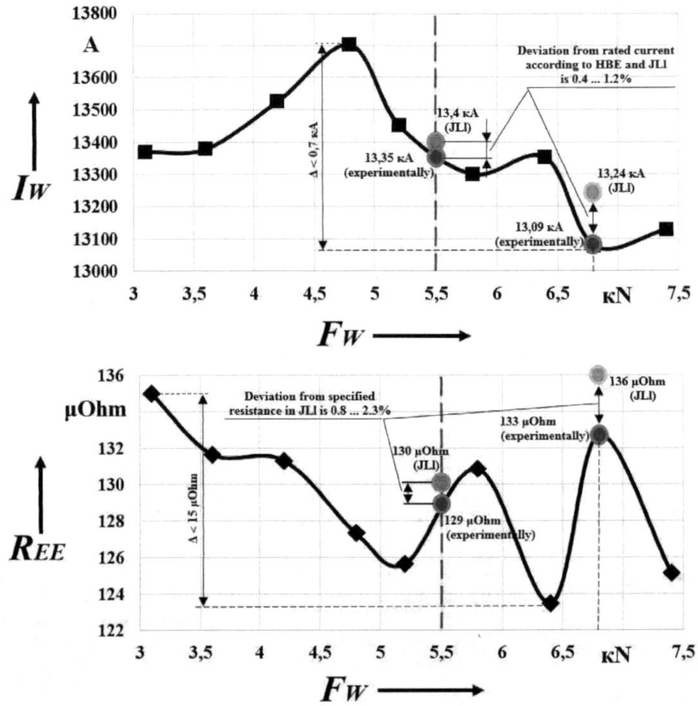

Figure 4. Deviations of experimental values I_W and R_{EE} from calculated ones for marked values of parameter F_W.

These data are necessary to substantiate the efficiency of the developed RPW processes with reduced energy intensity and two types of strong welded joints obtained: a) with mutual metal melting in parts being welded; b) with solid-phase zones in joints.

In this experiment, after performing welding and analyzing the process parameters, strength tests of joints by static shear loading were carried out (with the RGM-1000 machine). The value of the failure load P_D (kN) and the mode of the failure of the test specimens (brittle or ductile weld rupture in shear, base metal rupture) were estimated.

The tests showed that strong joints are formed regardless of the presence of molten metal splashes from the interelectrode space and cavities over the entire range of variation of the parameter F_W when performing RPW (Figure 5). The spread in the values of the average failure force P_D of specimens does not exceed 8% (2 kN).

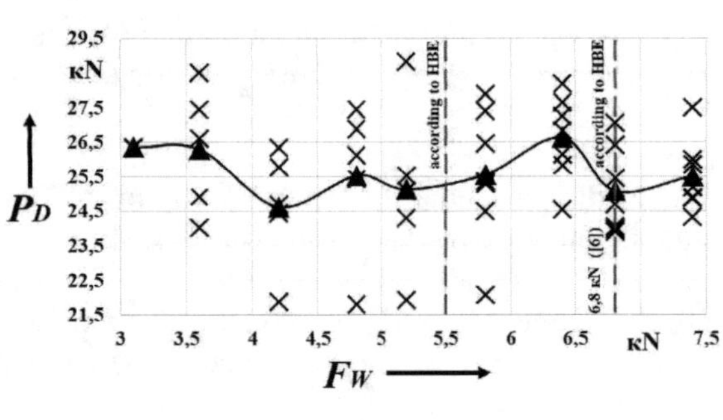

Figure 5. Change in average failure force of specimens obtained by welding with different values of parameter F_W.

IOP Conf. Series: Materials Science and Engineering 1061 (2021) 012034 doi:10.1088/1757-899X/1061/1/012034

In this regard, for the value of the parameter τ_W(0.36 s) taken to calculate the HBE, the RPW process with the variation in the parameter F_W ensures the formation of joints both with a common zone of mutual metal melting in the form of a weld nugget with a diameter d_{CD} (3.6 and 4.8 кN) and without it (7.4; 6.8; 6.4; 5.8; 5.2; 4.2; 3.1). In the second case, an annular solid-state joint with an activation zone of diameter d_{AZ} is formed (the width of the ring – b_{SFR}).

For welded joints produced with RPW corresponding to section I of the variation of the parameter F_W (up to 5.5 kN) (Figure 3), which lead to obtaining larger values Q_{EE} with respect to the calculated value, the formation of a weld nugget is possible (Figure 6). However, for these values of the parameter F_W, the requirement to fill the gap between the parts is not met. This is confirmed by incomplete deformation of the projection depression. The projection depression does not come into contact with the upper movable electrode until the current is switched off.

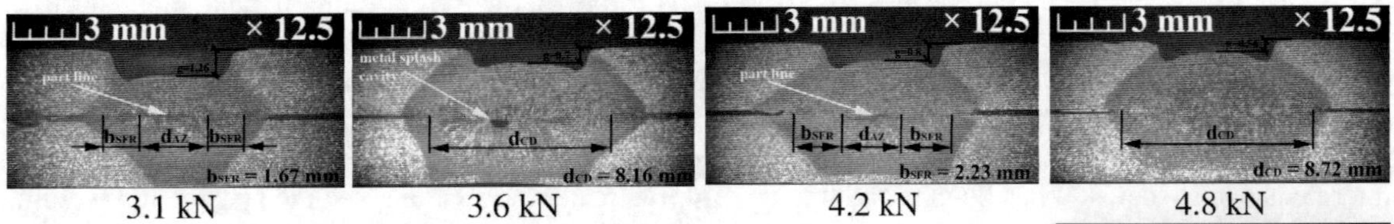

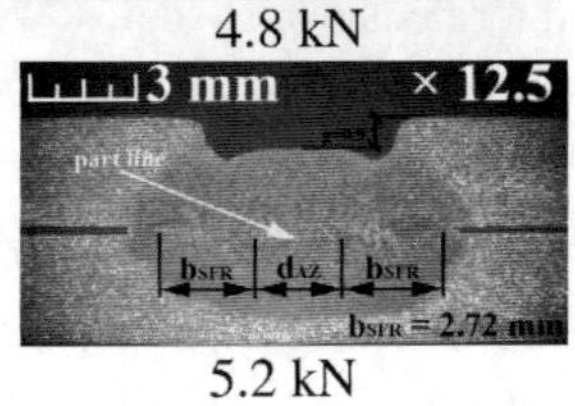

Figure 6. Structure of joints produced by RPW with increased Q_{EE} relative to the value calculated using the HBE (with F_W up to 5.5 kN).

For welded joints corresponding to section II of variation of the parameter F_W (greater than 5.5 kN) (Figure 3), which allow the values Q_{EE} not greater than the calculated value to be obtained for RPW, the weld nugget is not formed (Figure 7). In this case, the gap between the parts decreases and the projection depression is deformed by a greater amount (g ≤ 0.42 mm).

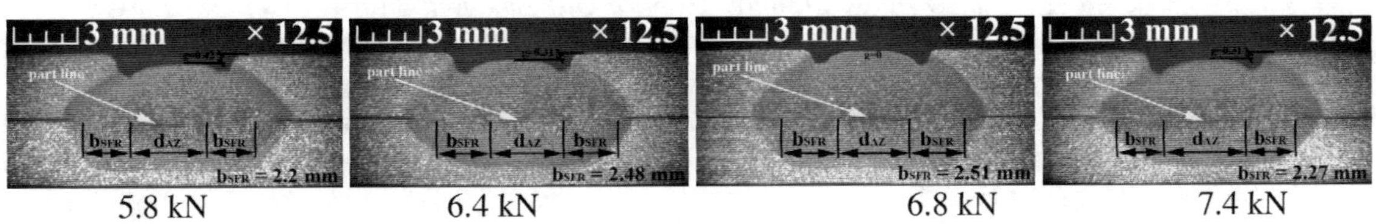

Figure 7. Structure of joints produced by RPW with value Q_{EE} not greater than the value calculated using HBE (with F_W greater than 5.5 kN).

Failure of 49 out of 50 projection joints obtained by varying the parameter F_W occurred as ductile weld rupture in shear along the annular region or in the weld nugget (Figure 8, a).

From this, we can conclude that the annular regions of the joints (solid-phase) contain zones of mutual metal melting, and during the RPW process with alternating current a weld nugget can start to form precisely from these regions.

The analysis of the rupture of the joints confirms their structure on macrosections. Three characteristic regions are clearly distinguished: I – annular joint (including the activation zone); II – a ring of extruded metal; III - the outer part of the area of heat transfer to the metal of the part (Figure 8, b).

a) b)

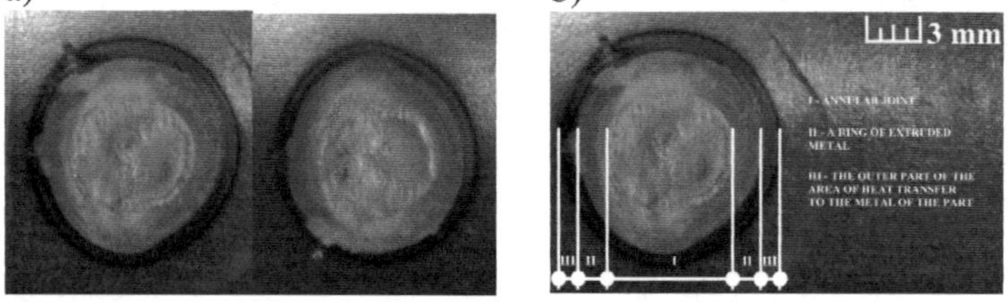

Figure 8. (a) Joint failures (on the left - part with projection, on the right - counter part) and characteristic regions of ruptures (b) of annular projection welded joint (a part with projection).

4. Conclusion

The research has shown that: 1) Using the example of resistance projection welding used for producing overlap joints of plates 2+2 mm thick with an embossed round projection with a truncated conical shape, the regularity of the effect of the electrode pressing force of on the amount of heat input into the interelectrode space with a constant current of the welding machine and a dynamic variation in the resistance of the interelectrode space was established; 2) The resulting regularity, with reference to the Joule-Lenz law, rather accurately reflects the relationship of the following parameters of the welding process: the length of the time the current is supplied; the welding current; the resistance of the interelectrode space; 3) It has been found that it is impractical to apply the resistance projection welding of the joints under consideration in the modes corresponding to the obtained regularity when electrode pressing forces are below 5.5 kN. Although welded joints with sufficient strength are produced (a weld nugget is formed), the requirement to fill the gap between the parts is not met, which is unacceptable. With electrode pressing forces of above 5.5 kN, the resistance projection welding allows us to obtain strong annular joints presumably with areas of mutual metal melting; 4) Further development and comparison of similar regularities in relation to the process of resistance projection welding of overlap joints of plates with similar thicknesses with reduced values of the time of the current pulse (in harder welding modes) are required to obtain joints with a weld nugget.

References

[1] Posh G, Holzinger V and Bychkovsky S L 2014 Railway cars welding: tasks in the field of materials, processes and automation *Svarochnoe proizvodstvo* 3 pp 10–16

[2] Berezienko V P, Melnikov S F and Furmanov S M 2009 Pressure welding technology (Mogilev: Belarusian-Russian University Publ.)

[3] Berezienko V P 1997 Theoretical and technological fundamentals of increasing the bearing capacity of joints made by spot and projection welding, regulation of their stress-strain state *Doctoral dissertation* (Mogilev)

[4] Paliakou A Yu 2015 Projection welding of multilayer welded structures with automatic regulation of mode parameters *PhD dissertation* (Mogilev)

[5] Polyakov A Yu, Furmanov S M, Bendik T I and Kurlenkov A M 2017 Development of an energy-saving method of resistance projection welding of multilayer welded structures *Welding International* **8** pp 868–873

[6] Gulyaev A I 1985 *Technology and equipment for resistance welding* (Moscow: Mashinostroenie)

[7] Bendik T I 2009 Projection welding of T-shaped joints with directional elastoplastic deformation of the metal. *PhD dissertation* (Mogilev)

[8] Mikno Z 2016 Projection welding with pneumatic and servomechanical electrode operating force systems. *Welding Journal Publ.* 8 pp 286–299

[9] Paliakou A Yu 2019 Reducing energy intensity of resistance projection welding processes. Mogilev. *Belarusian-Russian University Publ.*

[10] Paliakou A Yu, Kulikau V P and Stsiapanau A A Resistance projection welding of sheet metal without formation of a mutual melting zone in the form of a cast nugget. *International Conference on Aviamechanical Engineering and Transport*. Atlantis Press. Series: Advances in Engineering Research. November 2019 Vol 188 p 264–270

A model for predicting the temperature of a machine tool structure by a neural network using the sliding window method

V V Pozevalkin[1], A N Polyakov[1]

[1] Orenburg State University, Ave. Pobedy 13, Orenburg, Russia

E-mail: pozevalkinvv@mail.ru

Abstract. The paper studies a model for predicting the surface temperature of structural elements of a machine tool using artificial neural networks. A method of forming a training sample by the sliding window method for solving the problem of retrospective forecasting is presented. As applied to a neural network, the sliding window method is an algorithm for forming a training set from an initial set of experimental data necessary to build a forecasting model. Research was carried out for various types of neural networks, namely, generalized regression neural network, radial basis function network and feed forward network. Extrapolation was performed using multistep prediction, in which the predictive system uses the data obtained at the output of the neural network to predict subsequent values. The efficiency and practical suitability of neural network models for predicting the temperature of key heat sources located in certain areas of the machine structure was verified.

1. Introduction

A modern numerical control machine tool, as a typical technological equipment, is an important component in most mechanical engineering enterprises. Therefore, the quality indicators of machine tools largely determine the competitiveness of enterprises in the world market of high technology products. One of the quality criteria of modern machine tools is their accuracy. In this case, thermal deformations of the structural elements of the machine tool, which form thermal errors, can lead to a significant loss of machining accuracy. According to various sources [1-3], thermal errors are approximately 70 % of the total machining error.

Thermal errors can be reduced by making changes to the machine tool structure using advanced design methods and engineering solutions. For example, such as the implementation of structural symmetry or the use of cooling systems. In this case, methods of numerical analysis are used, such as the finite element method, the finite difference method [4, 5]. Extensive studies are also carried out in the field of thermal error compensation [6]. This approach is based on the correlation between measured temperature values and the resulting displacement of the machine tool functional point, for example, the tool center point (TCP). This allows you to evaluate the change in the relative position of the tool and workpiece. While all of these methods can provide acceptable results for some machine tool tests, improvements in predictive accuracy are still relevant. An accurate and reliable predictive model is the most important part of studying thermal processes in machine tools.

In recent years, it has been shown that temperature values can be successfully predicted using empirical modeling methods such as multiple regression analysis, various types of artificial neural networks, and a combination of several modeling methods [7]. The combination of various numerical calculation methods and the advantages of an artificial neural network is the most reliable and effective tool for modeling thermal processes.

2. Materials and methods

Prediction of experimental values of temperatures is an important component of the study of thermal processes in machine tools. This information is very useful both at the design stage of the machine tool structure by refining thermal models, and at the operation stage, by increasing the accuracy of predicting the thermal error for compensation systems.

There are currently no absolutely reliable methods capable of predicting thermal characteristics, temperature field and thermal errors with extremely high accuracy. Therefore, many different methods are used to improve the temperature characteristics of the machine tool, which are most often divided into three main groups.

The first group is conventionally called the constructive method, which involves making changes to the design of the machine tool, for example, the use of heat pipes, cooling systems, various materials that are insusceptible to temperature changes. This approach is very expensive and ineffective in reducing the thermal error.

The second group of methods is based on compensation using a thermal error model embedded in a numerical control system, and is more efficient and economical. The model can be obtained using linear or multiple regression, artificial neural network, etc. [8] with a sufficient amount of experimental data.

The third group is based on finite element modeling, which allows thermal performance to be evaluated early in the design of a machine tool. This approach is the most economical since it does not require continuous full-scale experiments.

However, all of these methods assume an accurate and reliable mathematical model for predicting temperature. The implementation of the principle of extrapolation by the sliding window method allows obtaining predicted temperature values for a pre-emptive time period.

2.1. Using the sliding window method

As applied to a neural network, the sliding window method [9] is an algorithm for forming a training set from an initial set of experimental data necessary to build a model for predicting the thermal characteristics of a machine tool.

In this case, a window is understood as a time interval containing a set of values required to form a training example. During the operation of the algorithm, the window is shifted in the original sequence by one observation unit, and each position of the window forms one example.

A time series $X(t)$ is a set of values of a certain quantity $x(t)$ at successive times:

$$X(t) = \{x(t_0 - m\delta), ..., x(t_0 - \delta), x(t_0), x(t_0 + \delta), ..., x(t_0 + k\delta)\}, \quad (1)$$

where t_0 is current time; δ is prediction interval; m is immersion depth; k is forecast horizon.

The prediction task for a time series $X(t)$ on an interval $t = \overline{1, n}$, is to find the continuation of the time series on an unknown interval. This means that it is necessary to determine $x(n+1), x(n+2), x(n+3)...x(n+k)$. The set of known values of the time series forms a training sample, the dimension of which is characterized by the value n and k the forecasting horizon. In addition, the method is also characterized by the width of the sliding window with a dimension m equal to the number of time series elements simultaneously fed to the input of the neural network.

To build a sliding window, a certain segment of the time series is selected and several observations are selected from it, which will represent the input vector (Inputs) of the neural network. The desired output (Targets) in the tutorial example will be the following observation in order.

2.2. Neural network training

The sliding window method assumes the presence of two windows W_i and W_o with dimensions m and s, respectively, which move along the time sequence of the initial experimental data with a certain sliding step δ.

Thus, the sequence of pairs W_i and W_o forms the training sample, which is used to train the neural network in forecasting. After training, the network must predict a time series for a forward period of time [10].

In this case, the training set can be represented as a matrix, the rows of which are vectors fed to the input of the neural network and the vector of target values:

$$W_i = \begin{bmatrix} x(1) & x(2) & \dots & x(n-m) \\ x(2) & x(3) & \dots & x(n-m+1) \\ \dots & \dots & \dots & \dots \\ x(m) & x(m+1) & \dots & x(n-1) \end{bmatrix}, \quad (2)$$

$$W_o = [x(m+1), x(m+2), \dots, x(n)], \quad (3)$$

where W_i – input data matrix; W_o – target vector; m – sliding window width; n – training sample size.

The proposition that the sliding window method is characterized by a window width equal to the number of time series elements simultaneously fed to the network input determines the structure of the neural network, which includes distributed neurons and one output neuron (figure 1).

2.3. Implementation of the extrapolation principle

Recently, it has been established that neural networks do a good job of approximating experimental data from thermal tests of machine tools [11]. But for forecasting, it is necessary to realize the possibility of continuing the time series for an anticipatory period of time, that is, the principle of extrapolation.

To implement the principle of extrapolation, the window is shifted one position in the direction of increasing time, and the process of forming the next pair of training sample is repeated. When constructing a sliding window, data is taken from the table of the relative change in the predicted value. At the next step, the obtained predicted value is added to the original sequence and a new input data matrix is formed.

Long-term forecasting is carried out using multi-step forecasting, in which the forecasting system uses the obtained forecast values (output data) to predict subsequent values.

The proposed forecasting model was implemented in the MATLAB system using the Neural Network Toolbox application package. The following is the sequence of execution of the forecasting algorithm, which is written as follows:

Step 1: formation of a set of initial experimental data N.

Step 2: forming a training sample P_i. MATLAB uses a cell array that defines a matrix of training sample sequences and is given by the following expression:

$$P\{i\} = [N(i : m+i)], \quad (4)$$

where i – cyclic variable, $i = 1,2,3,\dots, n-m$; n – training sample size; m – sliding window width.

The input array of the neural network X is the first m values of the matrix columns, and the target output vector T is the $m+1$ column. In this case, the values X and T are set by the following expressions:

$$X\{i\} = P\{i\}(1:m)'; T\{i\} = P\{i\}(:,m+1), i = 1,2,\dots,n-m, \quad (5)$$

IOP Conf. Series: Materials Science and Engineering 1061 (2021) 012035 doi:10.1088/1757-899X/1061/1/012035

Step 3: train the neural network. At this stage, the network is trained based on the training example formed in the previous step.

Step 4: determine the forecast values. Forecasting is carried out similarly to the process of forming a training sample. After training, the last of the known samples is fed to the input of the network, and one predicted value $x(t_0 + \delta)$ for the initial time series is formed at the output of the network. After that, they obtained predicted value is added to the sequence and the input data matrix is formed again. The procedure is repeated until the forecast horizon is reached k .

Step 5: forming a sequence of predicted values.

The presented algorithm for the formation of a training sample of a neural network and realizing the principle of extrapolation by the sliding window method allows predicting the values of the thermal characteristics of the machine tool for a pre-emptive period of time.

3. Experimental research

The experiments were carried out on vertical machine tool model 400V (Russia). To determine the temperature, a multichannel temperature meter was used, the sensors of which were located on the surface of the spindle assembly. As a result, a sample of temperature values in the volume of 406 elements was formed.

MATLAB functions were developed to analyze and predict temperature data near the front spindle bearing. These functions include the implementation of the algorithm for the formation of the training sample by the sliding window method and the implementation of the algorithm for extrapolation of experimental data.

Prediction was carried out using three types of neural networks, namely: Generalized Regression Neural Network, Radial Basis Function Network, and Feedforward Neural Network. Figure 1 shows the block diagrams of neural networks implemented in the MATLAB system.

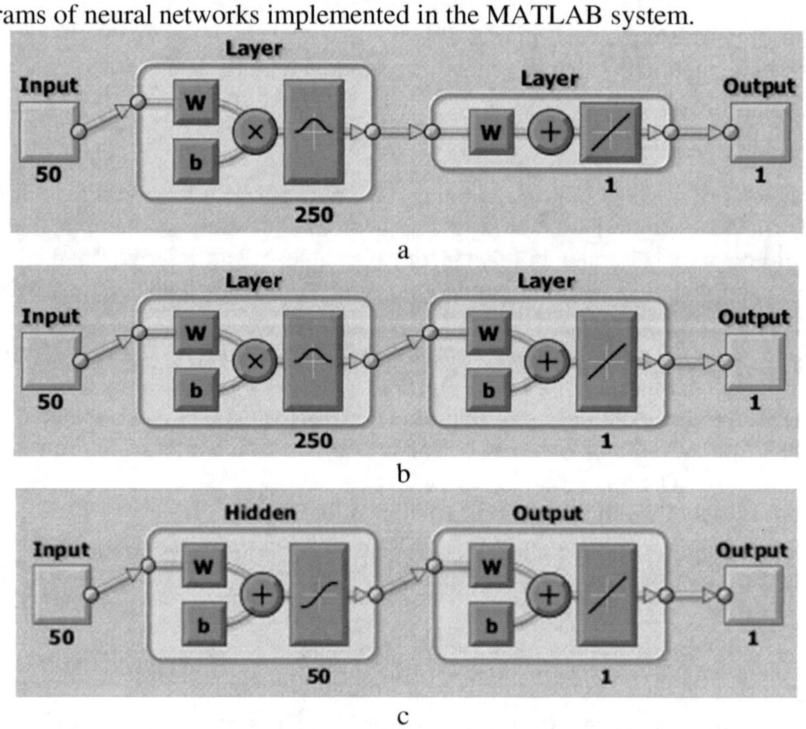

Figure 1. Block diagrams of neural networks: a – generalized regression neural network; b – radial basis function network; c – feedforward neural network.

For all neural networks, the same parameters were set, namely: the volume of the initial experimental data $n = 300$, the width of the sliding window $m = 50$, the forecasting horizon $k = 106$. The number of neurons in the hidden layer for the feedforward network was set to 50 neurons, which was trained using a standard learning algorithm.

It is proposed to evaluate the accuracy of the forecasting model using retrospective forecast. In this case, forecasting is carried out for a certain moment in the past, for which the actual data are already known.

To do this, we divide the initial experimental data into two parts. The first one will cover earlier data (300 nodes), with their help we will form a training sample for artificial neural networks. Based on the later data (106 nodes), we will form a test sample designed to obtain retrospective forecast values and check the correctness of the networks.

Figure 2 shows the results of the retrospective forecasting procedure.

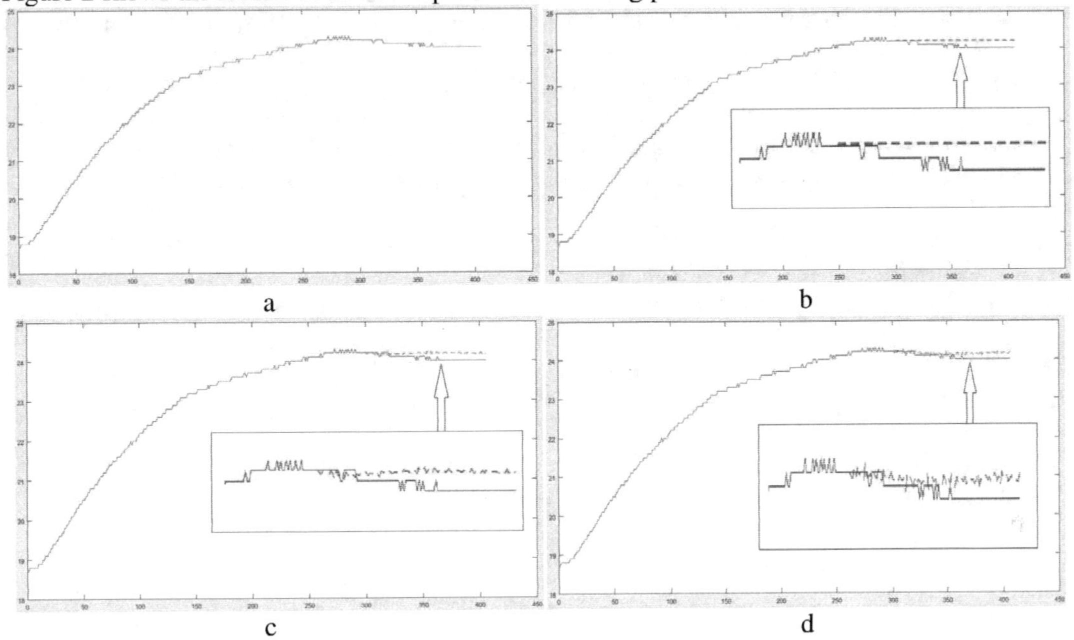

Figure 2. Forecasting results: a – initial experimental data; b – generalized regression neural network; c – radial basis function network; d – feedforward neural network.

On the graphs (figure 2), the abscissa shows the node numbers, and the ordinate shows the temperature values. The solid line denotes the initial experimental data; the dashed line indicates the predicted values of thermal characteristics.

The accuracy of the forecast can be judged by the magnitude of the error between the predicted and actual values of the experimental data [12].

4. Conclusion

Thus, the proposed method for forming a training sample allows you to create data arrays for training a model based on the mathematical apparatus of artificial neural networks, as well as to predict thermal characteristics beyond the set of initial experimental data obtained at the stage of machine tool operation.

5. Acknowledgments

The reported study was funded by RFBR according to the research project No. 20-38-90045.

References

[1] Mayr J, et al. 2012 Thermal issues in machine tools *CIRP Annals - Manufacturing Technology* **61(2)** 771-791

[2] Ramesh R, Mannan M A and Poo A N 2000 Error compensation in machine tools — a review: Part II: thermal errors. *International Journal of Machine Tools and Manufacture* **40(9)** 1257–1284

[3] Oomen T 2018 Advanced motion control for precision mechatronics: control Identification, and Learning of Complex Systems. *IEEJ J Ind Appl* **7(2)** 127–140

[4] Mian N S, et al. 2011 Efficient thermal error prediction in a machine tool using finite element analysis *Meas. Sci. Technol.* **22(08)** 510-517

[5] Li Y, et al. 2017 Boundary Conditions Optimization of Spindle Thermal Error Analysis and Thermal Key Points Selection Based on Inverse Heat Conduction *Int. J. Adv. Manuf. Technol.* **90(9-12)** 2803-2812

[6] Li J, et al. 2009 Thermal-error modeling for complex physical systems: the-state-of-arts review *Int. J. Adv. Manuf. Technol.* **42** 168–179

[7] Han J, et al. 2012 Thermal error modeling of machine tool based on fuzzy c-means cluster analysis and minimal-resource allocating networks *Int. J. Adv. Manuf. Technol.* **60** 463–472

[8] Eskandari S, et al. 2013 Positional, geometrical, and thermal errors compensation by tool path modification using three methods of regression, neural networks, and fuzzy logic *Int. J. Adv. Manuf. Technol.* 1–15

[9] Gao X, Mei J, Chen H et al. 2014 Approximate inference for dynamic Bayesian networks: sliding window approach *Appl Intell* **40** 575–591

[10] Ferreira P M, Ruano A E 2009 Online sliding-window methods for process model adaptation *IEEE Trans. Instrum. Measur.* **58** 3012–3020

[11] Pozevalkin V V, Parfenov I V, Polyakov A N 2019 Approximation of machine tool experimental thermal characteristics by neural network *Journal of Physics: Conference Series* 1399 044018

[12] Alberg D, Last M 2018 Short-term load forecasting in smart meters with sliding window-based ARIMA algorithms *Vietnam J Comput Sci* **5** 241–249

IOP Conf. Series: Materials Science and Engineering 1061 (2021) 012036 doi:10.1088/1757-899X/1061/1/012036

The concept of creating aerostatic architecture facilities for sustainable development of promising remote areas

N A Saprykina

The Moscow Architectural Institute (State Academy), 11/4, building 1, p. 4, Rozhdestvenka Street, Moscow, 107031, Russia

E-mail: nas@markhi.ru

Abstract. The article focuses on the topical problem of the formation of aerostatic architecture facilities for the sustainable development of remote areas, as well as the prospect of the use of guided flying devices in various fields of activity. Historical and modern examples of their creation and use in the context of the development of flying habitable objects are considered. The concept of the formation of aerostatic architecture objects as a resource for the development of an alternative space is presented. There are precedents of the future in connection with the general environmental orientation of society to address the important challenges of sustainable architecture. As a result of the study, innovative approaches to the formation of aerostatic architecture objects as a resource of the development of alternative space have been defined. The materials of the article can be useful for the theory and practice of the formation of the habitat space, as it opens up completely new possibilities in architectural activity.

1. Introduction

The role of architecture in a dynamic society is constantly transforming. Since the middle of the last century there has been a strong growth of scientific knowledge, rapid development of science and technology. This allows us to form the contours of a completely new society and its logistical equipment. In the theory of the formation of architectural space, we see the use of the ideas of recent achievements in other areas of knowledge related to the current problem of the formation of objects of aerostatic architecture for the sustainable development of remote areas [1].

The prospect of the use of guided flying devices in various fields of activity is quite obvious, despite the fact that the era of passenger pneumatic-airships was not continued in its time. However, in reality, inventors, engineers and entrepreneurs have tried a lot of times to bring back the era of mass flights on pneumatic-airships. Economic and technological thinking, science and technology have become the determining factors of the historical process. The introduction of the design and the construction of aerostatic architectural objects have been made possible by recent developments in electronic technology, as well as the close relationship between architecture and technical achievements.

2. Materials and methods

2.1. Creation of guided flying devices for the development of remote areas: the historical and modern objects

In remote areas of Russian Siberia and the Arctic, air transport plays a crucial role, allowing them to master their territorial length. In this regard, the hybrid pneumatic-airships, which can carry passengers and cargo, are interesting and promising today. This is due to the fact that at certain costs for operation of pneumatic-airships they remain a relatively cheap and reliable means, besides they can land in almost anywhere.

The main advantages of such guided flying devices are the smoothness and speed of movement, as well as the reliability of the flight - pneumatic-airships have demonstrated a greater survival of passengers in case of accidents than aircraft and helicopters. In contrast from them, pneumatic-airships do not create noise discomfort and do not pollute the atmosphere, and do not harm the surface of the tundra. The cost of ground equipment for berths of pneumatic-airships at unloading sites is incomparably lower than the cost of building airfields [2].

The heyday construction of the pneumatic-airships falls in the 1930s. New technologies have made these ships as attractive as possible in terms of comfort and environmental protection. Turning to the history of the origin and operation of guided pneumatic-airships in the early twentieth century, we can assume the level of comfort of staying at such facilities. This is evidenced by the interiors of the first passenger guided pneumatic-airships "*Graf Zeppelin*" (1928) or "*Hindenburg*" (1936).

Studies of the use of cargo airships in the North in the 1970s of the last century have allowed establishing that aerostatic flying machines can be used in remote areas of Russia with high efficiency to solve of many economic problems. For example, the delivery and installation of drilling equipment to the workplace, the transportation of natural gas from remote and hard-to-reach mining sites, the transportation of forest to the processing sites, and the patrolling and protection of forests and deer pastures from fires [2].

Further development of constructive and technological solutions of aircraft were carried out in 1979-1983, when hybrid airships were created ("aeroletets") to transport blocks of drilling and compressor stations, power lines in the collection, bridge and pipeline structures for some promising remote areas of Russia. Modern aerostat technologies are developing in three directions: light airships of small and medium volume; transport airships of large and super-large cargo capacity ("aeroletes"); stratospheric remotely controlled vehicles are lighter than air. They operate at altitudes of 18 to 25 km and serve as an element of telecommunications infrastructure, as well as for monitoring the earth's surface and atmospheric layers [2]. Domestic airships at aerospace and defense enterprises are created within the framework of conversion programs. Among them are the joint Russian-French project airship "*Voliris-900*", the Russian multifunctional modular airship "*MD-900*" with a capacity of 3.5 tons, as well as the airship "*DC-N1*" [3].

Aeronautics has found use in various fields of activity, but the most confidently entrenched in the field of tourism and sports - there are some examples of the use of airships in the role of platforms for surveillance and filming, advertising carriers and tourist "taxis". For example, the tour airship "*Zeppelin NT L-N07*" (Germany) makes pleasure excursions in the sky over Stuttgart. In the pneumatic-airships "*Skyship 600*" (Germany) thanks to two engines, the device easily develops sufficient speed. The most common airship built and operated in the United States is the "*A-60*". More than 20 airships of this type are used in South America, Africa, Australia, Asia, as well as in many European countries [3].

In the United States, China and several other countries, patrol airships are also used to protect borders, particularly in the fight against smuggling. In this regard, the patrol airship "*Au-12*" is equipped with the most modern equipment and is able to monitor vast areas, and the airship «*Mine seeker*», developed by the research agencies DERA and TLG, is designed to search for mines [3].

In addition, the use of aircraft can allow monitoring of surrounding airspace, as well as serve as a weather probe, telecommunication transmitter and dispatcher. Thus, the solar airship "*HALE-D*", according to the company-developer Lockheed Martin (USA), will be able to replace satellites. The surface of the airship is covered with a thin layer of solar panels that produce energy directly during the flight [4].

The launch and operation of the airship costs ten times less than the comparable in functionality of the satellites. From large ships can be distinguished pneumatic-airships to carry massive and large cargo - "*Aeroscraft*" company Worldwide Aeros (USA) - can lift up to 20 tons of cargo into the sky and move at speeds of up to 200 kilometers per hour. The idea of creating promising guided pneumatic-airships for tourism organization was reflected in the creativity of many design engineers. "*Aeroscraft*" Cruise Hotel, travelling by air, will be able to accommodate 250 passengers and has the opportunity to provide them with a lot of amenities, including casinos, restaurants and comfortable cabins [5].

The design and creation of aircraft is developing in a new concept of habitation - flying house, office, workshop, etc. Flying house "*Wolke 7*" (Swiss designer Timon Shacher) has all the essentials for permanent residence: living room, luxurious bedroom and even entertainment center [6].

Designed by artists Tino Shedler and Michael Brown, "*Strato Cruiser*" is designed for recreation - it will house a restaurant, a spa and even a swimming pool. This aircraft is filled with helium, and photovoltaic elements are used to generate electricity. The "Strato Cruiser" flight cruiser is a conceptuality model of a large-size flying palace, where people can experience a new kind of tourism - air cruises around the globe. In order to maintain the proper level of comfort and allow guests to carefully consider the landscapes over which they fly, there is a deliberately slow speed of the airship [7].

Flying Hotel "*Aircruise*" by company of Seymourpowell is a giant aeronautics facility designed to move travelers over long distances in luxurious conditions. It will house apartments for living, a bar, a restaurant, in addition, a large area of airships with glass floors will be allocated for viewing platforms. The use of solar panels and hydrogen fuel cells, which additionally serve as a function of providing the flying hotel with clean water, gives the air hotel "Aircruise" properties of ecology. To reduce the risk of fire are designed to be advanced technology, as well as the design of the shell, which is made of self-tightening in case of damage to the material [8].

In project by company of Geotectura proposes a campus concept for the Faculty of Architecture of the University of Delft (Netherlands) in the form of a flying structure "*AIRchitecture*". In the campus has created a flying workspace in the form of a zeppelin-airship, which houses rooms for workshops with ready-to-work classrooms. It is assumed that the structure of the zeppelin will be built of carbon material and aluminum, equipped with photovoltaic elements all over the surface [9].

Further development of the trend of wireless information transmission finds in various areas of human activity. Such technologies have become a leading trend in rapid air delivery of various services to meet the emerging demand for advanced unmanned technologies. In recent years, there has been an increase in the development of unmanned technologies to perform high-speed services in everyday life [10].

As an example, American architects have developed a project called "*Skyscraper for the Landing of Unmanned Drones*", which aims to create a terminal to accommodate personal or commercial drones. To ensure a safe landing, drones are attached parallel to the facade, and for maximum increase in surface area are offered two overlapping layers - external and internal. The modules on the "living facade" of the skyscraper are designed for nine different types of drones [11].

By defining a new approach to habitat formation, presented projects and development of aerostatic architecture can be applied in many areas of activity for different climatic conditions. The design and creation of next-generation special airships for development of remote areas is conditioned by their need to provide effective ways of moving people and transporting goods.

2.2. The formation of aerostatic architecture as a resource for the development of alternative space: precedents of the future

The development of the concept of mobile architecture in the context of the creation of flying objects is of great interest to modern architects due to the overall environmental orientation of society to solve the important problems of sustainable development of architecture [12].

Thus, architect Vincent Calbo proposes the project of the pneumatic-airships "*Hydrogenase*" flying on biofuels. He will be able to move at an altitude of about two kilometers at a speed of about 175 km/h, both vertically and horizontally. This flying facility provides living quarters, offices, scientific laboratories and production centers, as well as gardens and plantations of green microalgae, which will produce bio-hydrogen as the main fuel for the ship [13].

According to the architect's plan, the flying object will be semi-autonomous, which will require refueling stations, which will be floating structures, which are equipped with solar and photovoltaic batteries on top, and below the water level are equipped with turbines that convert the energy of sea tides. These stations will house plantations of green algae, which will act as a treatment plant by recycling waste into the bio-hydrogen, which will be used by the ship as fuel. It is expected that the stations will supply the flying ships and surrounding settlements with hydrogen, as well as will simultaneously absorb carbon dioxide and recycle organic waste. Electricity on the ship will supply fuel cells and flexible photovoltaic batteries that are able to recycle organic waste [14].

In order to improve the ecological situation on the planet and to provide the necessary resources to the inhabitants of nearby settlements, the architects of the Zhi Zheng, Hongchuan Zhao and Dongbai Song proposed the "*Himalaya Water Towers*" project, which should be located in the high-altitude Asian Himalayas and located at the level of the clouds. Their main task is to turn the clouds into water, which will then fall into the water systems of settlements located and near the complex, and far from it. Wind turbines are expected to be built into the towers to produce electricity for the needs of residents of China's highlands [15]

The project of the city with the name "*Heaven and Earth*" was designed by the modern Chinese architect Wei Zhao, guided by the fact that at the current rate of population growth of the Earth soon on the planet will not be free space for the life of new generations. In this regard, it was proposed to create above the ground flying self-sufficient cities, where it is possible to grow food, extract "green" energy and even produce consumer products. According to the author's proposal, the flying city will not produce carbon dioxide and will be stay in the air thanks to the same technology as the trains on an electromagnetic pillow. The flying city "Heaven and Earth" will not move from place to place, and will hang over one of the points of the earth for a long time [16].

The acute problem of environmental care is vital for the future of the planet and humanity, as over the past century a huge number of green zones have been used destroyed, which have ousted cities. Today, humanity is trying to return nature to cities wherever possible [17]. The most incredible ways are used, including vertical landscaping or so-called flying green gardens, which can take any form that mimics the environment as dynamic, three-dimensional ecological aerial objects.

Thus, the flying park of architects Ting Xu and Yiming Chen "*Light Park*" for Beijing residents is held by a large balloon with helium and operated by propellers that work on solar energy. The design collects rainwater for watering plants, which in turn clean the suffocating air of the city, absorbing carbon dioxide. The outlines of the flying park are completely combined with classical Chinese architecture [18].

It is expected that the gardens will float in the air on large remotely operated airships, each of which would house thousands of green oxygen plants. A group of airships can migrate to the city, moving to an area with strong gas content. Flying gardens will constantly move through the metropolis, bringing oxygen to places where there are no green spaces.

Company Rael San Fratello has proposed a project that in the future will help improve the ecology of megacities - flying jellyfish-gardens. They are a cluster of airships filled with hydrogen, which will house thousands of plants. They are provided to control airships remotely (both manually and autopilot). During the day, jellyfish gardens will move through areas of the city with a strong gas content and thanks to special sensors to follow for the weather, noise, movement, pollution level and other indicators in real time. In the evening, the flying gardens will return to the city's "base" to refuel, at the same time gardeners can take care of the plants. All instruments of flying platforms thanks to photovoltaic coating will work not on hydrogen, but on solar energy. Flying jellyfish-gardens will constantly move over the city, delivering oxygen and clearing air to areas where there is no green space. In addition, flying gardens can also be used for advertising purposes [19].

Similar to the previous proposal, flying skyscraper "*PH Conditioner*" project by Chinese specialists (Hao Tian, Huang Haiyang, Shi Jianwei) aims to collect acid pollutants and waste caused by large-scale industrial production in cities and turn them into useful substances (such as reclaimed water and chemical fertilizers) for the Chongqing region (China). The facility is installed at an altitude of 200-300 m, where acid pollutants are collected. The porous membranes of the object allow absorbing, collecting and placing pollution in the main purifier, where they are neutralized by alkaline substances. Once acid pollutants have been neutralized, they can be converted into a neutral liquid that will be absorbed by containers and used as a nutrient for green plants [20].

The project of the city "*Passing Cloud*", designed by design architect Tiago Barros, presents a group of spheres floating in the air, similar to the airship chamber. The flying spheres are made of plastic but durable nylon fabric, stretched on a light metal frame. The whole structure moves by air only under the influence of gusts of wind, so the flying city lacks the means to control the direction and speed of flight and, accordingly, a stable flight path [21]. This idea has found modern development in a unique project of flying capsule hotels, designed by architects of HKS Ins. These transparent spheres move around skyscrapers with the help of cables and roller system [22].

British specialists Lemire Abdul Halim Chehab, Suraj Ramkumar Suthar and Swapnil Sanjay Gawande have developed the "*Skyscraper Aakash*" project, which offers the placement of floating clusters in the sky over Mumbai (India) - one of the most densely populated cities in the world. The complex as a whole consists of a structure that branches into wide, floating modules connected to create urban flying buildings equipped with cloud-shaped helium balloons. This structure can grow indefinitely in all directions with the formation of intermediate green terraces available for public use at a fairly high level. The outer shells of helium balloons are covered with photovoltaic cells, which allows to make a structure with autonomous energy supply by using solar energy. The structure can grow over any clusters of the city [23].

The city "*Cloud Nine*" by Buckminster Fuller is also a project of a flying city in the form of a ball, a mile in diameter and covered with transparent plastic film, skipping rays of sunlight that heat the air inside the balloon. Several Cloud Nine balloons can be combined to create large air cities. These cities are free to float in the air or stand on anchor in a specific location. By controlling the penetration of sunlight into the balls, can be manage control the temperature of the air inside them and adjust the altitude of the flight [24].

As in the space-related projects of the early 20th century, the concept "*Stratosphere Network of Skyscrapers*" is a modern extension in a number of projects. Architects from China (Mingxuan Dong, Yuchen Xiang, Aiwen Xie, Xu Han) propose to make a populated layer in the stratosphere in the form of a network of skyscrapers at an altitude of several tens of kilometers [25]. The first skyscrapers are supposed to be built on supports, and when the network will cover all the land, the need for supports will disappear, as the connected buildings will be kept near the ground due because of the power of its attraction. The project proposal is connected with an attempt to solve the problems of catastrophic consequences of the ecological crisis and the growth of the earth's population. The horizontal mesh habitable layer in the stratosphere, according to the authors, may be the only platform for ensuring the continuity of human civilization [26].

An even more fantastic idea is proposed by American experts (Richard Porter, Chris Allen, Cam Helland, Stephen Phillips), proposing to expand the urban development space as part of NASA's innovative research project "*City in Space*". The space object is performed in the form of a habitable biomimetical network in a geostationary orbit tied to the Earth's atmosphere. As a "space elevator" offered a system, proposed a structure from woven fibers carbon nanotubes. The vital purpose of transport communications consists of local elevators with mass transit lines and hubs, which pass through the entire orbital network connecting several megacities. According to the authors, such a solution of the system of urban megacities in space can serve as an outlet in the inevitable situation in the post-human period [27].

The creation of an artificial habitable environment in space gives rise to the philosophy of understanding a new direction - *aerostatic architecture*. Active development of alternative space in space can lead to the emergence of spatial formations with a new maintenance infrastructure. It is expected that such settlements will be built over existing cities within the troposphere and above the ground, where space shuttles will be able to fly and to reach the space elevators, and settlements located in near-Earth space will become a kind of buffer between the earth and the space being developed [28].

3. Results and Discussion

The concept of building aerostatic architectural objects for the sustainable development of promising remote areas is outlined in this review in the context of the following focuses discussed in the article:

1. Creation of guided flying devices for the development of remote areas: the historical and modern objects. In the context of underdeveloped transport routes in remote areas, air transport, allowing them to master their territorial length, plays a crucial role. Aerostatic pneumatic-airships have found use in various areas of activity, as they can be used in remote areas with high efficiency to solve many of the national economic problems. For example, the delivery of drilling equipment to the place of work, its installation on land and offshore, in the transport of natural gas from remote and hard-to-reach mining sites, transportation of forest to the processing sites, as well as patrolling and protection of forests and deer pastures from fires.

There are isolated examples of the use of guided pneumatic-airships as platforms for surveillance and filming, elements of telecommunications infrastructure, as well as for monitoring the earth's surface and atmospheric layers. They can also be used as a carrier of advertising and tourist "taxis", in the field of tourism and sports, as they can land in almost any area. The design and creation of new generation special guided flying devices for the development of Russian remote areas is conditioned by the need to provide them with effective ways of moving people and transporting goods. The practical interest is unmanned technologies for high-speed services in promising remote areas.

These precedents of the use of aerostatic objects prove that the creation of such devices is possible as a result of the joint creativity of architects, engineers, mathematicians and programmers.

2. The formation of aerostatic architecture as a resource for the development of alternative space: precedents of the future. The development of the concept of aerostatic architecture in the context of the creation of flying objects is widely applied in connection with the general environmental orientation of society to address the important challenges of sustainable development of promising remote areas.

To solve environmental problems, solutions are proposed for the creation: flying jellyfish-gardens, which will house plantations of green algae, acting as treatment plants by recycling waste into bio-hydrogen; flying parks collecting rainwater for watering plants, which in turn clean the suffocating air of the city, absorbing carbon dioxide; flying jellyfish-gardens, which will move through areas of the city with a strong gas content and, thanks to special sensors, monitor the weather, noise, movement, pollution level and other indicators in real time; flying water towers in the air to turn clouds into water, which will be then directed into the water systems of settlements located and near the complex, and far from it.

Another direction is the creation of aerostatic objects: self-sufficient cities flying above the ground, where it will be possible to grow food, extract green energy and even produce consumer products; flying skyscrapers aimed at collecting acid pollutants and waste caused by large-scale industrial production in cities and turning them into useful substances; placing floating clusters in the sky with the formation of intermediate green terraces, available for public use at a fairly high level.

Fantastic ideas related to the cosmic orientation are of scientific and technical interest: a flying city in the form of a ball, a mile diameter and covered with transparent plastic film, passing rays of sunlight that heat the air inside the balloon; inhabited layer in the stratosphere in the form of a network of skyscrapers at an altitude of several tens of kilometers - the project proposal is connected with an attempt to solve the problems of catastrophic consequences of the ecological crisis and the growth of the Earth's population; he space object is carried out in the form of a habitable biomimetic network in a geostationary orbit tied to the Earth's atmosphere.

The above solutions of urban megacities systems in space can serve as an outlet from the inevitable situation in the post-human period. Active development of alternative space in space can lead to the emergence of spatial formations with a new maintenance infrastructure.

4. Conclusion

The designated topical direction of use of aerostatic objects and the organization of comfortable living space for specialists and travelers causes the emergence of unconventional methods and approaches to the architectural and technological formation of architectural objects. Here the line between technology and architecture is blurred, which expands creative possibilities and generates a new direction in architecture - *technological architecture*, which becomes a customer for new advances and innovative ideas [29].

An important factor influencing the formation of aerostatic architecture will be the active development of mass air transport. In this regard, there will be a need to design buildings equipped with numerous airfields, transfer hubs. The emerging conceptual precedents illustrate the possibility of creating multi-level mega-structures, united by a web of hinged transport arteries. There will be no need to connect them to engineering networks, as traditional engineering equipment will be superfluous, and its functions will take over devices that will work on completely different principles [30].

Presented projects of aerostatic architecture, equipped with special devices, opening one of the possible directions in architecture, which from the point of view of common sense are fantastic. Nevertheless, these new ideas reveal boundless possibilities for creative search, and the projects considered show the most decisive changes in this direction.

References

[1] Saprykina N A 2016 Developing the concept of mobile architecture: flying objects *«Architecture and Modern Information Technologies»* **1 (34)** 1-18 http://www.marhi.ru/AMIT/2016/1kvart16/saprikina/abstract.php
[2] Doves S N Air Rovers for the Russian North 27.08.2008 *Independent newspaper* (Moscow) http://www.ng.ru/science/2008-12-24/11_zeppelin.html
[3] Bendine S Airship Market Takes a New Breath 21.11.2002 "*CNews*" (Moscow). http://www.cnews.ru/articles/rynok_dirizhablestroeniya_obretaet_novoe
[4] HALE-D – solar airship that will replace satellites 30.11.2007 *Novate* (USA) http://www.novate.ru/blogs/240614/26786/
[5] Cruise Hotel "Aeroscraft" by Company Worldwide Aeros http://astronom.forumbb.ru/viewtopic.php?id=131
[6] Flying house "Wolke 7" http://billionnews.ru/arhi/1150-letayushhij-dom-wolke-7-14-foto.html
[7] Flying Hotel Palace "Strato Cruiser" http://lifeglobe.net/entry/5715
[8] Seymourpowell's Flying Hotel "Aircruise" http://alt-energy.org.ua/nazad-v-budushhee-letayushhie-sady-oteli-i-goroda/
[9] Aerial architecture: Flying buildings http://www.infuture.ru/article/1872
[10] Gagulina O, Matovnikov S 2018 Spatially organized "vertical city" as a synthesis of tall buildings and airships *E3S Web of Conferences* **33** 01002 https://doi.org/10.1051/e3sconf/20183301002
[11] The Hive: Drone Skyscraper http://www.evolo.us/competition/the-hive-drone-skyscraper/#more-34984
[12] Saprykina N A, Saprykin I A 2018 Sustainable development of spatial habitat Environment as a challenge to civilization *IOP Conference Series: Materials Science and Engineering (MSE)* Vol 463 pp 1-6 DOI: 10.1088/1757-899X/463/2/022052
[13] The project of the garden-pneumatic-airships "Hydrogenase". http://polezreniya.ru/vne-kategorii/doma-budushhego-primery-futuristicheskoj-arxitektury/
[14] Kizilova S A Form and functional features of modular floating structures *E3S Web of Conferences 2019* Vol 91 p 05013 (1-7).
[15] The project "Himalaya Water Towers" http://itw66.ru/blog/future_architect/610.html#cut
[16] Project of the city "Heaven and Earth" http://solium.ru/forum/showthread.php?t=6673<center
[17] Kasyanov N 2020 Research on the Similarities of Morphogenesis in Architecture and Nature / Advances in Social Science, Education and Humanities Research, volume **471** Proceedings of the 2nd International Conference on Architecture: Heritage, Traditions and Innovations (AHTI 2020) https://doi.org/10.2991/assehr.k.200923.045
[18] The project "Light Park" for Beijing http://ecosupport.ru/ecoarchitecture
[15] Flying jellyfish-gardens by Company Rael San Fratello http://ecosupport.ru/ecoarchitecture
[20] Flying skyscraper "PH air conditioning" for the Chongqing region (China) http://www.evolo.us/competition/ph-conditioner-skyscraper/
[21] Project of the city "Passing Cloud" http://www.dailytechinfo.org/np/2987-passing-cloud-proekt-goroda-letayuschego-v-nebesah.html
[22] Flying hotels capsules - innovation in the world of construction http://www.ooo-eurostroy.ru/novosti/648-oteli-kapsuly.html
[23] The project "Skyscraper Aakash" - floating clusters in the sky over Mumbai (India) https://www.evolo.us/aakash-skyscraper/

[24] The project «Cloud Nine» by Buckminster Fuller http://supercoolpics.com/samye-udivitelnye-nerealizovannye-proekty-gorodov-budushhego/
[25] Stratosphere Network of Skyscrapers 2013 http://www.evolo.us/competition/stratosphere-network-of-skyscrapers/
[26] Gogolkina O 2020 The Use of Shells in the Design of Recreational *IOP Conf. Series: Materials Science and Engineering* **753** (2020) 022049 https://doi:10.1088/1757-899X/753/2/022049
[27] The project "City in Space" 2007 http://www.evolo.us/competition/space-scraper/
[28] Cosmonautics and sustainable development of society (concepts, problems, solutions). Sat. Articles http://www.ihst.ru/~akm/35t9.htm
[29] Saprykina N A Formation of architectural objects for extreme habitat conditions in the context of innovative paradigms *IOP Conference Series: Materials Science and Engineering* 2019 Vol 675 p 012017 (1-10). DOI: 10.1088/1757-899X/675/1/012017
[30] Houses of the Future: Examples of Futuristic Architecture http://polezreniya.ru/vne-kategorii/doma-budushhego-primery-futuristicheskoj-arxitektury/

The method of aviation systems diagnostics according to the admissible level of non-failure operation probability

N I Sidnyaev[1], **Iu I Butenko**[1], **E E Bolotova**[1]

[1] Bauman Moscow State Technical University, 5/1, 2nd Baumanskaya str., Moscow, 105005, Russia

E-mail: Sidnyaev@yandex.ru

Abstract. The paper presents the methods of forecasting reliability of the created devices, design of devices with the given reliability and comparison of reliability of various aviation systems. Calculated relations at various types of redundancy in tests, prolongation of service life of blocks of aircraft, the basis of which is methodology of estimation of residual resource are resulted. The existing methods of reliability analysis are systematized and studied. The problems of uncertainty of information on input data at calculations by classical methods are stated. It is postulated that the principle of comparability of results of quantitative estimations of technical systems assumes the use of uniform composition of initial data for a certain reporting period of time, unity of generalized and private indicators used as well as unity of assumptions and limitations accepted. The application of general and elementary formulation of the state estimation of complex mechanisms with the use of diagnostic matrix and recognition theory to determine the state of technical objects is shown. Stages of the general process of object diagnostics are described, which consist in determination of parameter changes at influence on object elements and further processing of the obtained results using synthesis and analysis methods.

1. Introduction

Assessment of the technical condition of aircraft complexes (AC) is an important task for the enterprise-developer (manufacturer) in the analysis and evaluation of the technical condition of aircraft systems (AS) and its components development and release reports on the technical condition of the AK at the stages of production and operation.

The existing methods establish the basic principles and methods of analysis, a unified system of indicators of the technical condition of the AC, as well as the procedure for analysis and evaluation of the technical condition, the formation of conclusions and recommendations to improve the existing indicators of the technical condition of the AC and their components. The AC product includes: system, equipment, unit, device, unit, unit, electronic component base product, component element, software product, which are part of the aircraft or any of its structure. The set of properties, characterizing its ability to provide in the process of functioning, obtaining the output effect specified in the Technical Specifications under the specified conditions and operating modes. The aircraft failure is interpreted as an event, which consists in the fact that in the process of aircraft functioning the specified level of output effect is not achieved. The important role is played by experimental testing, which implies a set of works defined by a complex program of experimental testing and performed during tests on models, layouts, prototypes in order to verify compliance of product characteristics with the requirements of the Technical Specifications, ensuring the operation of products, determining the reserves of their performance under conditions close to real operating conditions.What is important is the technical state, which is interpreted as a set of quantitative and/or qualitative parameters, subject to change, characterizing the ability of the complex (product) to perform under certain conditions at a given moment or period of time the functions inherent in it at the level established in the regulatory documentation, taking into account the period of active existence, namely the calendar period of time from the moment of withdrawal (or acceptance into operation as part of the orbital system of the aircraft) to the moment of termination of use of the aircraft for its intended purpose. The period of active existence of an aircraft, when the limit state is reached due to degradation failures or exhaustion of consumed components of aircraft on-board systems, in the absence of design, production and operation failures. Mathematical model reproducing real properties, connections and interaction of aircraft products and external conditions in the process of AS functioning and allowing to investigate different AS properties. The main objectives of the analysis of the technical condition of the AS and its components are the assessment and determination of trends in changes over a certain period of time of the AS and its components, the timely development of recommendations and preventive measures aimed at ensuring the required level of the reliability, maintenance or improvement of this level (in serial production), as well as control of the implementation of developed measures [1-5].

In the analysis and assessment of the technical condition of the aircraftshould follow to the following basic principles of comparability of the results of quantitative assessment of the technical condition of the aircraft and its products, the completeness and reliability of information on failures and malfunctions identified during the reporting period, a systematic approach to the use of indicators of the technical systems of the aircraft and its constituent parts and main products, as well as the choice of the sequence of stages of analysis and assessment of the technical condition.

The principle of comparability of the results of quantitative assessments of the technical systems implies the use of a uniform composition of initial data for a certain reporting period of time, the unity of generalized and private indicators used, as well as the unity of the assumptions and limitations adopted. As a result, the technical condition indicators can be extended not only to different modifications, types, types of products (components) of the AS, but also to different complexes, calculated for the relevant reporting periods and should be comparable with each other. The principle of completeness and reliability of information about failures and malfunctions detected during the reporting period assumes that the initial data used for analysis and evaluation of technical systems fully and reliably characterize the events that took place and provide objective assessments of the AS technical condition. All failures and malfunctions revealed on the complex, regardless of whether they have affected the test results or not, should be taken into account, information documents should be drawn up on them (fault reports or claim acts, study acts, confirmed by study protocols). The initial data include all failures and malfunctions revealed on the finished products during the reporting period of time (according to the date of fault or failure detection) [6-8].

2. Diagnostic evaluation of aircraft systems

Diagnostics, i.e. an opinion on the technical condition of an aircraft, aims to determine its suitability for current and future use before the next maintenance. It follows from this that during the scheduled diagnosis, the statement of diagnosis contains elements of predicting the service life of the AS. To reduce operating costs, two types of diagnosis are used that differ in depth: general and elemental. Diagnosis of unit, system, aircraft in general is called general, and a detailed diagnosis, which determines the causes of performance degradation - elementary. Both general and elementary diagnostics are used to control technological processes and quality of aircraft service. According to the results of the general diagnosis, the AS is subjected to elemental diagnostics or sent for operation, and according to the results of elemental diagnostics, repair or maintenance is performed. Diagnosis of unit, system, aircraft in general is called general, and a detailed diagnosis, which determines the causes of performance degradation - elementary. Both general and elementary diagnostics are used to control technological processes and quality of aircraft service. According to the results of the general diagnosis, the AS is subjected to elemental diagnostics or sent for operation, and according to the results of elemental diagnostics, repair or maintenance is performed. In addition, the final diagnosis is used to control aircraft maintenance and repair. Diagnosing the state of relatively simple mechanisms, when you have to use one diagnostic parameter, does not meet any particular methodological difficulties. It practically comes down to measuring the value of the diagnostic parameter compared to the norm [9-11].

In this case, for mechanisms with a known continuous (increasing) pattern of change in the technical state can be three diagnosis options: $S > S_n; S_y < S_n; S < S_y$. The first option requires repair to correct the fault, the second one requires preventive maintenance due to insufficient resource, and the third one does not require restoration until the next scheduled diagnosis. For aircraft diagnosed with discrete diagnostic parameters, when it is not expedient to trace individual changes in the implementation of the technical condition, there are only two options for diagnosis: corrected and defective, i.e. $S > S_m; S < S_m$. The first option requires maintenance, while the second one does not require maintenance - the object is in good order. In this case, failure-free operation in the upcoming flight is guaranteed by early tightening of the diagnostic standardS_m.

Setting the elemental diagnosis of complex AS, when you have to use several diagnostic parameters, is significantly complicated.

The point is that each diagnostic parameter can be associated with a number of structural parameters. Therefore, a certain value of each of them may indicate a malfunction of the diagnostic object. It means that if the number of diagnostic parameters used is n, the possible number of technical states of the mechanism being diagnosed will be 2^n.

Theoretically, the diagnosis is to select one of the many possible states of the block to be diagnosed. Therefore, the task of diagnosing many diagnostic parameters is to reveal multiple links between structural parameters $X_1, X_2, ..., X_n$ and the corresponding diagnostic parameters $S_1, S_2, ...S_m$. For this purpose, diagnostic matrices are most commonly used in the practice of spacecraft components diagnostics.

The Diagnostic Matrix (Table 1) is a two-digit logical model describing the links between structural and diagnostic parameters that have reached an acceptable value. The horizontal rows of the matrix correspond to the applied diagnostic parameters. When a diagnostic parameter reaches a normative value, one or more malfunctions of the diagnostic object may occur. Vertical rows correspond to the malfunctions of the object, i.e. its structural parameters, which have reached the allowable value. A unit at the intersection of the horizontal and vertical rows means that a fault may exist. Zero indicates that there is no such possibility.

Table 1. Matrix model of links between diagnostic and structural parameters.

Normative values of structural parameters (faults)						
Normative values of diagnostic parameters		X_1	X_2	X_3	X_4	X_5
	S_1	0	0	1	0	0
	S_2	1	1	0	1	1
	S_3	0	0	0	0	1
	S_4	0	1	1	0	1
	S_5	0	1	0	0	0

When compiling the matrix, they strive to apply the minimum number of diagnostic parameters. Only the most sensitive and informative parameters are used for this purpose. At the same time, the number of diagnostic parameters should be sufficient to obtain a reliable unambiguous diagnosis. The analytical relationship between structural and diagnostic parameters can be written in general terms by the following equations:

$$\begin{cases} X_1 = f_1(S_1, S_2, ..., S_m); \\ X_2 = f_1(S_1, S_2, ..., S_m); \\ .... \\ .... \\ X_n = f_1(S_1, S_2, ..., S_m). \end{cases}$$

Here, the system of n equations (where n is the number of structural parameters X, each of which can be associated with m measured diagnostic parameters) describes all possible states of a diagnostic object expressed by diagnostic parameters S. For practical compilation of these equations, it is necessary to know the list of characteristic reactor malfunctions to be detected and the structural and consequence diagram of the corresponding diagnostic mechanism. The list of characteristic malfunctions of the mechanism is made on the basis of statistical indicators of its reliability [12-15].

Using such a scheme, based on the engineering study of the object of diagnosis, it is possible to establish the initial list of diagnostic parameters and relationships between the two. However, the logical equations compiled in this way, each of which represents a horizontal line of the matrix, may be insufficiently informative, because the structural and consequence diagrams do not contain a quantitative assessment of diagnostic parameters. In order to evaluate the significance of diagnostic parameters, one should analyze their qualities: sensitivity, informativity by probabilities of their occurrence in the presence of a given malfunction or by probabilistic links with the main parameters of the diagnostic object performance. At relatively small probabilities, the parameter is considered insignificant and equals zero. Otherwise, it is equated to one. The diagnostic parameters selected in this way allow filling the corresponding matrix cells with zeros and units.

Such a matrix makes it possible to solve the problem of localization of the diagnosed mechanism failure by the presence of the corresponding set of diagnostic parameters that have reached the normative value.

The physical essence of the solution to this problem is to exclude malfunctions incompatible with the existence of a certain combination of measured diagnostic parameters. The process of fault detection can be considered as a reduction of entropy (the degree of uncertainty of the technical state of the mechanism being diagnosed) by means of sequential introduction of information doses into the diagnostic matrix which are incommensurate with diagnostic parameters that have reached the normative value. For practical formulation of the diagnosis, the matrix is performed as an electronic device to which electrical signals corresponding to the measured diagnostic parameters are fed. With sufficient force, these signals pass through a threshold device (trigger) into its electrical circuit. Along the circuit, the signal is sent to the matching element of the control lamp X_1,X_2, etc., corresponding to the fault in which this diagnostic parameter may exist. If the lamp does not light up, it means that the dose of information obtained from this diagnostic parameter is insufficient to localize the existing malfunction and that additional information from other diagnostic parameters is required to make the diagnosis. For example, the fault signal X_1 will only be received after entering the signals from the two parameters S_1 and S_4, fault signal X_2 - after entering S_1 and S_3, etc. Consequently, the functioning algorithm of this logic device is$X_1 = S_2S_4$;$X_2 = S_1S_3$; $S_3 = S_1S_2S_4$; $X_4 = S_2S_3$; $X_5 = S_1S_3S_4$.

Such devices allow localizing faults and displaying information via appropriate channels to the central control station or to the operator's display. If it is necessary to enter corrected diagnostic indicators, the threshold devices are adjusted. Availability of switches makes it possible to adjust the diagnostic matrix to other diagnostic parameters. The logical matrix of the specified type can be the basis for synthesis of logical automata to determine the technical state of the diagnostic object by automatic fault location. In cases when fault localization by means of diagnostic matrix is difficult, for example, due to a small number of measured diagnostic parameters and strong intersection of state classes, theory and technical means of image recognition can be used for making a diagnosis.

The essence of making a diagnosis using image recognition theory is as follows. Complexes of implementations of diagnostic parameters for any mechanism (i.e., their values fixed in any coordinate system) are some sets with characteristic properties. These properties are manifested in the fact that familiarization of people or a machine (recognizing device) with the realizations of these sets makes it possible to recognize further belonging of a given realization to any of these sets. The sets of this type are called images.

A totality of different technical states of a diagnostic object is denoted in the case of the so-called "alphabet" of failures: $A=A_0,A_1,A_2,...,A_j,...,A_M$, where A_0 is a state of a technically sound object; A_j is an object state class expressed by the set of implementations of a particular failure; M is a number of classes of possible object states.

The task of fault detection is to attribute the implementation of the diagnostic parameter, which expresses the state of the mechanism to be diagnosed, to the appropriate classA_j. In this case, the implementations with common properties belong to one class.

If you measure diagnostic parameters of several identical, functional mechanisms, the parameter implementations obtained during diagnostics form the corresponding area. In doing so, they will be grouped relatively to some distribution center called A_0 reference. Occurrence of any particular malfunction in the same mechanisms will cause shift of the implementation region and its A_{1r}reference to another place. The presence of another fault will again change the location of the implementation area of diagnostic parameters and its reference A_{0r}, etc. If the location of the mechanism's characteristic fault areas is known, an unknown fault can be determined knowing the implementation obtained during diagnostics by its correspondence to a certain class of the object's states.

Therefore, the recognition of the technical state of this mechanism can be understood as establishing the fact that the A_{jx} image is similar to the reference A_{jr}image, considering that the degree of similarity is determined by the distance between the implementation of A_{jx} and the reference A_{jr}. There are two ways of detecting malfunctions: by including the resulting implementation (image) inside the malfunctioning state area and by the distance between its distribution centerA_{jx} and the reference image A_{jr}. In the first case, the material on which the resulting implementation (called a tutorial sequence) is formed is expressed by the boundaries between the areas of characteristic faults. In the second case, recognition is reduced to determining the distance between A_{jx} and A_{jr}:

$$\rho(A_{jx}, A_{jr}) = \sqrt{\sum_{K=1}^{N} (X_{kx} - X_{kr})^2},$$

where A_{kx} and A_{kr} are the values of diagnostic parameters (implementations) presented for recognition; N is the number of diagnostic parameters characterizing the fault alphabet.
By presenting the values of diagnostic parameters, for example, in the coordinates of y - wear time, x - movement of the device, you can get an area of serviceable states, similar to area 0, and area 1-5 typical aircraft malfunctions. Similar faults can be detected with relatively inexpensive electronic devices.

The described method of applying image recognition theory to AS diagnosis can be used for other mechanisms as well. It makes it possible to determine the technical condition of an aircraft directly during operation.

3. Generaldiagnosticprocess

The general process of diagnosing the technical state of the aircraft includes: test action on the object, measurement of diagnostic parameters, processing of the received information and making the diagnosis according to the specified standard.

Test impact on the object of diagnosis is carried out either during the operation of the aircraft itself at the given load, speed, thermal conditions, or using appropriate devices (stands with loading devices, rolling and portable devices, etc.). Test actions shall provide maximum information about the technical condition of the aircraft at optimal labor and material costs.

Diagnostic parameters are measured with sensors. The types and types of sensors correspond to the physical nature of the diagnostic parameters. When measuring vibration parameters piezo-sensors are used, in thermal diagnostics - thermometers, in electrical measurements - induction current collectors, etc. are used. A distinction is made between easily removable and built-in sensors. The former are installed on the object for the time of diagnosis (magnetic, hinged, on the clamps, threaded, etc.), and the latter are standard accessories of the aircraft. Built-in sensors can be connected to control devices for permanent monitoring or to centralized plug connectors.

When using multichannel narrow information, several different types of sensors can be used, and when using wide information, one sensor can be used. On the way to the measuring instrument, the information obtained from the sensors is processed by the corresponding images. Processing consists in amplification of the received signal, removal of interferences, analysis and filtering of the signal by size and phase.

Diagnosis setting consists in comparing the obtained one or more processed diagnostic parameters with the specified standards. In the simplest case, when a single diagnostic parameter is used, the diagnostic procedure ends there.

Excess of $S_y$means the need for technical impact of the set volume, and the absence of excess - the ability to operate until the next control.

When used for the diagnosis of a large number of parameters are used methods of synthesis and analysis of the processing of captured information. The process of diagnostics by synthesis of differentiated information obtained from several sensors and by analysis of generalized information recorded by one sensor are shown in the circuit diagram.

The essence of the process of diagnosing a complex object by synthesizing differentiated information is as follows. The mechanism to be diagnosed is usually subjected to test action. The sensors $D_1,D_2,\dots, D_n$, fix the values of structural parameters $X_1,X_2,\dots, X_n$each of the elements of the object $E_1,E_2,\dots, E_n$. Values of structural parameters expressed by electric signals $S_1,S_2,\dots, S_n$ are received by the gain unit and after processing into the threshold device. The latter skips the signals exceeding the allowed values of S_y. These signals $S_1^{''}, S_2^{''},\dots, S_m^{''}$ come to the logical device (diagnostic matrix). The matrix synthesizes the information from several sensors and generates a diagnosis. The diagnosis can be refined by differentiating the required effects on specific types: warning or repair [16].

The process of diagnostics by analysis of generalized information differs from the described one by the fact that changes of structural parameters $X_1,X_2,\dots, X_n$ are fixed integrally by groups of object elements with the help of one sensor (for example, vibration parameter piezo-sensor). After amplification, the signal containing generalized information about the technical state of the object should be analyzed, highlighting the most characteristic and useful components. These components are then filtered by value using a threshold device and diagnosed. When diagnosing simple objects, the process described above is simplified by exclusion of separate stages.

4. Conclusion

One of the main factors ensuring a high level of technical condition of aviation systems, aircraft and other aerospace equipment products is the availability of a developed regulatory framework that establishes a unified approach to the order of their creation, production and operation, which is based on the latest achievements in science and technology. One of the most important elements of this regulatory framework is a set of regulatory documents that establish requirements for ensuring the necessary level of technical condition.

An indicator of the effectiveness of these activities was the fact that despite a number of accidents, including those related to the loss of aircraft and payloads, significant financial losses were avoided, while large-scale negative consequences associated with harm to personnel, the population and the environment were avoided.

At the same time there is a tendency to decrease the level of technical condition of aviation equipment products. To the main reasons of appearance of the specified tendencies it is necessary to carry: consequences of underfunding of organizations of aviation industry for the last decades which have appeared in deterioration of quality of processes of creation, manufacture and operation of products of AT and accumulation of some problems; considerable quantity of again accepted legislative and other by-laws and as consequence occurrence of the new requirements, incompletely harmonized with requirements of branch normative documents on maintenance of technical condition;

incompletely harmonized with requirements of branch normative documents on maintenance of technical condition of products of aircraft.

References

[1] Pavlov I V 1982Statistical methods for assessing the reliability of complex systems. *Moscow, RadioCommunication* p 168

[2] Sidnyaev N I 2018 A Study of the Destruction of Spacecraft Surfaces at Contact Interactions with Microparticles of the Space Environment *Cosmic Research* Vol 56 (3) pp 213–222

[3] Gecha V Ya, Barbul R N, Sidnyaev N I and Butenko Yu I 2019 Methodology for assessing the reliability of spacecraft in the design and engineering study *Nadezhnost.* vol 2 pp 3-8

[4] Gnedenko B V 1965 Mathematical methods in the theory of reliability. *Moscow, Nauka* p 524

[5] Sidnyaev N I Stolbova VA 2019 Mathematical modeling of nitrogen flow around hypersonic bodies *AIP Conference Proceedings* Vol. 2171 Art.no 130015

[6] Morozov D B and Chermoshentsev S F 2019 Technique for improving the reliability of the control system of an unmanned aerial vehicle in flight in case of a failure in on-board control equipment. *Nadezhnost.* vol 19 (1) pp30–35

[7] Sidnyaev N I Klimova N S 2020 Changes in the Surface and Volume Properties of a Spacecraft during Absorption and Recombination of Oxygen and Nitrogen Atoms *Cosmic Research* Vol 56 (3) pp213-222.

[8] Morris S F 1990 Use and application of MIL–HDBK–217. *Solid Slate Technology. Vol33 (6) pp 65–69.*

[9] Sidnyaev N I 2020 Study of heat transfer in the boundary layer in a nitrogen flow past a catalytic graphite wall *Thermophysics and Aeromechanics* Vol 27(2) pp 193–203

[10] Brennom T R 1988 Should us MIL–HDBK–217 be 8888. *IEEE Trans. Reliab.* vol 37 (5) pp 474–475

[11] Sidnyaev N I 2019 On a theorem of impulse and energy to determine parameters of the electric jet engine KA *Journal of Physics: Conference Series* Vol. 1392 Art.no 012024

[12] Barlow R 1969Mathematical theory of reliability. ed RBarlow, F Proshan. *Moscow: Soviet radio* p 488

[13] Lavrischeva E M, Zelenov S V and Pakulin N V 2019 Methods for assessing the reliability of software and technical systems. Proceedings of the Institute for System Programming. vol 31 (5) pp 95–108

[14] Sidnyaev N I 2016 Space weather factors affecting onboard elements of low-orbit spacecrafts. ed N I Sidnyaev, L A Makridenko, V Ya Gecha and V N Onufriev 2016 Electromechanics issues. VNIIEM Proceedings. Proceedings of the Fourth International Scientific and Technical Conference "Actual Problems of Creation of Space Systems for Remote Sensing of the Earth". *Moscow: JSC "Corporation" VNIIEM "* pp 90–10

[15] Pokhabov Yu P 2018 What to understand by calculating the reliability of unique highly responsible systems in relation to the mechanisms of one-time actuation of spacecraft. *Reliability* Vol 4 pp 28–35

[16] Antonov S G and Klimov S M 2017 Methodology for assessing the risks of violation of the stability of the functioning of software and hardware systems in the conditions of information and technical influences. *Nadezhnost* Vol 17 (1) pp 32–39

Theoretical background of experimental methods for studying the influence of technical conditions of shock absorbers on lateral tire reactions

D A Tikhov-Tinnikov[1], A I Fedotov[2], N Batzhargal[3]

[1] East Siberian State University of Technology and Management, 40B/1 Klyuchevskaya St., Ulan-Ude, 670013, Russia
[2] Irkutsk National Research Technical University, 83 Lermontov St., Irkutsk, 664074, Russian Federation
[3] Erdenet Institute of Technology, 1 Sh.Otgonbileg St., Erdenet, Mongolia.

E-mail: dm_tt@mail.ru

Abstract. The article studies the technical condition of shock absorbers which influences the occurrence of lateral reactions when the vehicle is passing through a single irregularity under the influence of lateral forces. The lateral reactions determine the properties of the directional and trajectory stability of the car. These properties allow us to consider lateral reactions as diagnostic parameters for the method of operational control of the technical condition of the suspension according to the stability criteria of the car. A mathematical model developed by the authors was applied. The model is based on the dynamics equations of the sprung and non-sprung masses, taking into account the technical condition of the shock absorbers. To describe the longitudinal and lateral reactions, the normalized slip function is used. The process of car movement around a circle was simulated with various parameters of the technical condition of the shock absorbers. Simulation results were processed using approximation and correlation analysis methods. The dependences of changes in the maximum value of lateral reactions on each wheel depending on the technical condition of shock absorbers were obtained. The presence of a strong correlation between the magnitude of the lateral reaction and the technical condition of its own shock absorber is established. Changing the technical condition of improper shock absorbers to a lesser extent affects the lateral reaction, but at the same time reduces the quality of diagnosis. Foundations of the method for conducting a road experiment were developed.

1. Introduction

Course stability and trajectory stability are one of the main properties ensuring safety of vehicles. It is well known that the technical condition of the suspension system has a significant impact on vehicle stability. At the same time, its technical condition is not assessed by stability properties. One of the reasons is the lack of scientific knowledge about the impact of changes in technical conditions of the suspension system on vehicle stability.

There is a road method for monitoring the technical condition of shock absorbers by analyzing kinematic parameters of movement, in particular the angle of vehicle rotation relative to the vertical axis [1]. The method is used for functional diagnostics of the damping system along the vehicle axes. The method cannot be used for the elemental determination of technical conditions due to the impossibility of identifying a diagnostic parameter characterizing each shock absorber. The differential diagnosis deals with lateral reactions that occur when the vehicle moves along the curved path. They are force parameters of interaction of an elastic tire with a road surface. Values of lateral reactions depend on many factors: a normal load on the wheel, a lateral adhesion coefficient, an angle of withdrawal, and a wheel rolling mode. The nature of lateral reactions is well known for stationary modes of suspension [2], when shock absorbers do not work, and lateral reactions depend on characteristics of the elastic suspension elements rather than on technical conditions of the dampers. Thus, in order to assess the influence of technical conditions of the suspension system on stability properties, it is necessary to study the vehicle movement under unsteady suspension operation. The study can be analytical and experimental. Road experiments are costly and require planning and preparation. The study was carried out using mathematical simulation methods. The result is a theoretical justification of the road experiment

2. Materials and methods

We applied a mathematical model used to describe a M1 vehicle moving along the circle with a radius of 15 meters and passing through a single square roughness with a side of 5 centimeters. This configuration forms scheme 1 of the road experiment simulation (Fig. 1, a). The mathematical apparatus includes [3] equations of dynamic equilibrium of the sprung mass for six degrees of freedom; equations of dynamic equilibrium of unsprung masses for one degree of freedom; calculation of longitudinal and lateral reactions using the normalized slip function [4] and taking into account a smoothing ability of the tire and an unsteady suspension operation mode [5]. The road surface was a surface with a zero vertical coordinate, except for the location of a single irregularity, where the coordinate was determined by the disturbance height.

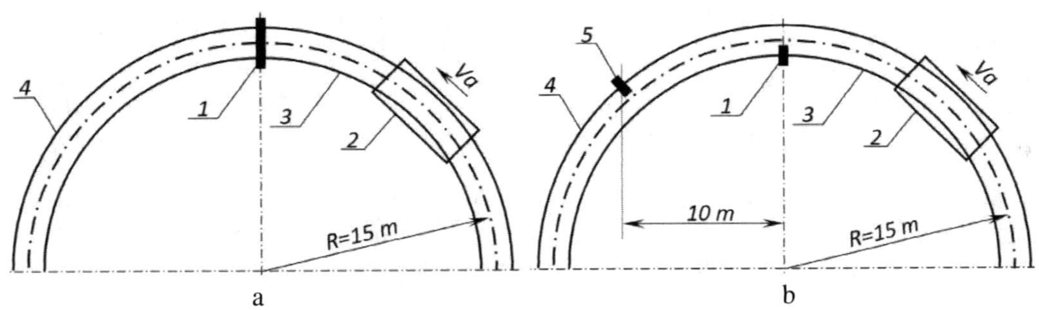

Figure 1. Road experiment schemes: a – a single irregularity is located in one section of the road (scheme 1); b - two single irregularities are located along the sides of a vehicle (scheme 2); *1*, *5* - single irregularities; *2* - tested vehicle; *3* - the middle line of a path of the inner wheels; *4* - the middle line of a path of the external wheels.

The program code implemented in the SciLab computer environment provides for calculations using the Euler method of numerical integration with a step of 0.001 seconds. There is a satisfactory convergence with results of other road experiments [6], which allows the use of their mathematical apparatus.

The program allows the calculation of vehicle movement parameters with an independent disturbance of the suspension of internal and external wheels. The model contains coordinates of single irregularities displaced along the tracks. Such a configuration of parameters determines scheme 2 of the road experiment simulation (Fig. 1, b).

The efforts of shock absorbers, which are part of the dynamic equilibrium equations, are empirical dependencies on the speed of rod movement. When changing the technical condition, the force is calculated as a product of the function of a working shock absorber by a correction factor *A*, which characterizes a decrease in damping properties, and varies from 0 to 1.0 corresponds to a shock absorber which has no damping properties; 1 does not change the force function. The values of coefficient *A* were

0.1, 0.2, 0.3, 0.55, 0.8, 1.0 for the front shock absorbers; and 0.1, 0.25, 0.4, 0.6, 0.8, 1.0 for the rear ones. The values were set in four cycles, which allowed us to perform calculations for 1296 combinations of conditions of the shock absorber.

3. Results

The dependence of lateral reactions on time was obtained for various values of coefficient *A*. Figure 2 shows the time dependence of lateral reaction R_y of the front inner wheel according to scheme 1, when the vehicle is travelling at 10 m/s. The free run starts one second before the wheel passes through the irregularity. The dependences of changes in lateral reactions of the front external and rear wheels, as well as the simulation results according to scheme 2 are similar and not presented graphically. In Figure 2, the solid line denotes lateral reactions for the sound shock absorbers, the dotted line denotes lateral reactions for the defective ones.

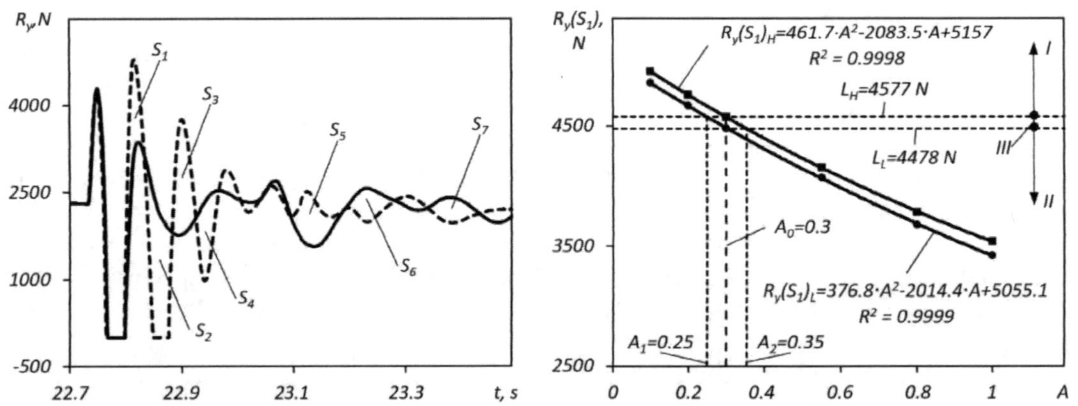

Figure 2. Characteristics of lateral reaction R_y of the front inner wheel: a - areas of local diagnoses (S_1-S_7); b – boundaries of the diagnostic parameter $R_y(S_1)$.

The simulation shows that a decrease in damping properties of shock absorbers changes temporal characteristics of lateral reactions. The changes form areas S_1–S_4, when the front wheels pass though the irregularities and areas S_5–S_7 - when the rear wheels pass through the irregularities.

The simulation results were processed. For S_1, dependences (Fig. 2, b) that describe the boundaries of changes in the maximum value of lateral reaction $R_y(S_1)$ depending on coefficient *A* were obtained. The upper $R_y(S_1)_H$ and lower $R_y(S_1)_L$ boundary levels were approximated by a rational second-order function:

$$R_{yij}(S_1)_{H,L}=a(k)\cdot A^2+b(k)\cdot A+c(k), \quad (1)$$

where $a(k)$, $b(k)$ are coefficients determining the degree of dependence $R_y(S_1)_H$ and $R_y(S_1)_L$ on A; $c(k)$ is the function value at $A=0$; k is the index corresponding to the road experiment design number; i, j is the wheel index ($i=1$ is the front axle, $i=2$ is the rear axle, $j=1$ – right, $j=2$ – left). The numerical values of the parameters and determination coefficients $R^2(k)$ are presented in Table 1.

Table 1. Function Parameters $R_{yij}(S_1)_H$ and $R_{yij}(S_1)_L$

$R_{yij}(S1)$	$a(1)$	$b(1)$	$c(1)$	$R^2(1)$	$a(2)$	$b(2)$	$c(2)$	$R^2(2)$
$R_{y11}(S_1)_H$	462	-2084	5157	0.999	609	-2456	5469	0.999
$R_{y11}(S_1)_L$	377	-2014	5055	0.999	741	-2614	5376	0.999
$R_{y12}(S_1)_H$	215	-1697	4947	0.999	419	-2130	6140	0.999

$R_{y12}(S_1)_L$	207	-1570	4598	0.999	274	-1900	5937	0.999
$R_{y21}(S_1)_H$	1902	-3176	5057	0.997	2837	-4462	5718	0.999
$R_{y21}(S_1)_L$	1568	-2819	4224	0.998	2238	-3864	5244	0.999
$R_{y22}(S_1)_H$	185	-205	5151	0.754	2581	-4358	6178	0.999
$R_{y22}(S_1)_L$	-94	311	3445	0.999	2265	-4137	5937	0.999

A correlation analysis was performed for maximum values of lateral reactions $R_{yij}(S_1)$ and absorber damping reduction coefficients $A_{ij}(k)$ (Table 2).

Table 2. Correlation coefficients between $R_{yij}(S_1)$ and $A_{ij}(k)$

$R_{yij}(S_1)$	$A_{11}(1)$	$A_{12}(1)$	$A_{21}(1)$	$A_{22}(1)$	$A_{11}(2)$	$A_{12}(2)$	$A_{21}(2)$	$A_{22}(2)$
$R_{y11}(S_1)$	-0.997	-0.004	0.005	0.042	-0.994	0.001	0.004	0.065
$R_{y12}(S_1)$	0.147	-0.986	0.048	-0.019	0.019	-0.997	0.045	-0.013
$R_{y21}(S_1)$	-0.09	-0.098	-0.86	0.363	0.009	-0.126	-0.882	0.043
$R_{y22}(S_1)$	-0.03	-0.09	0.019	-0.958	-0.04	-0.042	-0.035	-0.928

The results can be used for further studies.

4. Discussion and conclusion

Previous studies [7] show that lateral reactions R_y in S_1–S_3 (Fig. 2, a) can be used as diagnostic parameters. These areas are characterized by the greatest level of sensitivity of the parameter to changes in shock absorbers. The nature of the process is unique. Given the similarity of reaction properties, further studies were carried out in S_1. Moreover, the maximum value of lateral reaction $R_{yij}(S_1)$ was used as a diagnostic parameter.

The correlation analysis (Table 2) shows that the maximum value of a lateral reaction depends on the technical condition of a wheel's shock absorber. The effect of shock absorbers of other wheels on the lateral reaction is insignificant. However, they change the lateral reaction even under the constant technical condition of own shock absorbers. This circumstance does not allow for the unambiguous identification of technical conditions of shock absorbers by the value of the lateral reaction. The higher the ambiguity of a diagnosis, the greater the value of the lateral reaction due to the influence of other shock absorbers on technical conditions. Figure 2 shows the line with square markers which indicate the upper boundary of diagnostic parameter $R_y(S_1)$, and the line with circle markers which indicates the lower border. According to [1], the front shock absorbers are serviceable with a damping force of at least 30 % ($A_0 = 0.3$) of the nominal level. For rear shock absorbers, this indicator is 40 % ($A_0 = 0.4$). The front inner wheel (Fig. 2, b) with $A_0 = 0.3$ has a zone of uncertainty of diagnosis *III* between L_L=4478 N and L_H=4575 N. If the maximum lateral reaction falls within this range, the own shock absorber can be in both good and bad conditions. If the reaction is in zone *I*, at $R_y(S_1) > L_H$, the absorber is defective; in zone *II* at $Ry(S1) < L_L$, it is not defective.

Let us determine the range of coefficient A determined the influence of technical conditions of other shock absorbers on the diagnostic parameter. This assessment can be graphical and analytical. Using the graphical method, it is necessary to build a perpendicular to the abscissa axis from the intersection points (boundary lines $R_y(S_1)_H$ and $R_y(S_1)_L$ and lines L_L and L_H (Fig. 2, b). The range of coefficient A is determined by values A_1=0.25 and A_2=0.35. The analytical method is used to solve quadratic equation (1) with $R_y(S_1)_H=L_L$ and $R_y(S_1)_L=L_H$. The roots of the equations and calculation parameters for the first and second experimental simulation schemes are presented in Table 3.

Table 3. Parameters and roots of quadratic equations

$R_{yij}(S_1)$	A_0	$L_H(1)$, H	$L_L(1)$, H	$A_1(1)$	$A_2(1)$	$L_H(2)$, H	$L_L(2)$, H	$A_1(2)$	$A_2(2)$
$R_{y11}(S_1)$	0.3	4577	4478	0.25	0.35	4796	4670	0.24	0.36
$R_{y12}(S_1)$	0.3	4460	4150	0.09	0.5	5540	5394	0.22	0.38
$R_{y21}(S_1)$	0.4	4063	3464	0.1	19.47	4370	4040	0.27	0.62
$R_{y22}(S_1)$	0.4	5083	3553	-	17.19	4836	4630	0.32	0.51

An analysis of the range of coefficient A as a difference between $A_2(k)$ and $A_1(k)$, allowed us to compare the diagnostics modes. The larger the range, the higher the probability of first and second order errors. Simulation according to scheme 1 indicates the impossibility of diagnosing the rear external wheel. In this case, the root of quadratic equation $A_1(1)$ is a complex number, the second root is $A_2(1)=17.19$ (Table 3), which falls outside the range of acceptable values of coefficient A. Thus, the zone of diagnosis uncertainty covers the range of the coefficient of decreasing damping properties of the shock absorber (Fig. 3, a).

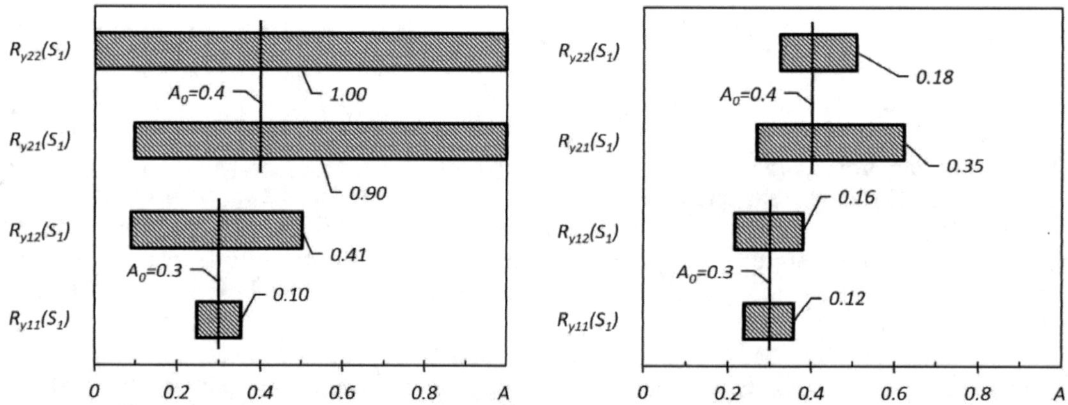

a b

Figure 3. The ranges for coefficient A: a - according to scheme 1; b - according to scheme 2.

Diagnostic parameter $R_{y21}(S_1)$ of the rear inner wheel is of little use. The boundaries of the uncertainty zone are from 0.1 to 1.0. Unambiguously, it is possible to identify the condition of only those shock absorbers whose damping force does not exceed 10 % of the nominal value.

The shock absorbers of the front wheels have lower uncertainty, for the external wheel, the range of coefficient A is 0.41; for the inner one, it is 0.1.

According to scheme 2, the simulation of movement is characterized by a smaller variation of the diagnostic parameter (Fig. 3, b). Uncertainty zones are reduced. For the rear internal and external shock absorbers, they are 0.35 and 0.18, respectively. Diagnosis uncertainty for the front external shock absorber is 0.16. A slight increase in uncertainty for the front internal shock absorber is due to the simulation of movement at a constant speed (without rolling).

According to scheme 2, smaller values of the uncertainty diagnosis zone make it more preferable to use as a basis for developing a power method for diagnosing the suspension by the value of lateral reactions when the vehicle moves along a curved path passing through single irregularities.

Uncertainty can be removed by including additional operations of in-depth studies of shock absorbers, for example, using specialized stands [8]. In this case, shock absorbers will be divided into three groups. The first group will include shock absorbers with diagnostic parameter $R_{yij}(S_1)$ which is less than L_L. These shock absorbers are workable. The second group includes shock absorbers whose diagnostic parameter $R_{yij}(S_1)$ falls into the uncertainty zone in the range from L_L to L_H. For such shock absorbers, additional diagnostics is required. The third group includes defective shock absorbers with a diagnostic parameter whose value is higher than L_H.

Another approach is to determine the maximum permissible value of the diagnostic parameter which will be lower than L_L. Shock absorbers of the second group are decommissioned without additional diagnostics. The effectiveness of the approaches is evaluated using the economic and probabilistic method for determining standards of technical operation of vehicles [9].

Despite the purely theoretical nature of the results, they allow us to formulate the basics of the method for conducting a road experiment. At the initial stage, scheme 2 can be simplified as an exception from experiment 5 with a single irregularity (Fig. 1, b), and the car travels clockwise and counterclockwise. In this case, the internal wheels will interact with a single irregularity. By shifting the trajectory to the center of rotation, the external wheels are tested clockwise (right ones) and counterclockwise (left ones). The data obtained at the initial stage make it possible to verify the theoretical results and serve as the basis for planning and conducting road experiments according to scheme 2.

Taking into account that S_1–S_4 are formed for less than 0.1 second, and the vehicle speed does not exceed 10 m/s, the lateral reaction should be measured on a road section of no longer than one meter. To fix the parameters, it is advisable to use road measuring plates equipped with strain gauges for measuring lateral reactions [10]. By changing the distance from a single irregularity to the slab, it is possible to determine the lateral reaction for any area of local diagnoses.

The theoretical studies allow us to draw the following conclusions. There is a strong correlation of the lateral reaction and the technical condition of the own shock absorber under unsteady suspension operation. The technical condition of other shock absorbers affects the lateral reaction to a much lesser extent. However, the presence of this effect makes it difficult to use a lateral reaction as a diagnostic parameter due to the diagnosis uncertainty. The uncertainty value depends on the scheme of single irregularities which should be taken into account when conducting road experiments.

References

[1] Lysenko A V 2019 *Road Method for Monitoring the Technical Condition of Shock Absorbers under Operating Conditions* (Irkutsk dis. cand. tech. sci.)

[2] Hachaturov A A, Afanas'ev V L, Vasil'ev V S, Gol'din G V, Dodonov B M, Zhigarev V P, Kol'cov V I., Jurik V S. and Jakovlev E I 1976 *Dynamics of a System Road - Tire - Car – Driver* (Moscow: Mashinostroenie)

[3] Fedotov A I, Tikhov-Tinnikov D A, Ovchinnikova N I, Lysenko A V 2017 *Earth Environ. Sci.* **87** 082018

[4] Dik A B 1988 *Calculation of Stationary and Non-Stationary Characteristics of the Brake Wheel when Driving with a Slip Angle* (Omsk dis. cand. tech. sci.)

[5] Kuznecov N Ju 2019 *Monitoring the Technical Condition of Automobile Shock Absorbers Based on Tire Grip Characteristics* (Irkutsk dis. cand. tech. sci.)

[6] Kuznecov N Ju, Lysenko A V and Zedgenizov V G. *Proceedings of Irkutsk State Technical University* **130(11)** 191–198

[7] Tikhov-Tinnikov D A, Fedotov A I, Alekseev A V *International Journal of Advanced Studies* **10(1)** 81–93

[8] Ryabov I M, Pozdeev A V, Erontaev V V, Danilov S V, Omarova Z K, Kovalev A M and Silchenkov D D 2019 *Mater. Sci. Eng.* **632** 012060

[9] Kuznecov E S, Voronov V P, Boldin A P, Lukin V P, Chernenko V A, Vlasov V M, Frolov Ju N, Kramarenko G V, Janchevskij V A and Pankratov N P *Technical Maintenance of Cars: Textbook for High Schools* (Moscow: Transport)

[10] Golomidov A M 1986 *Operational Properties of a Car with Front-Wheel Drive* (Moscow: Mashinostroenie)

Guidance dynamics of the optical load of a multielement manipulator

N E Tochilkin[1], A I Kimyaev[1] and V V Sheval[1]

[1] Moscow Aviation Institute, 4, Volokolamskoe shosse, Moscow, 125993, Russia

Abstract. The dynamic equations of a flat motion of the angular-type multielement manipulator are considered; the manipulator has the payload as an image formation device of a mobile observed object. Dynamics constraints of a manipulator-driven load via stepper motors are investigated. Nonlinear control algorithms are suggested for the tracking of an observed object via feedback actuators associations operating as one multicoupling control feedback actuator.

tochilkin.77@mail.ru

1. Introduction

Tasks of the mobile objects tracking by the visual fields of the carrier-mounted optical devices for an image formation are appeared in many fields of technology. Main quality operation conditions of the control systems (CS) of the observation complexes (OC) with movable carrier are providing constant location of an observed object (OO) in the visual field (VF) of the optical device and clearness of a received image. In control terms, the conditions define the strict requirements to quality indexes (QI) of CSoOCs, such as speed of response, dynamic accuracy in quasi-steady-state modes, and smoothness in the responding to the creep speed. Nontriviality of the assurance of mentioned QIs is that the assurance methods of every QI contradict to ones of other QIs.

One solution of the control problem is considered in [1] for the movable carrier providing the direction of the VF of the onboard device for an image formation. In [1], error signal in the closed-loop system of the mobile OO tracking is automatically generated according to the results of an adjacent frames analysis.

Flat two-element manipulators are used as a mechanical suspension for receiving more informative foreshortenings of the location of the onboard device for an image formation [2], the requirements being the same for the QIs of the VF direction control as for above-mentioned ones.

Also, such tasks of the mobile object observation are solved in the CS with multielement manipulators, in which video camera (VC) is used as an operation tool [3]. Range of tasks solved with such manipulators is wide and can include work with radioactive substances in the isolated chambers, maintenance of the numerical control machines, carrying out high precision three-dimensional measurements, etc.

In synthesizing the CSoOCs of such complex commercial-grade items (OC mounted on movable and static carriers), designer must consider their features:

- combination of relative and bulk motion velocities of the mechanical movements;
- multicoupling of the mechanical part of a controlled element;
- nonstationarity of the parameters of a controlled element;
- nonlinear character of the relations in a controlled element;
- operation of the several feedback actuators (FA) with the same load;
- presence of a delay in the error signal discriminator based on technical analysis of received images.

Choice of the actuation motors significantly influences on the possibility of assurance of the required dynamic characteristics. In many robotics problems, stepper motors (SM) are widely used having noticeable advantages in solving the steady-state control tasks with absence of the additional curve requirements. Primarily, these advantages are provided by absence of feedback in setting only initial and end placement points of a payload. However, one can occur loss of synchronization of set and current steps [4] in operation of the SM with a load; it is unacceptable in CSoOCs where important QIs are dynamic accuracy and speed of response and it consequently leads to introduction the output signals feedbacks of the local FAs. Well-known SM closed-loop control methods are steps verification and registration of the back EMF in the exciting windings [5]. Disadvantages of steps verification are lack of shaft speed control and impossibility of compensation of the steady-state error results from active external forces and torques. The last disadvantage also refers to back EMF registration.

In this paper, one of the new options of path tracking controller design is investigated. The controller provides the following requirements:

- high speed of response (due to criticality of the OC using item even to minimal images losses);
- low error of tracking for OO elements having small-scale structure even in the foreshortening change process (due to criticality of OC using item even to minimal image quality degradation).

2. Dynamic equations

In this section, mechanical subsystem of the OC based on a five-degree-of-freedom manipulator (further, controlled element (CE)) with VC as an operating tool is considered. Physical analog of the subsystem is shown in Fig. 1, where $Pr_j, j = 0, 1, 2, 3, 4$ are controlling electric actuators; $o_a x_a y_a z_a$ is the fixed (further, absolute) axes system (AS); $o_0 x_0 y_0 z_0$ is the local AS related with actuator Pr_0. Origins of $o_a x_a y_a z_a$ and $o_0 x_0 y_0 z_0$ ASs are coincided with the OO, and their vertical axes are coincided with the spin axis of the actuator Pr_0 and the local vertical. The vertical motion plane of the CE and $o_0 x_0 y_0$ plane are coincided. Payloads driving by corresponding electric actuators are marked with dotted line contours: I, II, III contours are the loads driving by actuators Pr_1, Pr_2, Pr_3, respectively; the load of the actuator Pr_0 is the whole moving part of the manipulator, which is marked with rectangular dotted line contour.

As mentioned above, the fidelity of a nonstationary attitude of the VC is especially important in the continuous tracking. Since cross couplings appeared as a result from cooperation of the local FAs [6] significantly influence to the dynamic accuracy in high precision tracking tasks, it is advisable to analyze this impact and to design the correction algorithms for the actuators if it is necessary. Significant importance in the design problems hierarchy is considering the motion of the CE in $o_0 x_0 y_0$ plane. Further, the motion CS of the OC in $o_0 x_0 y_0$ plane will be called a subsystem of the CSoOC (SCSoOC).

Physical analog of the flat representation of the CE is shown in Fig. 2, where $o_i x_i y_i, i = 1, 2, 3$ are the local ASs related with corresponding electric actuators Pr_i; φ_i, φ_i^a are the relative and absolute rotation angles of the manipulator elements, respectively; φ_d is the angular foreshortening of the optical axis of the VC; R_d is the slope distance between the VC and the OO.

We will use Lagrange equations of the second kind with relative rotation angles and angular velocities of the elements as generalized coordinates for formulation the equations of the plane motion of the CE:

$$\frac{d}{dt}\left(\frac{\partial (T_\Sigma - U_\Sigma)}{\partial \dot{q}_i}\right) = \frac{\partial (T_\Sigma - U_\Sigma)}{\partial q_i}, \tag{1}$$

where T_Σ is the kinetic energy of the CE, U_Σ is the force function, $q_i = \varphi_i$ is the relative rotation angles of the manipulator elements, $\dot{q}_i = \dot{\varphi}_i = \omega_i$ is the relative angular velocities of the manipulator elements.

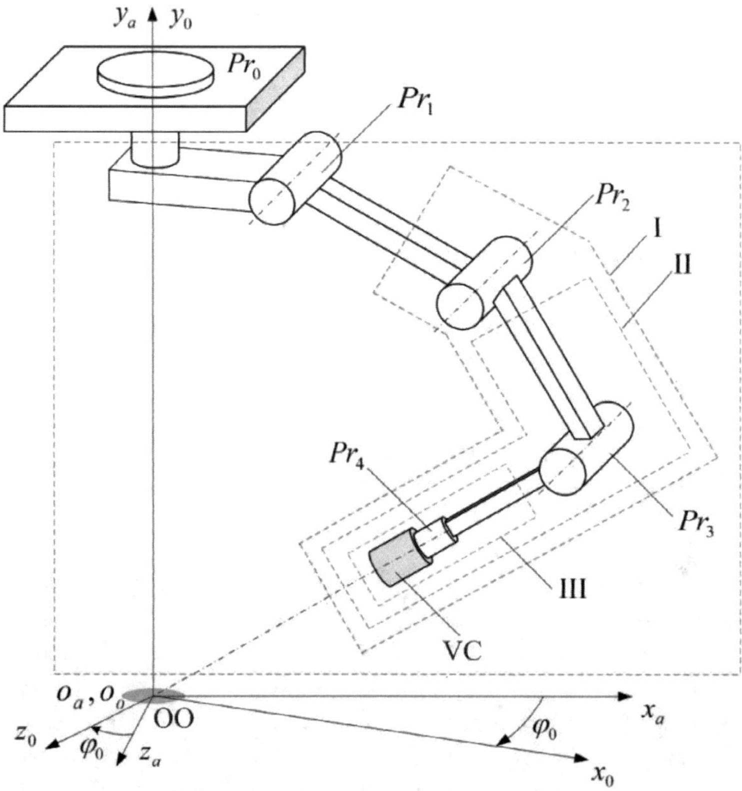

Figure 1. Physical analog of the mechanical subsystem

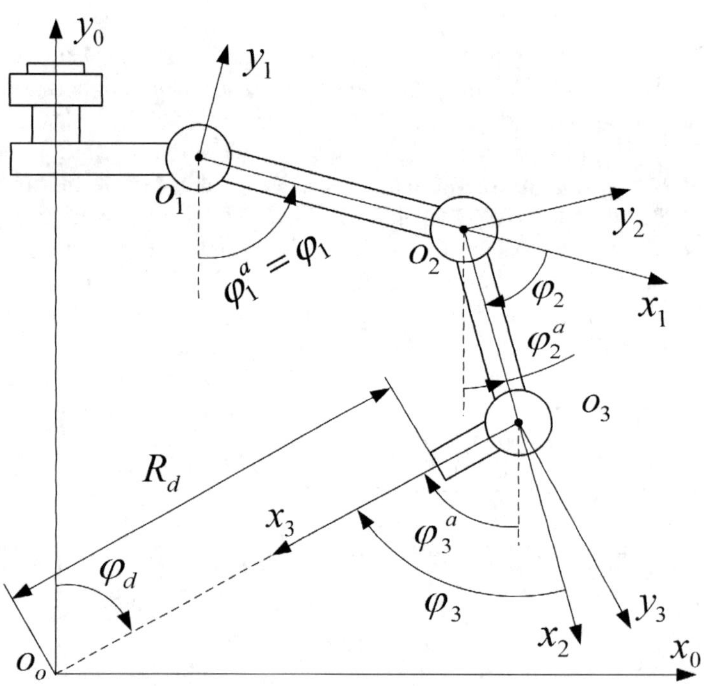

Figure 2. Physical analog of the plane mechanical suspension of the CE

3. Kinetic energy

The kinetic energy of the CE is defined as the sum of the kinetic energies of the payloads of the corresponding actuators:

$$T_{\Sigma} = T_{\Sigma 1} + T_{\Sigma 2} + T_{\Sigma 3}, \tag{2}$$

where $T_{\Sigma 1}, T_{\Sigma 2}, T_{\Sigma 3}$ are the kinetic energies of the payloads of actuators Pr_1, Pr_2, and Pr_3, respectively.

Kinetic energy $T_{\Sigma 1}$ is defined by the rotational motion of the payload of the actuator Pr_1 around the axis o_1:

$$T_{\Sigma 1} = \frac{1}{2}\boldsymbol{\Omega}_1^T J_{\Sigma 1}\, \boldsymbol{\Omega}_1, \tag{3}$$

where $\boldsymbol{\Omega}_1 = \boldsymbol{\omega}_1$ is the absolute rotational velocity of the load, $J_{\Sigma 1} = f(\varphi_2, \varphi_3)$ is the moment of inertia of the load around the axis o_1.

Load driving by actuator Pr_2 is involved into complex motion so it is necessary to consider its relative and bulk motion velocities for obtaining the kinetic energy. The velocities are shown in Fig. 3a, where $\mathrm{CM}_{\Sigma k}, k = 2,3$ are the mass centers of the payload of the actuators Pr_k; $\boldsymbol{R}_{\mathrm{CM}_{\Sigma k}}$ is the vector between the spin axis of the actuator Pr_k and $\mathrm{CM}_{\Sigma k}$; $\boldsymbol{V}_{o_k}$ is the linear velocity of AS $o_k x_k y_k$; $\mathbf{V}^r_{\mathrm{CM}_{\Sigma k}}$ is the linear velocity of $\mathrm{CM}_{\Sigma k}$ movement relative to the AS $o_k x_k y_k$; $\boldsymbol{\Omega}_k$ is the absolute rotational velocity of AS $o_k x_k y_k$ movement.

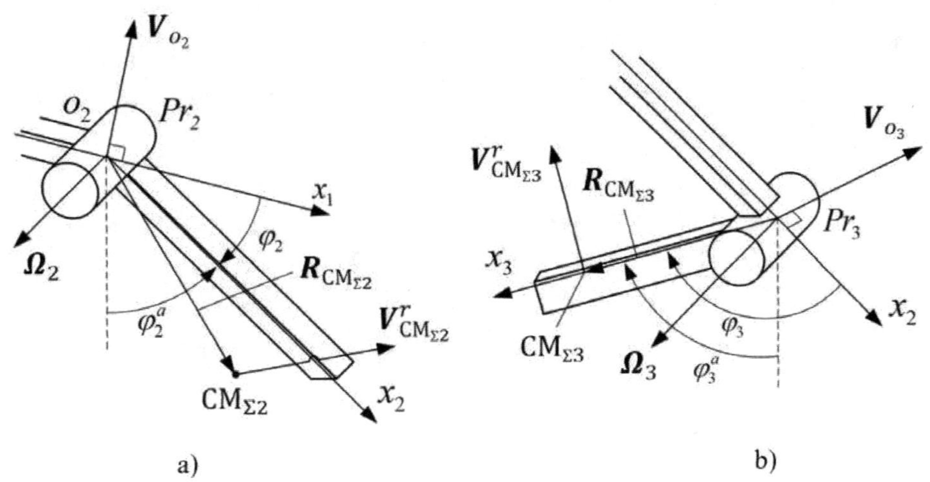

Figure 3. Velocity vectors

The kinetic energy of the payload of the actuator Pr_2 is defined by:

$$T_{\Sigma 2} = \frac{1}{2} m_{\Sigma 2}\left|\boldsymbol{V}_{o_2} + \boldsymbol{V}^r_{\mathrm{CM}_{\Sigma 2}}\right|^2 = \frac{1}{2}\boldsymbol{V}^T_{o_2} m_{\Sigma 2} \boldsymbol{V}_{o_2} + m_{\Sigma 2}\left(\boldsymbol{V}_{o_2} \cdot \boldsymbol{V}^r_{\mathrm{CM}_{\Sigma 2}}\right) + \frac{1}{2}\boldsymbol{\Omega}_2^T J_{\Sigma 2}\, \boldsymbol{\Omega}_2, \tag{4}$$

where $m_{\Sigma 2}$ is the load mass, $\boldsymbol{\Omega}_2 = \boldsymbol{\omega}_1 + \boldsymbol{\omega}_2$ is the absolute rotational velocity of AS $o_2 x_2 y_2$ movement, $J_{\Sigma 2} = f(\varphi_3)$ is the moment of inertia of the load around the axis o_2.

Analogically, the kinetic energy of the payload of the actuator Pr_3, which is also involved in complex motion (Fig. 3b), is defined by:

$$T_{\Sigma 3} = \frac{1}{2} m_{\Sigma 3} \left| \boldsymbol{V}_{o_3} + \boldsymbol{V}^r_{\mathrm{CM}_{\Sigma 3}} \right|^2 = \frac{1}{2} \boldsymbol{V}^T_{o_3} m_{\Sigma 3} \boldsymbol{V}_{o_3} + m_{\Sigma 3} \left(\boldsymbol{V}_{o_3} \cdot \boldsymbol{V}^r_{\mathrm{CM}_{\Sigma 3}} \right) + \frac{1}{2} \boldsymbol{\Omega}^T_3 J_{\Sigma 3} \boldsymbol{\Omega}_3, \quad (5)$$

where $m_{\Sigma 3}$ is the load mass, $\boldsymbol{\Omega}_3 = \boldsymbol{\omega}_1 + \boldsymbol{\omega}_2 + \boldsymbol{\omega}_3$ is the absolute rotational velocity of AS $o_3 x_3 y_3$ movement, ($J_{\Sigma 3} = const$) is the moment of inertia of the load around the axis o_3.

Force function

The force function of the CE is defined as the sum of the elementary works done by the active moments of forces driving the corresponding payloads:

$$U_\Sigma = \int dA_\Sigma = \int dA_{\Sigma 1} + \int dA_{\Sigma 2} + \int dA_{\Sigma 3}, \quad (6)$$

where $dA_{\Sigma i}$ is elementary work done by forces driving the payloads of the actuator Pr_i.

Among active moments acting to the CE, we will highlight the torques of the electric actuators of the SCSoOC and the torques produced by gravitational forces acting to payloads. Forces and torques mechanism is shown in Fig. 4, where $\boldsymbol{G}_{\Sigma i}$ is the vector of gravitational force acting in point $\mathrm{CM}_{\Sigma i}$, M_i is the torque of the actuator Pr_i.

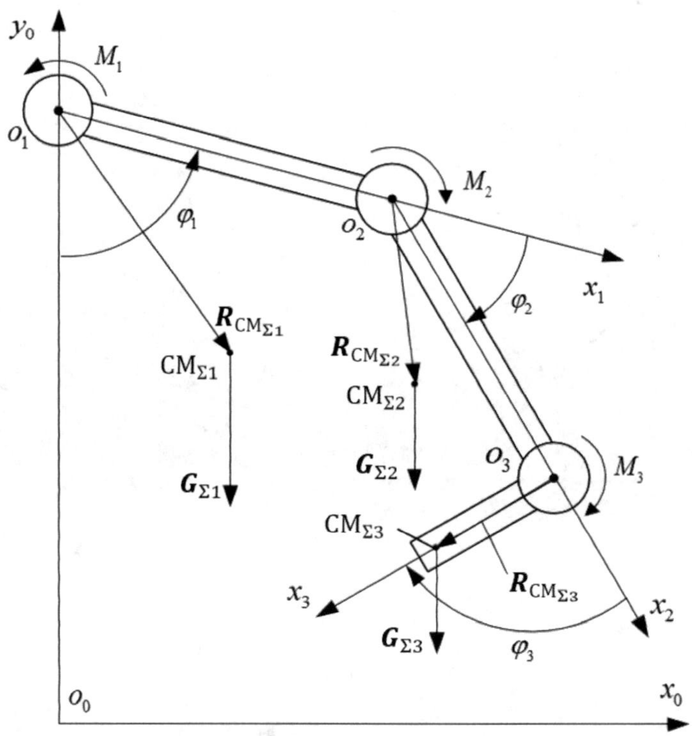

Figure 4. Forces in the subsystem

Relations for works of the active moments are formulated as:

$$dA_{\Sigma 1} = \left((\boldsymbol{M}_1 + \boldsymbol{M}_{G_{\Sigma 1}}) \cdot d\boldsymbol{\varphi}_1 \right) = \left((\boldsymbol{M}_1 + [\boldsymbol{G}_{\Sigma 1} \times \boldsymbol{R}_{\mathrm{CM}_{\Sigma 1}}]) \cdot d\boldsymbol{\varphi}_1 \right) \quad (7)$$

$$dA_{\Sigma 2} = \left((\boldsymbol{M}_2 + \boldsymbol{M}_{G_{\Sigma 2}}) \cdot d\boldsymbol{\varphi}_2 \right) = \left((\boldsymbol{M}_2 + [\boldsymbol{G}_{\Sigma 2} \times \boldsymbol{R}_{\mathrm{CM}_{\Sigma 2}}]) \cdot d\boldsymbol{\varphi}_2 \right) \quad (8)$$

$$dA_{\Sigma 3} = \left((\boldsymbol{M}_3 + \overline{\boldsymbol{M}}_{G_{\Sigma 3}}) \cdot d\boldsymbol{\varphi}_3\right) = \left((\boldsymbol{M}_3 + [\boldsymbol{G}_{\Sigma 3} \times \boldsymbol{R}_{\mathrm{CM}_{\Sigma 3}}]) \cdot d\boldsymbol{\varphi}_3\right) \tag{9}$$

Lagrange equations of the second kind

In this research, computation of the mass centers of the payloads of the SCSoOC actuators, the moments of inertia of the loads, kinetic energies, and force functions according to (2) – (9) is implemented with the Wolfram Mathematica notebook document. In the document, the Lagrangian, its derivatives with respect to generalized coordinates and velocities are also computed, and the formation of the Lagrange equations of the second kind for actuators Pr_1, Pr_2, Pr_3 of the manipulator are formed. Further, solution of these equations relative to angular accelerations $\varepsilon_i = \ddot{\varphi}_i$ of the actuators will be called the dynamic equations of the CE motion, the relation $|\boldsymbol{R}| = R$ being correct for all vectors.

Since the CE has three degrees of freedom, three dynamic equations are defined for each of them. Dynamic equation for the first degree of freedom is described by:

$$\begin{aligned}
\varepsilon_1 = \{&M_1 + g \sin\varphi_1 m_1 R_{c1} - (J_{\Sigma 2} + J_{\Sigma 3})\varepsilon_2 - J_{\Sigma 3}\varepsilon_3 + \\
&+m_2 g[\, R_1 \sin\varphi_1 + 2g\, R_{c2} \sin(\varphi_1 + \varphi_2) + \\
&+R_{o2}(R_{\mathrm{CM}_{\Sigma 2}}(\omega_1^2 \sin\varphi_1 - \varepsilon_2 \cos\varphi_1) - \\
&-(2\omega_1 + \omega_2)\dot{R}_{\mathrm{CM}_{\Sigma 2}} \cos\varphi_1)] + \\
&+m_3[g\, R_1 \sin\varphi_1 + 2g\, R_2 \sin(\varphi_1 + \varphi_2) + \\
&+3g\, R_{c3} \sin(\varphi_1 + \varphi_2 + \varphi_3) - R_{\mathrm{CM}_{\Sigma 2}} R_{o2}\varepsilon_2 \cos\varphi_1 - \\
&-R_{o3}\left(R_{o3}\varepsilon_2 + R_{\mathrm{CM}_{\Sigma 3}}(2\varepsilon_2 + \varepsilon_3)\cos(\varphi_1 + \varphi_3)\right) + \\
&+R_{\mathrm{CM}_{\Sigma 2}} R_{o2}\omega_1^2 \sin\varphi_1 + R_{\mathrm{CM}_{\Sigma 3}} R_{o3}\omega_1^2 \sin(\varphi_1 + \varphi_2) - \\
&-R_{\mathrm{CM}_{\Sigma 3}} R_{o3}(\omega_2^2 + 2\omega_1\omega_3 + \omega_2\omega_3 + \omega_3^2)\sin(\varphi_1 + \varphi_3) - \\
&-2R_{o2}\omega_1 \dot{R}_{\mathrm{CM}_{\Sigma 2}}(\omega_1 + \omega_2)\cos\varphi_1 - (2R_{o3}(\omega_1 + \omega_2) + \\
&+R_{\mathrm{CM}_{\Sigma 3}}(2(\omega_1 + \omega_2) + \omega_3)\cos(\varphi_1 + \varphi_3))\dot{R}_{o3}]\}/ \\
&/\{J_{\Sigma 1} + J_{\Sigma 2} + J_{\Sigma 3} + \\
&+(m_2 + m_3)R_{o2}\left(2R_{\mathrm{CM}_{\Sigma 2}} \cos\varphi_1 + R_{o2}\right) + \\
&+2m_3 R_{\mathrm{CM}_{\Sigma 3}} R_{o3} \cos(\varphi_1 + \varphi_3) + m_3 R_{o3}^2\},
\end{aligned} \tag{10}$$

where g is the gravitational acceleration; m_i is the mass of the element i; R_{ci} is the distance between axis o_i and mass center of the element i; R_i is the length of the element i; R_{oi} is the distance between axes o_1 and o_i.

Dynamic equation for the second degree of freedom is described by:

$$\varepsilon_2 = \{M_2 - (J_{\Sigma 2} + J_{\Sigma 3})\varepsilon_1 - J_{\Sigma 3}\varepsilon_3 + m_2(2gR_{c2}\sin(\varphi_1 + \varphi_2) + \\ +R_{o2}(R_{CM_{\Sigma 2}}\omega_1^2 \sin\varphi_1 - (R_{CM_{\Sigma 2}}\varepsilon_1 + \omega_1\dot{R}_{CM_{\Sigma 2}})\cos\varphi_1)) + \\ +m_3[g(2R_2\sin(\varphi_1 + \varphi_2) + 3R_{c3}\sin(\varphi_1 + \varphi_2 + \varphi_3)) + \\ +R_{CM_{\Sigma 2}}R_{o2}(-\varepsilon_1\cos\varphi_1 + \omega_1^2\sin\varphi_1) - R_{o2}\omega_1\dot{R}_{CM_{\Sigma 2}}\cos\varphi_1 + \\ +R_{o3}\left(-R_{o3}\varepsilon_1 + (\omega_1 + \omega_2)\left(\frac{dR_{o3}}{d\varphi_2}(\omega_1 + \omega_2) - 2\dot{R}_{o3}\right)\right) + \\ +R_{CM_{\Sigma 3}}(R_{o3}(-(2\varepsilon_1 + \varepsilon_3)\cos(\varphi_1 + \varphi_3) + \\ +(\omega_1 + \omega_3)(2(\omega_1 + \omega_2) + \omega_3)\sin(\varphi_1 + \varphi_3)) + \\ +\left(\frac{dR_{o3}}{d\varphi_2}\omega_1(\omega_1 + \omega_2 + \omega_3) - (2(\omega_1 + \omega_2))\dot{R}_{o3}\right)\cos(\varphi_1 + \varphi_3))]\}/ \\ /\{J_{\Sigma 2} + J_{\Sigma 3} + m_3R_{o3}(2R_{CM_{\Sigma 3}}\cos(\varphi_1 + \varphi_3) + R_{o3})\}, \quad (11)$$

where

$$\frac{dR_{o3}}{d\varphi_2} = -\frac{R_1R_2\sin\varphi_2}{\sqrt{R_1{}^2 + R_2{}^2 + 2R_1R_2\cos\varphi_2}} \quad (12)$$

Dynamic equation for the third degree of freedom is described by:

$$\varepsilon_3 = \frac{1}{J_{\Sigma 3}}\left\{M_3 - J_{\Sigma 3}(\varepsilon_1 + \varepsilon_2) + \frac{dR_{CM_{\Sigma 2}}}{d\varphi_3}m_2R_{o2}\omega_1(\omega_1 + \omega_2)\cos\varphi_1 + \right. \\ +m_3(3gR_{c3}\sin(\varphi_1 + \varphi_2 + \varphi_3) + \frac{dR_{CM_{\Sigma 2}}}{d\varphi_3}R_{o2}\omega_1(\omega_1 + \omega_2)\cos\varphi_1 + \\ +R_{CM_{\Sigma 3}}(-R_{o3}(\cos(\varphi_1 + \varphi_3)(\varepsilon_1 + \varepsilon_2) + \omega_2(\omega_1 + \omega_2)\sin(\varphi_1 + \varphi_3)) - \\ \left. -(\omega_1 + \omega_2)\dot{R}_{o3}\cos(\varphi_1 + \varphi_3)))\right\}, \quad (13)$$

where

$$\frac{dR_{CM_{\Sigma 2}}}{d\phi_3} = -\frac{m_3(m_3R_2 + m_2R_{c2})R_{c3}\sin\varphi_3}{(m_2 + m_3)^2R_{CM_{\Sigma 2}}} \quad (14)$$

Simulation model

In this research, structure of the OC with the manipulator, which has bipolar SM-based electric actuators with absolute rotation angles feedback via tilt sensors, is considered. In flat task, position of the VC is definitely described by its distance to OO R_d and angular foreshortening of its optical axis φ_d, and besides $\varphi_d = \varphi_3^a$ (Fig. 2). Required coordinates of the VC must be transformed to the control values of absolute angles φ_1^c, φ_2^c, and φ_3^c by the solution of inverse kinematics for three feedback actuators according to the CE gears diagram. Admitting $R_d = \text{const}$, one can study the CE dynamics in feedback actuators association (FAA) during forced motion. Generalized structure of the SCSoOC can be represented as in Fig. 5, where *SA*SA_i are the stepper actuators; θ_i are the errors of FAs; θ_g is the guidance error. Blocks f_i are the dynamic equations obtained from (10), (11), and (13).

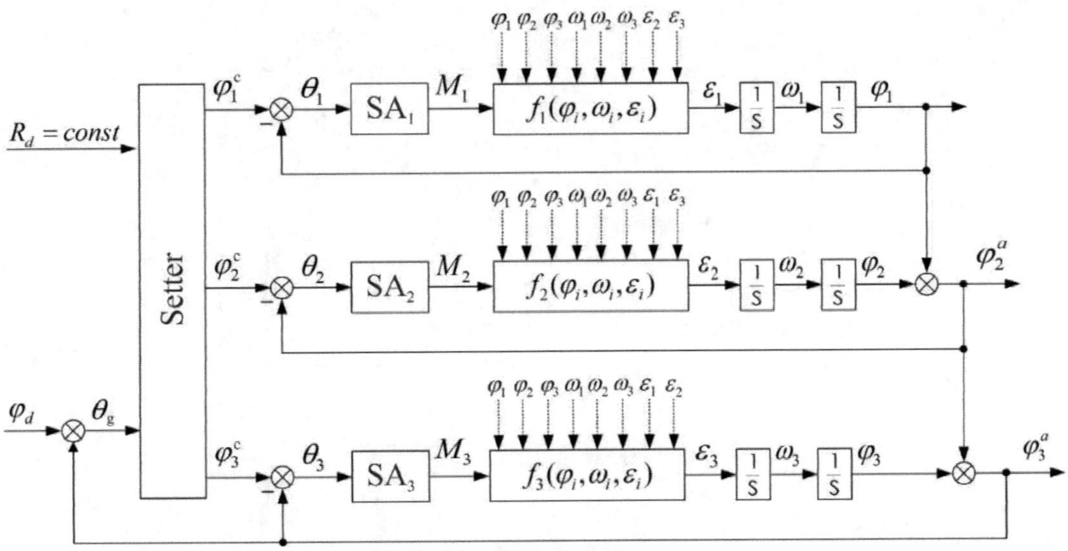

Figure 5. SCSoOC structure

Simulation model of the bipolar SM-based actuator uses the SM mathematical model included the system of non-linear equations [9]. Practically, SM is operated with driver generated the control voltagefor the exciting windings. Driver mathematical model supported microstep mode was designed for the imitation of the control voltages. The model is described by the following relations:

$$U_A = A\,sin\left[\frac{2\pi(n+D)}{4D}\right] \tag{15}$$

$$U_B = A\,sin\left[\frac{2\pi n}{4D}\right], \tag{16}$$

where U_A, U_B are the control voltages of the windings A and B, respectively; A is amplitude of the control voltage; D is the step divider value; n is the fixed-step control counter value corresponded to the current voltage:

$$n_{i+1} = n_i + 1 \tag{17}$$

Switching frequency of the counter defines the control voltages frequency, i.e. step frequency of SM rotor, which corresponds to its rotational speed. Thus, SA includes SM, driver, and control counter.

Principle of a position feedback system design is based on decreasing the error signal up to zero with its stability. It is effective to use the error including information both about position coordinate and its derivative as an input signal of the counter in the system under consideration. The voltages of SA generated by driver must be fixed on the last received values in decreasing the error. It means that the counter must be fixed on some last value in this position as well. Thus, the counter counts to some maximum value proportional to the control angle φ^c (or maximum error) in decreasing the error up to zero. Therefore, it is effective to design the control counter in accordance with

$$n = \left\lfloor K \int \theta\, dt \right\rfloor, \tag{18}$$

where K [Hz/V] is the proportionality constant (further, gain) between error expressed in voltage and counter frequency.

Simulation model of the SCSoOC built in accordance with the structure in Fig. 5 and given data using MATLAB Simulink is shown in Fig. 6.

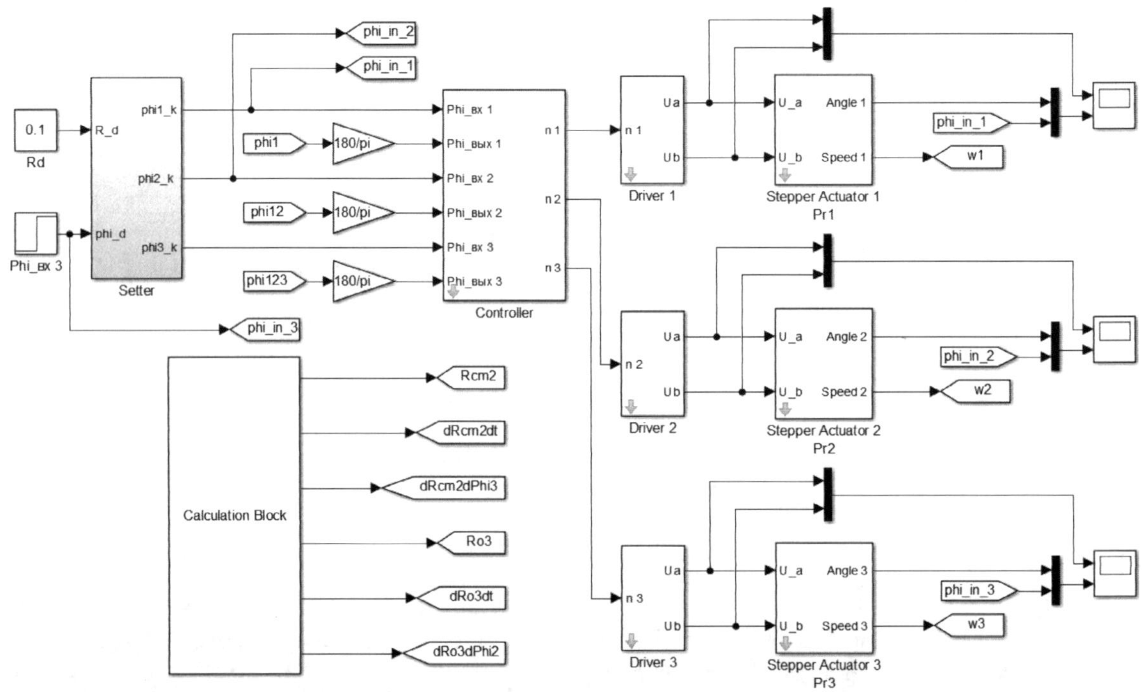

Figure 6. Simulation model of the SCSoOC

The solution of inverse kinematics is implemented in «Setter» block according to [8]; three control counters are generated in «Controller» block according to (18); the SA control voltages described by «Stepper Actuator» blocks are generated in «Driver» blocks according to (15) and (16); parameters of the dynamic equations are calculated in «Calculation Block». Dynamic equations (10) – (14) are considered in «Stepper Actuator» blocks in accordance with the SCSoOC structure (Fig. 5).

The characteristics of the American manipulator AR2 [11] are admitted as mass-dimensional parameters of the CE. The manipulator includes the following actuation motors:

- Pr_1 *– NEMA 23HS22 – 2804SS – HG50 (Stepper Actuator 1)*;
- Pr_2 *– NEMA 17HS15 – 1684S – HG50 (Stepper Actuator 2)*;
- Pr_3 *– NEMA 17LS19 – 1684E – 200G (Stepper Actuator 3).*

In simulation, distance between VC and OO is equal to 0.1 m.
Designer-defined gains K_i, in fact, define the quality factor of each local FA and can be selected so that a subsystem will have required QI. One of the simulation tasks was the research of the influence of QI to transient response of the SCSoOC.

Responses of the SCSoOC to step input with amplitude $\varphi_d = 5°$ and different QI values are shown in Fig. 7.

The curves obtained show that speed of response of the SCSoOC increases with increasing of the gains. Significant increase of a gain can lead to the motor automatic braking phenomenon that can be seen from the transient response with gains equal to 140 Hz/V. This phenomenon appears when switching frequency of the control voltages is higher than motor speed capabilities defined by its load [4].

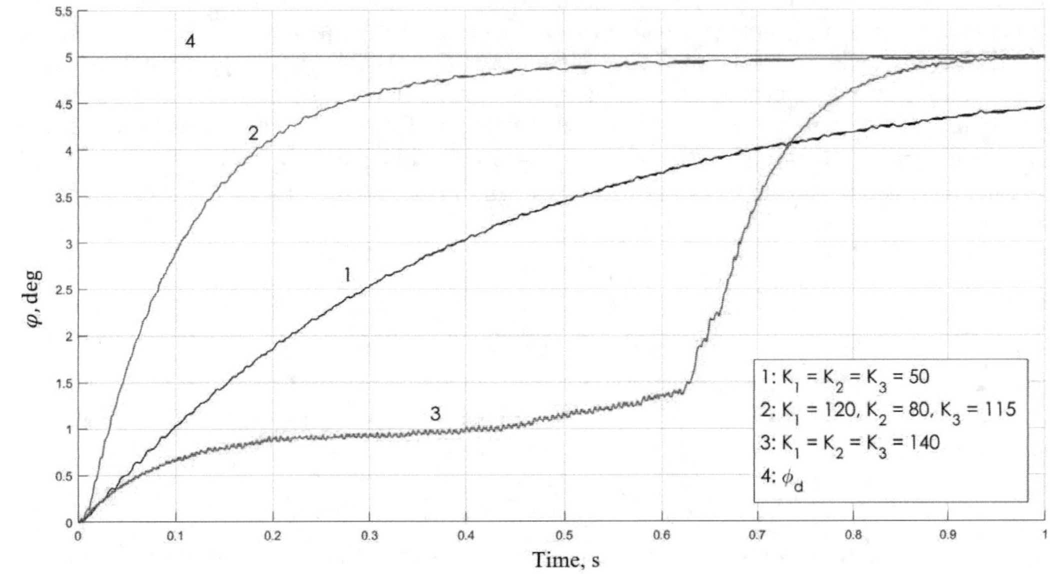

Figure 7. Response of the SCSoOC to step input

Rational (from out-of-non-synchronism-zone maximum speed of response perspective) gains values experimentally found are

$$\begin{cases} K_1 = 120\ \text{Hz/V}, \\ K_2 = 80\ \text{Hz/V}, \\ K_3 = 115\ \text{Hz/V}. \end{cases} \quad (19)$$

Response to the input signal, which changes its value in the range $\varphi_d = 5 \ldots 10°$, i.e. initial error is equal to 5°, is considered for the research of dynamic accuracy of the SCSoOC forced motion. In that case, the behavior of the dynamic accuracy as a composition of QI, such as speed of response and dynamic accuracy in quasi-steady-state modes is researched. Simulation results are shown in Fig. 8.

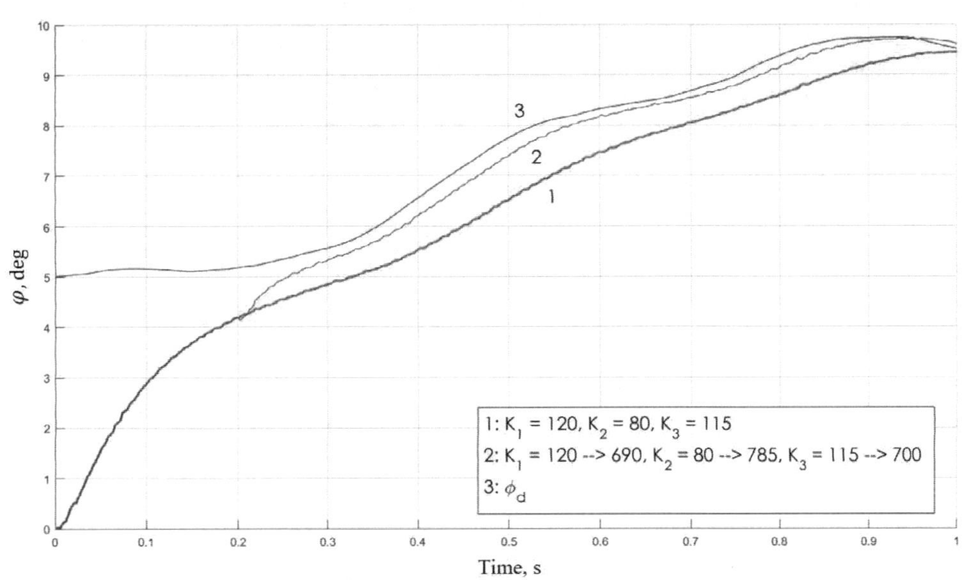

Figure 8. Response of the SCSoOC to nonstationary input

With above-admitted rational gains values of the SCSoOC, speed of response is limited by the requirement of the missing to nonsynchronism zone, but after that both error and its derivative become less than the determined threshold (approximately during 0.2 s). Possibility is appeared to change gains values in a relay way to higher ones increasing speed of response and dynamic accuracy in the quasi-steady-state mode in this error zone. Using graded structure [12], we will consider the response to input signal providing gains values are corresponded to ones in (19) in the high deviation area $(|\theta_g| \geq 1°)$ and to experimentally defined increased values in the small deviation area $(|\theta_g| < 1°)$, i.e. approximately after 0.2 s):

$$\begin{cases} K_1 = 690\ \text{Hz/V}, \\ K_2 = 785\ \text{Hz/V}, \\ K_3 = 700\ \text{Hz/V}. \end{cases} \quad (20)$$

Simulation results of the SCSoOC operation in gains switch mode are shown in Fig. 8 (curve 2).

Fig. 9 shows guidance errors of the SCSoOC defined as a difference between input and output signals:

$$\theta_g = \varphi_d - \varphi_3^a \quad (21)$$

Integral estimate of the dynamic error θ_g defined as an area under its curve is used for the quantitative estimation of the SCSoOC dynamic accuracy (Fig. 10).

$$\bar{\Delta}_\theta = \int_0^T |\theta_g|\, dt \quad (22)$$

Fig. 10 shows that relay switching of gains assists decreasing of dynamic error in the low error area. With speed of response of a SA limiting by threshold frequency of rotor steps, actuator operation area can be divided to more parts than two; in these parts, effective gains will be determined. Thus, SA will operate with maximum possible frequency on each part. It will allow to increase speed of response and dynamic accuracy of the SCSoOC.

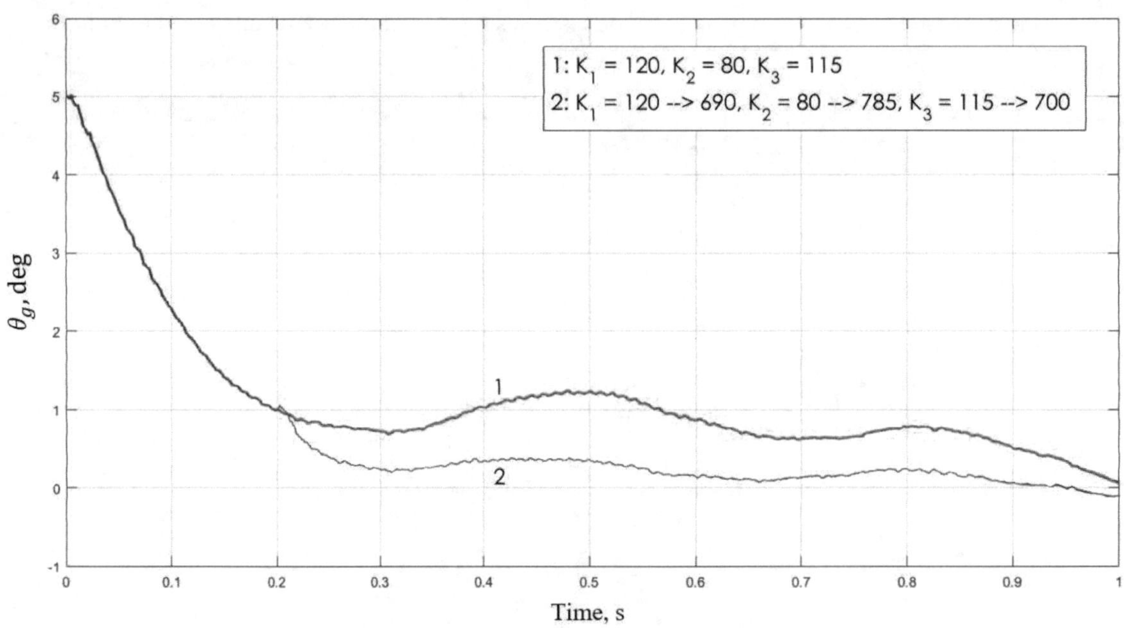

Figure 9. Guidance errors

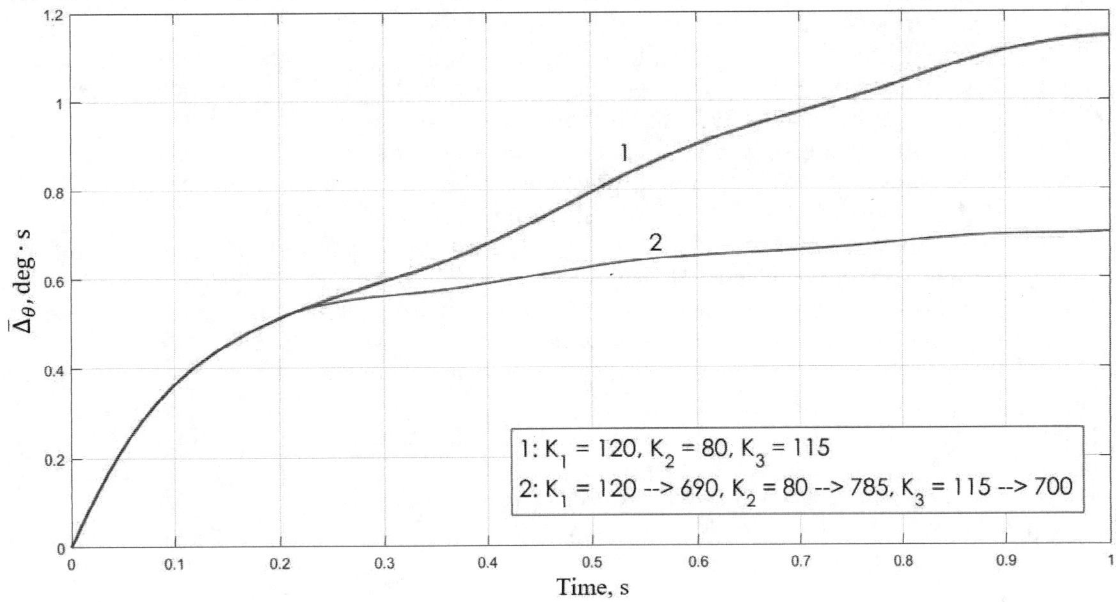

Figure 10. Integral estimations of the guidance errors

Results and conclusion

1. In this paper, research of the controlled elements of complex (in fact, multicoupling) system of the feedback actuators association is carried out. The research allowed to determine a description reduction methodology of the mechanical part of the system.
2. It is shown that stepper motors widely used in robotics can be constraint for speed of response of a system and thus decrease the overall dynamic accuracy in the mobile targets (observing object) tracking task.
3. Since gain selection criterion for each stepper-motor-based feedback actuator is a capability of achievement of the cutoff frequency of its operation, effective control option can be a partition of operation angular range of a stepper motor to the areas having their own gain. In high initial errors areas, maximum possible gain that will not be a reason of the motor automatic breaking phenomenon should be chosen. In low errors areas, the same gains influence to decreasing of both switching frequency of the control voltages and rotational speed of an actuator, and they should be increased in accordance with the requirements to given dynamic accuracy.

In conclusion, nonlinear control of a feedback actuators association is substantiated to decrease influence of the above-said constraints under defined conditions and thus to increase tracking dynamic accuracy of an observing object.

References

[1] Shon B V, Bushuev, et al 2015 Algorithms for controlling a flying robot when the robot is tracking a moving object. *Journal of Instrument Engineering* **58 (8)** pp 593–600

[2] Margun A A, Zimenko K A, et al 2014 Control system for unmanned aircraft equipped with robotics arm. *Scientific and Technical Journal of Information Technologies, Mechanics and Optics* **6** pp 54–62
[3] Universal Robotics A/S, available at: https://www.universal-robots.com/
[4] Metelev A S 2016 Resonance of stepper motors and controlling methods analysis *Student Scientific Spring 2016* (Bauman Moscow State Technical University)
[5] Bartos A J 2005 *Closed-Loop Stepper Motion Alternative*
[6] Osmolovskiy P F 1968 Iterative principle of design of the multi-channel control systems. *Transactions (Doklady) of the USSR Academy of Sciences.* **181 (1)**
[7] Yablonskiy A A 1966 *Course of the classical mechanics. Part II. Dynamics.* (Moscow: Vyshaya Shkola Publishing House)
[8] Musatov A E 2020 *Design and simulation research of the two feedback actuators of association of the videocamera spacial motion system* (Moscow: MAI, Master's thesis)
[9] Kenjo T 1987 *Stepping motors and their microprocessor controls.* (Moscow: Energoatomizdat)
[10] Fortuna V, Sergeev S Microstep mode of the stepper motor operation, available at: electroprivod.ru/microstepping_mode.htm
[11] Annin Robotics, available at: https://www.anninrobotics.com
[12] Sheval V V, Dorokhov E I, Zemtsov V I and Isakov S A 1984 *Dual-zone feedback systems.* (Moscow: Energoatomizdat)

Integral criterion of the state of physical parameters of the surface layer of machine parts

A V Totay, M N Nagorkin

Bryansk State Technical University, 7, 50 let Oktyabrya Blvd, Bryansk, 241035, Russia

E-mail: totai_av@mail.ru

Abstract. Exoelectronic emission, which is used as a non-destructive method for monitoring the parameters of the surface layer of parts, and the level of exoemission current from the machined surface, that is a complex criterion for the physical and chemical state of the metal have been studied. Correlations between the main parameters of the surface layer and exoemission are found.

1. Introduction

The reliability of technical systems is proved by the trouble-free performance of their parts that undergo the most significant loads of various types during operation. Achievement of a target product life requires a science-based approach to the design of route and operational technological processes ensuring an optimality state of physical parameters of the surface layer for the specified operating conditions.

In the function of the integral criterion for parameter estimation of the surface layer (PESL) it is desirable to use a non-destructive evaluation method that objectively describes the physics of the surface of the part after technological attacking. As such, the method of exoelectronic emission (EEE) can be used.

The term "exoelectronic emission" defines the phenomenon of non-stationary electron emission from the surface of a solid body in an excited state, under external thermal or light stimulating action with the energy below the threshold for the occurrence of stationary emission effects. By now the fact of relationship of emission parameters and unsoundness of materials is considered the proof of it. Technological methods of attacking the surface, accompanied by deformation and thermal phenomena cause complex physical and chemical processes that cause some imbalance in the electronic system of materials [1 – 4]. Despite the difference in the physical nature of the transformations that occur in the surface layer as a result of external influences, the emission reaction is reduced to a decrease in the energy threshold. It is also established that the EEE method is not sensitive to the type of surface defect, but only serves as its indicator. Therefore, its use for monitoring the state of the surface layer should be accompanied by studies of the relationship between EEE parameters and the regularities of the defect formation process in the surface layer. Wide practical use of the method is caused by a number of features:

–no restrictions related to the nature, structure and properties of materials;

– the ability of acquisition of information both in integral form from the entire controlled surface, and locally, from its individual sections;

IOP Conf. Series: Materials Science and Engineering 1061 (2021) 012040 doi:10.1088/1757-899X/1061/1/012040

–rapid response to surface defects, which makes possible to use it in dynamic testing;

–measurements can be carried out in a variety of medium, a wide temperature range, and after various external energy depositions on the surface.

As is commonly known, the real surface of a solid body is always covered with layers of adsorbates, an oxide film, that results in a modification of the energy spectrum of existing electron states. Due to the structural imperfection the surface is a source of dislocation generation, which inevitably affects its mechanical properties. Many researchers associate this phenomenon with the presence of grain boundaries and cristal twins, oxide films, cracks, etc.

In applied technical problems, there are mainly two types of EEE:

1) photon stimulated electronic emission (PSEE) – emission recorded during lighting at a constant temperature;

2) thermostimulated electronic emission (TSE) is an emission recorded when a controlled object is heated.

2. Results

The technological system (TS) of diamond burnishing (DB) of flat surfaces of gray cast iron (SCh20) parts, used in the manufacture of basic elements of equipment and technological equipment, after preliminary face milling (or flat grinding) was considered as an object of research.

In this paper, the PSEE was used to estimate the physical and chemical state of surfaces.

To avoid high-energy effects on the surface, an ultraviolet visible radiation probe is used in the measurement process. Photostimulation results in delocalization of electrons from traps, which take place due to defects in the surface layer of the material. Light with quantum energy hv ◘ E_{onm}, where E_{onm} – optical ionization energy of trapping center causes either ionization of center with further trap depletion or electron excitation. Although metals have a sufficient number of free electrons, EEE requires excitation and stimulation. For most metals and alloys, the work of electron liberation is 4–5 eV, so that the EEE of metals should occur if the excitation and stimulation reduce the surface potential barrier. In thin oxide layers on the surface of metals, there are structural defects, which, as a rule, are the centers of electron localization.Such traps can recapture electrons, which found themselves in in carrier band of oxide. But under the conditions of photostimulation they are immediately emptied in accordance with the PSEE mechanism.

The energy state of the surface layer, measured by the level of ecoemissions current depending on methods of processing, types of materal parts, tool materials and machining modes, was investigated in an experimental apparatus of the Moscow aviation Institute construction.

The vacuum system of the unit ensures stable pressure in the chamber 10^{-7} Pa, which after replacement of the next sample is achieved within approximately 30 minutes. For photostimulation of EEE a monochromator of spectrophotometer SPh-26 was used, that allows controlling the parameters of luminuous flux. The quartz optical system provides an adjustable diameter UV probe. The ion counter served as the register system IC-04, the range of detected currents – from $1\cdot10^{-13}$ to $2\cdot10^{-18}$ A.

The segments, exsected from cylindrical test-pieces by means of electroerosion method, served as a model. Scanning was performed along the cylinder's generatrix along a 12 – 16 mm long track. Before exoemission analysis, the samples were thoroughly washed with ethanol.

Scanned images taken from the sample surfaces made the obtaining of the average value of the EEE level from one surface possible, as well as identifying areas with an increased degree of deformation besides analyzing the heterogeneity of the energy properties of the surfaces.

Thus fig. 1 shows scans from the surfaces of samples made of harden steel 45, turned on a lathe and processed by circular grinding and diamond burnishing.

Analysis of the scans shows that in absolute terms, the instability of the energy properties of the studied surfaces is more typical for grinding and less for turning. In addition, there is a significant difference in the average level of exoemission flow I for these processing methods: for turning it is approximately 160 sec^{-1}, for grinding – 500 sec^{-1}, for diamond burnishing – 930 sec^{-1}. These data indicate a significant difference in the physical state of the sample surfaces due to the excellent conditions of plastic deformation in the surface layer expressed by individual force, speed, and heat factors for each processing method.

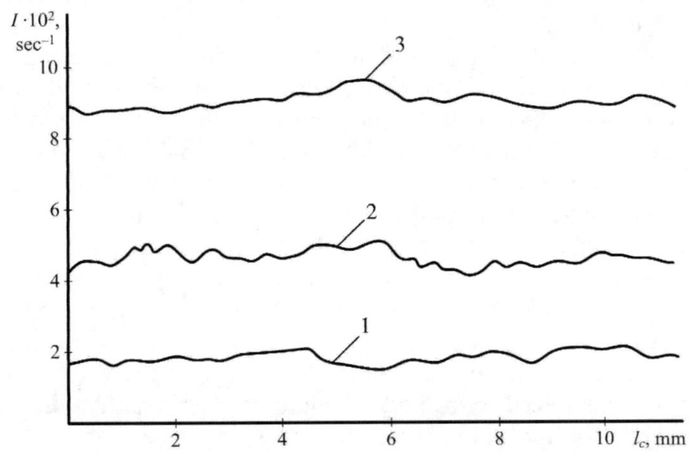

Figure 1. Standard scans of the treated surfaces of harden steel 45: 1 – fine turning with composite 10, t = 0.3 mm, S = 0.06 mm/rev, V = 180 m/min; 2 – plunge-cut cylindrical grinding (white electrocorundum) t = 0,005 mm, S_{np} = 800 mm/rev, V = 28 m/min; 3 – diamond burnishing (synthetic polycrystalline diamond), r = 1,5 mm, S = 0,05 mm/rev, V = 120 m/min

To study the stability of energy properties over the area of the machined surface, the end surface of a cylindrical sample made of harden steel 65G was taken (fig. 2). The decision to investigate the end surface was due to the need to exclude possible energy losses when a section of the cylindrical surface is excited consequent on various conditions owing to the reflections of the absorption of the light stream from the scanning spot.

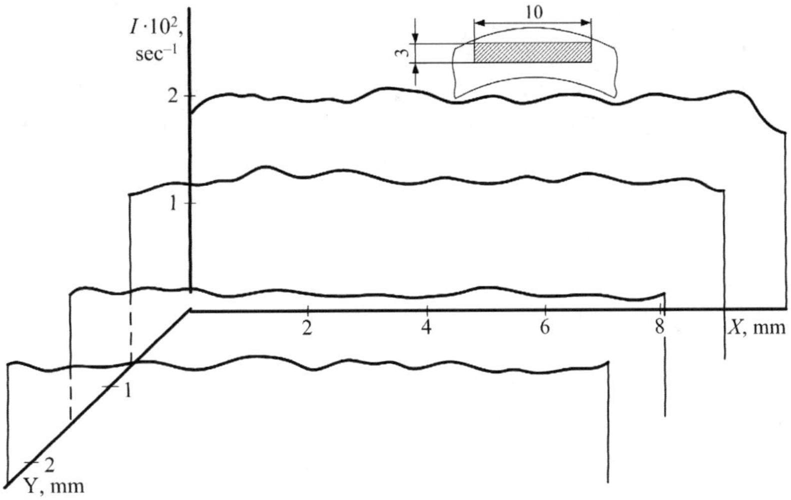

Figure 2. Exoemission topogram of the end surface section of a sample made of harden steel 65G, turned with a composite 10: t = 0,45 mm, S = 0,08 mm/rev, V = 210 m/min

The presented graphs of the dependence of the exoemission flow I on the section of the route (x, y), made every 1 mm of the surface area of 30 mm^2 indicate a high degree of uniformity of the exoemission currents, which is expressed in close average values of the fixed parameter (162, 170,

167, 174 sec^{-1}). This allowed concluding that the energy properties of the entire sample were adequately evaluated for the level of EEE taken from a single track of a section of the cylindrical surface. Experimental dependences of the differentiated influence of the main elements of the modes of the studied processing methods on EEE are shown in fig. 3 (the research was conducted in partnership with V. V. Shorin).

During the process of turning the depth, feed and the cutting speed have bearing on the volume of exoemmision current. With their increase the parameter I rises and in the studied ranges of changes in processing modes, these dependencies are monotonous. The growth of exoemission current with increasing cutting depth and feed is associated with an increase in deformation effects on the metal of the surface layer and, consequently, with an increase in the number of defects in the grains of the material. A cutting speed step-up is accompanied by the increase in the deformation temperature, which contributes to a more intensive process of localization of free electrons at the surfaces of newly formed grains and causes a decrease in the work function during further photostimulation. In terms of quality the effect of the depth and speed of plunge-cut cylindrical grinding on the EEE remains the same as during the process of turning.

The effect of longitudinal flow is reversed, i.e. by increasing it the value of exocurrent decreases and that is connected with a reduced probability of multieffect on one and the the same microscopic volume of metal surface layer of the grains of the circle.

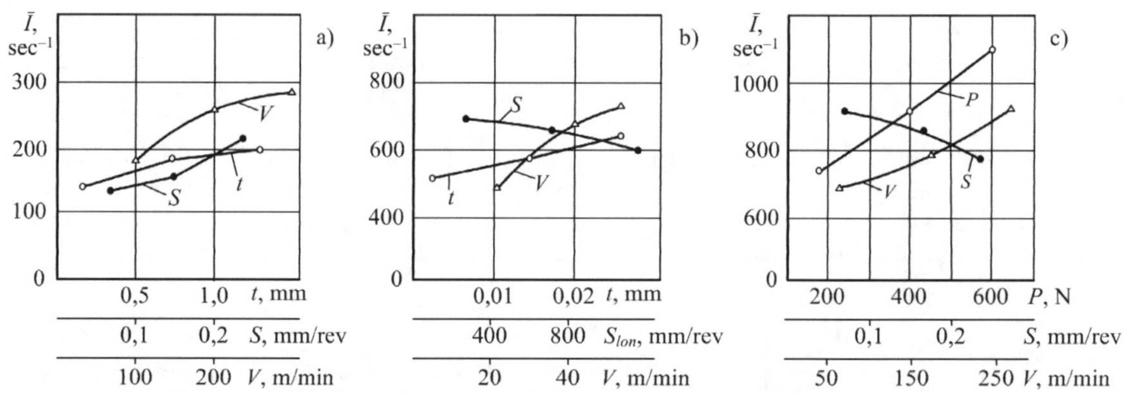

Figure 3. Influence of technological factors on the level of EEE in various methods of processing of harden steel 45: a – turning with composite 10); b – circular grinding (white electrocorundum); c – diamond burnishing

A similar effect of the line feed on the exoemission can be seen in the case of diamond burnishing. It is obvious that the effectiveness of the deformation affection on the metal of the surface layer decreases with increasing of line feed.

An increase in the smoothing force and speed in the studied ranges contributes to an increase in EEE. Planned experiments were carried out to identify the complex influence of machining factors for obtaining corresponding empirical dependencies. Steel 45 was used as a feedstock. This method extends the effect of the obtained empirical dependences on other materials by experimental refinement of b_0 coefficient of the standard model of the corresponding treatment process.

Table 1 shows the average EEE values for the studied processing methods.

As a result of processing the experimental results, the following dependencies were obtained:

– turning:

$$I = b_0 t^{0.18} S^{0.16} V^{0.28} r^{0.12}, \quad (1)$$

– circular grinding:

$$I = b_0 t^{0.21} S^{-0.18} V^{0.32}, \quad (2)$$

– plunge-cut cylindrical grinding:

$$I = b_0 S_p V^{0.23} Z^{0.12}, \quad (3)$$

– diamond burnishing:

$$I = b_0 P^{0.21} S^{-0.13} V^{0.14}. \quad (4)$$

Only statistically significant factors are included in dependencies (1) – (4). The values of the coefficients b_0 taking into account the processed material are given in table 2.

Table 1. Average values of exoelectronic emission depending on the processing method

№ experiment	Processing methods			
	Turning	Circular grinding	Plunge-cut cylindrical grinding	Diamond burnishing
1	110.6	529.6	466.0	575.6
2	185.2	651.4	688.9	874.3
3	180.7	416.3	536.5	503.3
4	175.2	511.8	801.4	762.4
5	201.8	658.3	509.4	670.2
6	196.3	812.4	761.3	1021.3
7	190.5	520.3	586.3	584.6
8	320.6	638.7	868.7	889.6

Table 2. The values of the coefficients b_0

Processing methods	Material			
	Normalized steel 45	Harden steel 45	Harden steel 65G	High-grade cast iron
turning	80,8	98,6	116,3	73,8
circular grinding	1.15·103	1.58·103	1.92·103	1.45·103
plunge-cut cylindrical grinding	362.6	471.6	546.4	395.5
diamond smoothing	54.8	94.5	86.8	69.7

The rank diagram obtained for turning shows that the critical factor in this process is cutting speed, which primarily affects the temperature and rate of plastic deformation. Factors mainly shaping the force pattern in the facing layer (*r* and *t*) are approximately 1.5 times lower if the degree of influence over EEE is meant. During the circular grinding the line feed designating the time of effect of the circular on the machinable surface comes first. By way of illustration, the rank of the "average grain size" factor is shown, which was not statistically significant.

As the plunge-cut cylindrical grinding rank chart shows, the greatest influence over EEE is made by radial feed (actually depth of cut) and speed, with the S_p factor in the first place. This suggests that with this method, which is accompanied by abundant cooling, the main factor designating the strength of the exocurrent from the surface is the deformation one.

Expectedly the indenter force factor came out on top in case of diamond burnishing. Correlation analysis was used to establish statistical relationships between the EEE intensity and the parameters of the surface layer state.

The following parameters were preferred:

– Ra, µm – arithmetic mean deviation of the profile,
– H_μ kg/mm^2 – surface microhardness,
– $U_н$ – coefficient of the degree of cold work,
– σ_0, MPa – technological tangential macrostresses on the sample surface,
– d – the number of grains of the metal surface layer,
– ρ, cm^{-2} – dislocation density.

The initial data for correlation analysis were the results of both single-factor and multi-factor experiments. The coefficients of pair correlation between all analyzed parameters were obtained (table 3).

Table 3. Coefficients of pair correlation between parameters of the surface layer state

Parameters	*Ra*	H_μ	$U_н$	σ_0	*d*	ρ	*I*
Ra	1.00	0.17* –0.22 –0.31	0.21 –0.19 –0.33	0.34 0.29 0.43	–0.58 –0.71 –0.63	0.16 –0.08 –0.11	0.17 0.36 –0.47
H_μ	0.17 –0.22 –0.11	1.00	0.92 0.84 0.96	0.16 0.22 0.27	0.78 0.83 0.91	0.91 0.84 0.89	0.86 0.88 0.92
$U_н$	0.21 –0.19 –0.33	0.92 0.84 0.96	1.00	0.18 –0.11 0.14	0.79 0.78 0.86	0.87 0.82 0.89	0.91 0.84 0.87
σ_0	0.34 0.29 0.43	0.16 0.22 0.27	0.18 –0.11 0.14	1.00	0.27 0.17 0.21	0.31 0.42 0.36	–0.11 0.17 0.08
d	–0.58 –0.71 –0.63	0.78 0.83 0.91	0.79 0.78 0.86	0.27 0.17 0.21	1.00	0.74 0.77 0.82	0.37 0.79 0.92
ρ	0.16 –0.08 –0.11	0.79 0.78 0.86	0.87 0.82 0.89	0.31 0.42 0.36	0.74 0.77 0.82	1.00	0.93 0.89 0.91
I	0.17 0.36 –0.47	0.86 0.88 0.92	0.91 0.84 0.87	–0.11 0.17 0.08	0.87 0.79 0.92	0.93 0.89 0.91	1.00

* – the top line shows the correlation coefficient for turning, the second line – for circular grinding, and the third line-for diamond smoothing.

3. Discussion of results

Considering the strength of the relationship of each parameter separately with the intensity of the exocurrent, first of all, it should be noted that for all three studied processing methods, there is a fairly good convergence of the values of the pair correlation coefficients.

If we conveniently classify microgeometric and physical parameters included into correlation analysis into two groups we can definitely say that surface microhardness, coefficient of the degree of cold work, the number of grains and dislocation density are allied to EEE, while arithmetic mean deviation of the profile and technological tangential macrostresses are loosely correlated with the studied parameter. At another point, if you look at the coefficients of pair correlation between H_μ, $U_н$, *d* и ρ, they all fall into the range of values 0.74 – 0.96, which indicates their strong mutual influence. Therefore, the relationship of the first group of parameters with EEE should be considered as a whole. Thus, with an increase in the surface microhardness and, consequently, coefficient of the degree of cold work, there is a decrease in the grain size (an increase in its number) and an increase in the dislocation density. This statistical analysis does not contradict the physics of the surface layer. Since an increase in hardness is nothing more than an increase in the metal's resistance to plastic deformation, in the vast majority of cases this is due to a decrease in the size of the metal grains of the surface layer and the barrier effect of grain boundaries [4, 8]. These last two factors significantly inhibit the movement of dislocations in the real crystal lattice of polycrystals, which is explained by the barrier effect of grain boundaries (the smaller the grain, the larger the boundaries) and the "locking" property of dislocation planes. This close correlation of all four factors and the amount of exoemission is explained by the sensitive reaction of this method to the material defect, the increase of which is accompanied by a significant increase in the number of exoelectrons in the surface.

The effect of the roughness parameter *Ra* on the EEE is rated by correlation coefficients from 0.47 for diamond burnishing to 0.36 for grinding, which can be considered rather weak from the point of view of mathematical statistics. From the point of view of physics phenomena theoretically it is difficult to say what will affect the growth of exoemission current mostly – the increase of surface area due to increasing *Ra* and thus there is higher probability of electron exit into the vacuum during the process of stimulating or power reduction of excitant light probe as a result of large losses when scanning the surface with orange-peel defect. Besides, depending on the methods and modes of fine machining and strengthening treatment the microrelief can be formed under the influence of relatively minor force influences (grinding with fine grain circles having small cutting depths), and at high specific loads (diamond burnishing). Such a situation can form a combination of low roughness in some cases with a relatively large surface grain and a small density of dislocations, and in others a similar microrelief will be on the surface with a smaller grain and a higher density of dislocations. It would be safe to assume that this physical model explains weak correlation between the heigh parameter of roughness and the value of EEE.

As for the coefficients of pair correlation between surface macrostresses and the energy characteristic of the surface, in table 3 they are bottom in absolute terms.

The statistical relationship between the grain size *d* (number) and the dislocation density *p* with EEE is of practical interest. Using data for correlation analysis, we obtain equations of the form:

$$d = C_d I + a_d, \tag{5}$$

$$\rho = c_p (I - a_p). \tag{6}$$

The values of the coefficients in dependencies (5) and (6) are shown in table 4.

Table 4. The values of the coefficients in dependencies (5) and (6)

Coefficient	Material		
	Steel 45 (normalized)	Steel 45 (harden)	Steel 65G (harden)
C_d	0.019	0.024	0.029
a_d	6.5	6.9	7.2
C_p	$2.2 \cdot 10^6$	$1.9 \cdot 10^7$	$2.9 \cdot 10^7$
a_p	22	15	84

4. Conclusion

According to the results of conducted research and obtained experimental data it can be concluded:

1. The level of exoemission current from the treated surface can serve as a complex criterion for the physical and chemical state of the metal of the surface layer.
2. EEE can be used as a non-destructive method for monitoring the quality indicators of parts surface, which allows providing the specified parameters of the surface layer state.

References

[1] Kortov V S, Slesarev A I and Rogov V V 1986 *Exoemission control of the surface after processing* (Kiev: Naukova dumka) p 176
[2] *New in the study of the surface of a solid body* 1977 (Moscow: Mir, vol. 2) p 315
[3] Totay A V, Akulich P P and Poplavsky A M 1982 *Proceedings 2-nd Int. Symp. "Exoemission and its application"* (Moscow state university) pp 93–94
[4] Totay A V and Akulich P P 2003 *Proceedings 3-rd Int. Symp "Technological quality management of machine parts surfaces"* (Kiev: ATMU) pp 185–89
[5] Buckley D 1981 *Surface effects in adhesion, friction, wear, and lubrication* (Amsterdam: Elsevier) p 359
[6] Akhmatov A S 1963 *Molecular physics of boundary burning* (Moscow: Fizmatgiz) p 472
[7] Vinogradov A A 1988 *Superhard material* **5** 44–49
[8] Kuznetsov V D 1977 *Physics of cutting and friction of metals and crystals. Selected works* (Moscow: Nauka) p 310
[9] Vainer L G 2013 *Russian engineering research* **33(3)** 140–143

Mathematical description of automobile headlight contamination when driving on a wet road

V G Vlasov[1], A I Fedotov[1], V O Gromalova[1], S M Gergenov[2]

[1] Irkutsk National Research Technical University,83 Lermontov st., Irkutsk, 664074, Russia
[2] Eastern Siberian State University of Technology and Management, 40v Klyuchevskaya st., Ulan-Ude, 670013, Russia

E-mail: fai@istu.edu

Abstract. The article describes the process of contamination of automobile headlights in order to calculate contamination speed using a mathematical model. The authors published results of the study aimed at solving the problem of negative application of the CAIM (chemical anti-icing materials) on winter roads. In particular, it was found that when driving vehicles on winter roads covered with CAIM, the headlights become contaminated. The headlight contamination with dirty droplets from under the wheels of a vehicle ahead was analyzed. The detachment of droplets from the moving wheel and the path of droplets from the wheel to the headlights of the vehicle behind were analyzed and described. Process is studied of contaminant film formation which decreases light transmission was carried out. Road experiments were carried out. Along with the physicomathematical description of the process under consideration, they give new knowledge about headlight contamination depending on the vehicle speed. The mathematical model supplemented by the experimental study allows us to predict the rate of headlight contamination by the wheels of vehicles ahead when moving along roads covered with chemical anti-icing reagents depending on the speed and the distance between the vehicles.

1. Introduction

Driving on a wet road is more dangerous than driving on a dry road. This is due to the deterioration of traction properties of the wheels [1,2] and poor view of the road situation. If in the daytime the view can be improved using good wipers, in the dark the illumination with headlights is significantly impaired due to their contamination. The headlights are contaminated due to a liquid dirty film formed by dirty droplets from under the wheels of the vehicle ahead [3].

2. Materials and methods

The article aims to describe the process of headlight contamination using a mathematical model developed to predict headlight contamination.

The headlight contamination process consists of three stages:
- detachment of droplets from the moving wheel;
- movement of droplets from the wheel to the headlights of the vehicle ahead;
- formation of a contaminant film on headlights.

Detachment of droplets from a moving wheel occurs if their centrifugal inertia exceeds the centripetal force of attraction to the wheel due to molecular wetting forces [4]. Wetting occurs if attraction forces between the molecules of liquid and solid bodies are greater than between the molecules of liquid bodies. At low speeds, the centripetal force is superior to the centrifugal force (there is no detachment of droplets), and the wetting film is relatively thick. With an increasing speed, the centrifugal force increases, droplets detach from the "thick" film, overcome the weaker attractive force between the liquid molecules, and the film thickness decreases. The latter means that with an increasing wheel speed, the centrifugal centripetal forces increase, since attractive forces between the molecules of a liquid body and a solid body (tire) play a crucial role. Thus, when droplets detach from the wheel, there is no trigger.

The above discussion allows us to draw an important conclusion: the number of droplets detaching from the wheel moving on the wet road increases with its speed.

Let us describe the movement of droplets from the wheel to the headlights of the vehicle ahead.

The movement equation is a system of three differential equations

$$m\frac{d^2\vec{S}}{dt^2} = \vec{F}_g + \vec{F}fr + \vec{F}res, \quad (1)$$

with the initial condition

$$\vec{S}(0) = \vec{0}, \quad \frac{d\vec{S}}{dt}(0) = \vec{V}_0. \quad (2)$$

In equations (1-2) $\vec{V}_0$ - speed of the droplet, m - mass of the droplet, $\vec{S} = (x, y, z)$ - coordinates of the droplet along its trajectory, its movement is determined by the forces acting on it, the first force is $\vec{F}_g = -m(0,0,g)$ gravity.

In (1), the second term is the viscous friction force described by the Stokes's formula

$$\vec{F}fr = -6\pi\eta r \cdot (\frac{d\vec{S}}{dt} - \vec{V}_в) \quad (3)$$

where $\eta = 1.8 \cdot 10^{-5} kgm^{-1}c^{-1}$ is the dynamic air viscosity coefficient at 20° C, r is the radius of the droplet, $\vec{V}_в$ is the air flow speed behind the car.

The third term is the force of air resistance described by the Newton's formula

$$\vec{F}res = -\frac{1}{4}\pi\rho r^2 \cdot (\frac{d\vec{S}}{dt} - \vec{V}_в)^2 \quad (4)$$

where $\rho = 1.2 kgm^{-3}$ is air density at 20°C.

Let us analyze the system of differential equations (1-2).

A) The system (1-2) has an exact solution at $\rho = \eta = 0$

$$z(t) = \frac{V_0^2 Sin^2\alpha}{2g} - \frac{g}{2}(t - \frac{V_0 Sin\alpha}{g})^2, x(t) = V_0 Cos\alpha \cdot t, \ t \in \left[0, \frac{2V_0 Sin\alpha}{g}\right] \quad (5)$$

$x(t)$ and $z(t)$ are coordinates of the droplet along and perpendicular to the surface, α is the angle of speed direction relative to the horizon when the droplet detaches from the wheel, V_0 is the modulus of speed of the droplet equal to the modulus of speed of the automobile.

Solution (5) means that the path of the droplet is a parabola in the XOZ plane.

Let us calculate the parameters of the droplet's path neglecting the forces of friction and air resistance. Let the vehicle speed be 72 km / h (20 m / s) and the droplet detachment angle be 30° to the horizon. The droplet will fly away in the horizontal direction by 35 meters, the maximum lifting height will be 5.1 meters, the time will be 2 seconds. Since for 2 seconds the automobile will drive away 40 meters from the place of detachment of the droplet, at a distance of 75 meters, the headlights could become dirty from the vehicle ahead if there was no influence of the air environment.

B) When analyzing the movement of the droplet in the air, the important criterion is the Reynolds number

$$Re = \frac{2\,r\,\rho\,V}{\eta}, \tag{6}$$

At large values of the Reynolds number ($Re \geq 500$), the force of viscous friction can be neglected. Let the radius of the droplet be $r = 10^{-3}m$, $V_0 = 20ms^{-1}$, then $Re = 2600$, i.e. the force of viscous friction is small.

The system (1-2) neglecting the forces of viscous friction and gravity at $\vec{V}_в = 0$ has a simple analytical solution:

$$\frac{dS}{dt} = \frac{V_0}{1 + aV_0 t}, \; S(t) = \frac{1}{a}\ln(1 + aV_0 t), \tag{7}$$

where $a = \frac{\pi\,\rho\,r^2}{4m} = 0.225m^{-1}$ for a spherical droplet with a density of $10^3\,kgm^{-3}$.

Accordingly, the droplet speed decreases from 20 m/s to 2 m/s, while its distance is 10 meters, and the distance between the droplet and the wheel is 50 meters. Thus, if the distance between the vehicles is 50 meters, the headlights of the second vehicle may become contaminated with droplets flying out from under the wheels of the first vehicle. If the radius of the droplet is 2 mm and its density is 2 times higher than density of water, for 2 seconds the droplet will be at a distance of 21 meters moving at an initial speed of 20 m/s, and the distance between the droplet and the wheel will be 61 meters.

It should be noted that accurate calculation of the path of the droplet requires knowledge of speed of air flows $V_в(x, y, z, t)$. It is impossible to calculate them. The speed of air flows is determined by the size, geometry and speed of the vehicle, atmospheric wind, air flows from moving vehicles, and the road landscape.

The processes of detachment and movement of droplets have been described above. An experimental study of these processes is expensive.

Let us describe the process of liquid contaminant film formation on the headlights. The liquid contaminant film decreases light transmission which can be measured. Thus, an experiment is required. It would be possible to install a measuring device next to the headlight. It would measure light transmission of the contaminant film.

However, it is difficult to install this device. A simpler way is to install screens on the headlights, whose light transmission is easy to measure in laboratory conditions. Previously, the authors of the article experiments on automobile lights contamination in wintertime were conducted under real road conditions. Variation in lights intensity due to contamination by deicing chemicals was studied [5,6]. This experiment was conducted when two cars moved at the same speed for five kilometers keeping the distance of 50 meters. Then the headlight screens were replaced with clean ones. The results of this experiment are presented as dots in Fig. 1.

In order to determine the dependence of headlight contamination on speed, it is required to analyze physics of this process. The value of headlight contamination is directly proportional to thickness of the contaminant liquid film g, i.e.

$$Y = N(1 - K_{trans})g. \tag{8}$$

K_{trans} is the transparency coefficient of the liquid film, depending on its chemical composition, for pure water, its value is close to 1, N is the normalization factor with a value which is the inverse length.

As has already been shown, the number of droplets detaching from the wheel should increase with an increasing speed, which means that at a higher speed, the thickness of the contaminant film is greater. The contaminant film is formed due to wetting forces determined by the attraction of liquid molecules to the headlight glass molecules. With an increasing film thickness, the outer layer of film molecules is no longer attracted to the glass. Forming droplets, they will slide off the headlight under the influence of gravity and air pressure. The above process can be described by the following differential equation

$$\frac{dg}{dV} = Q_{cont} - \frac{2}{V_{cont}} g \,. \tag{9}$$

This equation describes the dependence of thickness of the contaminant film per one kilometer on the vehicle speed.

$$g(V) = 0.5 Q_{cont} V_{cont} (1 - e^{-\frac{2V}{V_{cont}}}) \,, \tag{10}$$

where V_{cont} is speed at which thickness of the contaminant film is 87% of the limit equal to $0.5 \cdot Q_{cont} V_{cont}$, Q_{cont} is conditional time of film formation.

Combining equations (8-10), we have a theoretical formula for the value of headlight contamination at a five-kilometer distance depending on the speed

$$Y_{hi} = Y_0 (1 - e^{-\frac{2V}{V_{cont}}}) \,, \; V_{cont} = 80 km/h \,, \tag{11}$$

The parameters included in (11) were obtained by comparing the experimental data (blue dots in Fig. 1) with the contamination function curve. The limiting calculated degree of contamination turned out to be equal, i.e. under the experimental conditions, the transparency coefficient of liquid forming the film was very high $K_{trans} = 0.772$

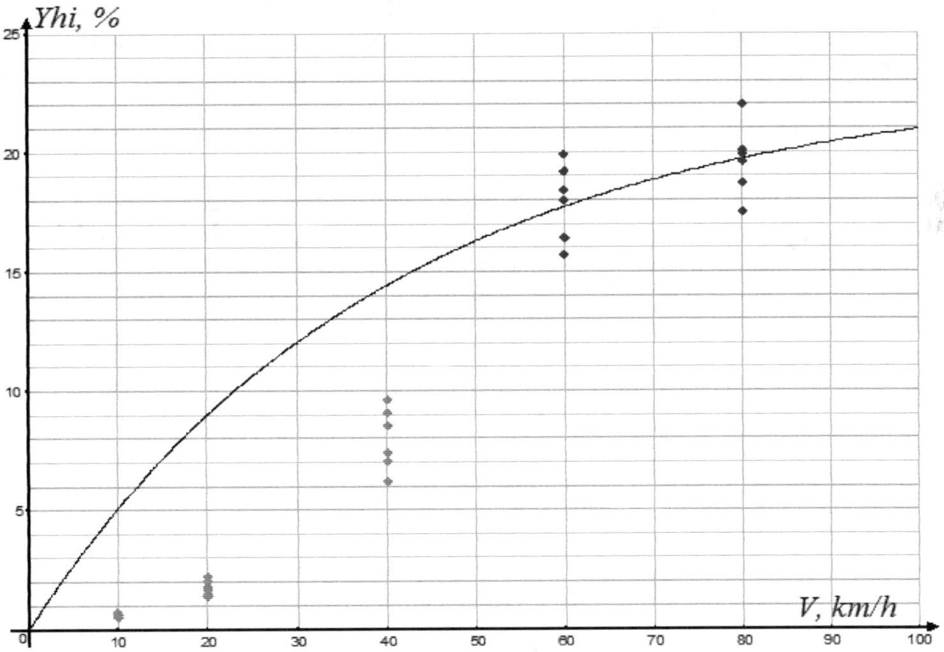

Figure 1. Contamination Y_{hl}, % of Nissan Qashqai headlights depending on its speed at a distance of 50 meters from the vehicle ahead on a road covered with chemical anti-icing reagents

Why is the value of headlight contamination so low (pink dots in Fig. 1) at 10 km/h and 20 km/h? To answer this question, let us use formula (5), according to which the distance between the wheel and the droplet is determined by formula

$$X = \frac{2V_0^2}{g} Sin\alpha(1 + Cos\alpha) \quad (12)$$

The maximum distance is reached by a droplet detaching at an angle of 60 degrees equal to

$$X_{max} = \frac{3\sqrt{3}V_0^2}{2g} [m/s] = 0.02V_0^2 [km/h]. \quad (13)$$

Thus, if the vehicles move at 60 km/h, the maximum distance at which the vehicle ahead contaminates the vehicle behind is 72 meters. With a decreasing speed, the distance between the vehicles at decreases; at 20 km/h, it becomes equal to 8 meters.

In the experiment described, the distance between the vehicles was 50 meters, and the headlights should be kept clean at 50 km/h and less. At 10 km/h and 20 km/h, slight contamination might be due to the droplets detaching from the wheels of overtaking vehicles.

At 40 km/h, the contamination value turned out to be unexpectedly high, which might be due to the fact that the distance between the vehicles was sometimes 32 meters.

3. Conclusion

The mathematical model supplemented by the experimental study allows us to predict the rate of headlight contamination by the wheels of vehicles ahead when moving along roads covered with chemical anti-icing reagents depending on the speed and the distance between the vehicles.

References

[1] Molev Yu I 2007 Ensuring road safety in winter *Thesis* VlSU Vladimir p 389

[2] Molev Yu I and Proshin D N 2019 On the issue of controllability loss when moving along a layer of loose snow (aqua planning on snow) *XIV International Scientific and Practical Conference* Orenburg pp 490-500

[3] ODM 218.4.005-2010. Industry road guidance document "Recommendations for ensuring traffic safety on roads"

[4] Ivlev L S and Dovgalyuk Yu A 1999 *Physics of atmospheric aerosol systems* RIE SPbSU p 194

[5] Fedotov A I, Kornyakov M V, Gromalova V O and Gergenov S M 2018 Influence Of Deicing Chemicals Contaminating Automobile Headlamps On Light *Intensity Proceedings of the International Conference "Aviamechanical engineering and transport"* pp 141-146

[6] Gromalova V O, Fedotov A I, Gergenov S M, Zedgenizov V G and Neskoromnykh V V 2019 On impact of vehicle headlights contamination with products of road chemical deicing agents upon pedestrian visibility on unlit roads *IOP Conf. Series: Materials Science and Engineering 632 012056 IOP Publishing doi:10.1088/1757-899X/632/1/012056* Volume 632

IOP Conf. Series: Materials Science and Engineering 1061 (2021) 012042 doi:10.1088/1757-899X/1061/1/012042

Mathematical modeling of the percussion mechanism with a single impact energy increase

V G Zedgenizov[1], D V Kokourov[1], T A Senotrusova[1]

[1] Irkutsk National Research Technical University, 83, Lermontova ave., Irkutsk, 664074, Russia

E-mail: vzedgenizov@bk.ru

Abstract: The present study aims to develop a mathematical model of the percussion mechanism with a single impact energy increase and determine the influence of the spring stiffness and the ratio of inertial and projectile masses on the main parameters of the destruction process. The method of mathematical modeling of the destruction process was applied. A mathematical model of the percussion mechanism with a single impact energy increase and a diagram for its solution in Matlab-Simulink have been developed; dependences of the main parameters of the desctrution process on the spring stiffness and the ratio of inertial and projectile masses have been obtained. The mathematical model of the percussion mechanism makes it possible to increase the energy of a single impact by three-five times in comparison with a freely falling projectile. The reactive component is not transmitted to the base machine.

1. Introduction

Currently, the least energy-consuming method is a mechanical destruction of durable materials. Table 1 compares impact fracture indicators with other technologies [1].

Table 1. Specific energy intensity of various rock destruction technologies

No	Rock destruction technology	Specific energy consumption, J / cm^3
1	Ultrasound destruction	10104
2	Water jet destruction	1684
3	Low-power destruction	253
4	Rolling cutter destruction	126-253
5	Powerful impact destruction (with an impact speed of up to 10 m/s)	17-25
6	Explosion destruction (with an impact speed of 700-900 m/s)	6

Hydraulic hammers as replaceable working equipment for single-bucket excavators have a wide range of applications and are used for destroying oversized rocks and reinforcing concrete and concrete structures, opening asphalt concrete pavements, and developing frozen soil [2].

The destruction efficiency increases with increasing energy of a single impact [3, 4]. However, its increase is restrained by the reactive component (recoil), which is transmitted to the base machine [5].

2. Materials and methods

The paper uses mathematical modeling methods, the tool-Matlab-Simulink.

3. The study of the structure of the modified lead-tin-base bronze

The aim of the present study is to develop a percussion mechanism with increased energy of a single impact, which is not transmitted to the base machine.

Figure 1 shows a design diagram of the mechanism.

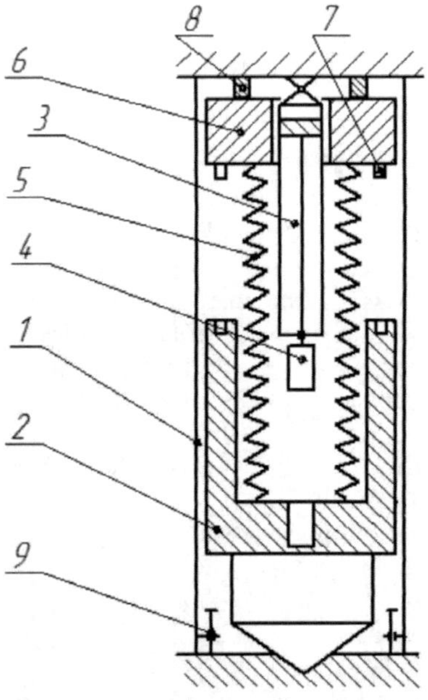

Figure 1. Design diagram of the percussion mechanism

The impact mechanism consists of barrel 1, projectile 2 coaxially placed in the barrel, hydraulic cylinder 3 pivotally connected with barrel 1, grip 4, spring 5 and inertial mass 6 with locks 7. Spring 5 is located between projectile 2 and inertial mass 6.

There are support 8 and safety elements 9.

In the non-working state, barrel 1 is installed vertically, projectile 2 is in a lower position and bumps into the destroyed surface. Inertial mass 6 and locks 7 are located in the upper part of barrel 1 and bumps into spring 5, which is in the unfolded state. The rod of hydraulic cylinder 3 is retracted and grip 4 is not connected with projectile 2.

When the working fluid is supplied to the piston cavity of hydraulic cylinder 3, the rod with grip 4 is mixed down until it contacts projectile 2. Then, the working fluid is directed into the rod cavity of hydraulic cylinder 3, projectile 3 is captured and moves along barrel 1.

As projectile 1 moves, it compresses spring 5, while inertial mass 6 is kept from moving by supports 8. When projectile 2 rests against inertial mass 6, locks 7 are closed, forming a single shock mass.

When the rod cavity of hydraulic cylinder 3 is connected to the drain, grip 4 releases projectile 2, and under gravity, the single impact mass moves downward. Having travelled a certain distance, locks 7 of inertial mass 6 open, and under released energy of compressed spring 5, projectile 2 acquires additional acceleration and strikes against the destroyed surface. Moving in the opposite direction, inertial mass 6 perceives the reactive component of spring energy, which is not transmitted to the base machine, and the mechanism returns to its original state.

Safety elements 9 protect the structure in case of failures of projectile 3.

The operations of the percussion mechanism can be described by the following system of differential equations:

$$m_1 \frac{d^2x_1}{dt^2} = c_2(x_2 - x_1) - G_1 \quad (1)$$

$$m_2 \frac{d^2x_2}{dt^2} = -c_2(x_2 - x_1) - G_2 \quad (2)$$

where m_1 – inertial mass, kg;
x_1 – displacement of the inertial mass, m;
c_2 - spring stiffness, N / m;
x_2 – projectile movement, m;
G_1 - weight of the inertial mass, N;
m_2 – projectile weight, kg;
x_2 – projectile movement, m;
G_2 – projectile weight, N.

Equation (1) determines the movement of the inertial mass, while equation (2) is used to determine the movement of the projectile, depending on the main parameters of the percussion mechanism.

When developing a mathematical model of the impact mechanism, the mass movement resistance was not taken into account.

To solve the system of equations in Matlab-Simulink, a diagram shown in Fig. 2 has been developed. Figure 3 shows the simulation results [6].

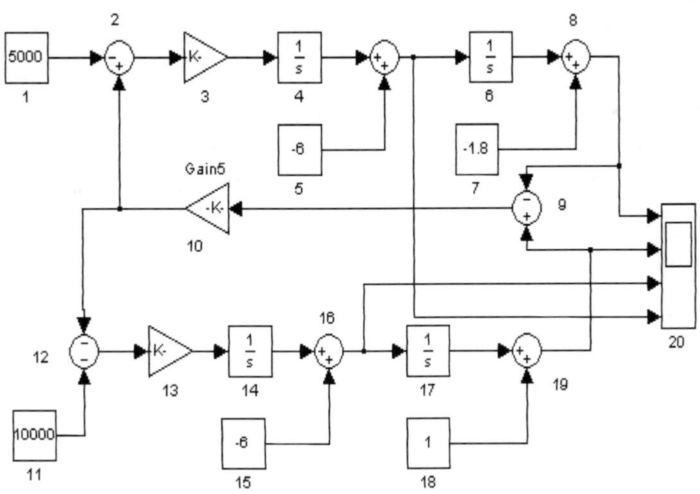

Figure 2. The diagram in Matlab-Simulink

The chain of blocks 1-10 simulates the solution of equation (1), blocks 11-19 solve equation (2), the results are output to block 20.

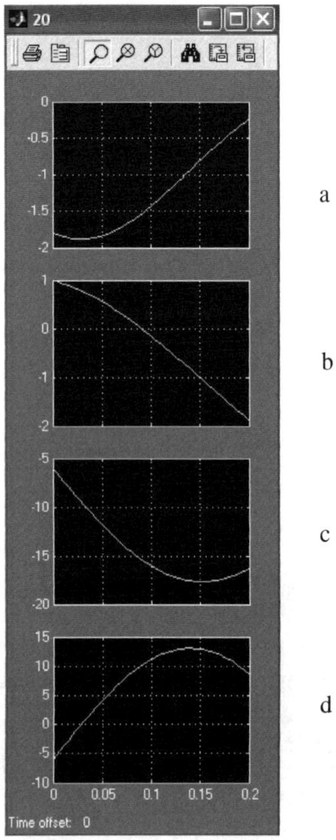

Figure 3. Simulation results:
a - displacement of the inertial mass (x1); b - movement of the projectile (x2); c - projectile speed; d - inertial mass speed

The initial conditions are as follows: projectile mass - 1000 kg; inertial mass - 500 kg; spring rate - 40 kN / m; the projectile, together with the inertial mass, is raised to a height of 2.8 m and dropped down. When the locks open at a height of 1 m, and the inertial mass continues to move down; however, as affected by the spring, it stops and begins to move in the opposite direction 0.025 seconds after (Fig. 2-a).

At a height of 1 m, the projectile acquires an additional impulse and strikes the destructible surface at the zero mark (Fig. 2-b). Moreover, its initial velocity increases from 6 to 16 m/s at the time of impact (Fig. 2-c). The maximum speed of the inertial mass is 13 m/s (Fig. 2-d).

Using the percussion mechanism model, the dependence of the main process parameters on spring stiffness was obtained (Fig. 4).

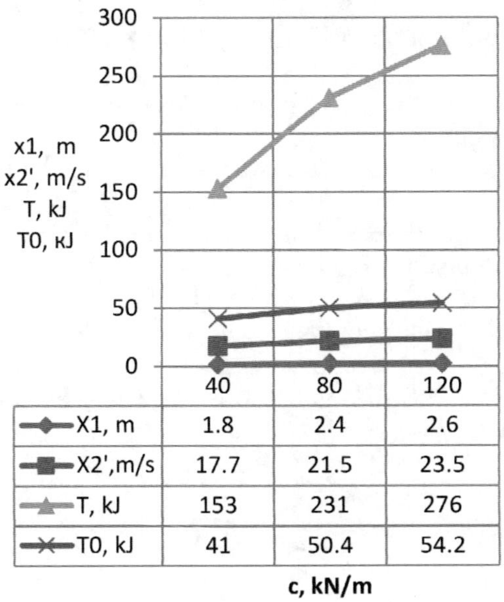

	40	80	120
X1, m	1.8	2.4	2.6
X2',m/s	17.7	21.5	23.5
T, kJ	153	231	276
T0, kJ	41	50.4	54.2

Figure 4. Dependence of the main process parameters on spring stiffness

With an increasing spring stiffness, all the indicators increase. For comparison, the graph shows T0 which is the impact energy during the free fall of a projectile weighing 1500 kg from a height equal to the sum of displacements x1 and x2. A 3-5-fold increase in impact energy can be observed.

Figure 4 shows the dependence of the main process parameters on the ratio of inertial mass and projectile mass.

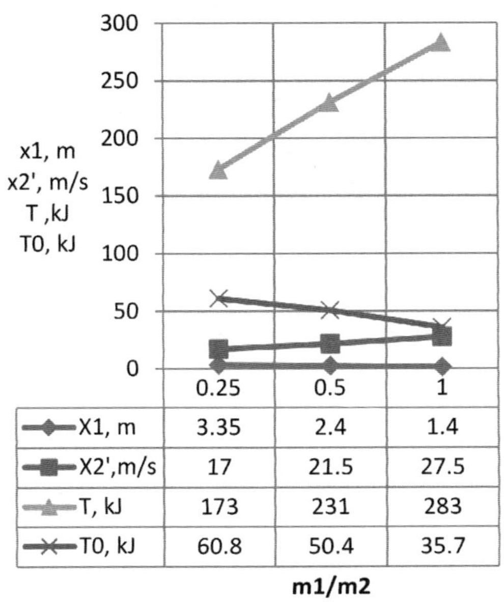

	0.25	0.5	1
X1, m	3.35	2.4	1.4
X2',m/s	17	21.5	27.5
T, kJ	173	231	283
T0, kJ	60.8	50.4	35.7

Figure 5. Dependence of the main process parameters on the ratio of inertial mass and projectile mass.

Conclusion

The model developed in the present study makes it possible to increase the energy of a single impact by 3-5 times in comparison with a freely falling projectile, while the reactive component is not transmitted to the base vehicle.

References

[1] Bazhal A I et al 2008 Essential analysis of technologies for the destruction of solids *Materials of the Ural mining decade* (April 14-23, 2008)

[2] SpecNavigator *Application of a hydraulic hammer based on an excavator* Available at: https: //specnavigator.ru/gidromolot/na-baze-ekskavatora.html

[3] Nabiulin R Sh 2006 A comparative analysis of tools for destroying oversize materials *Mining equipment and electromechanics* 1 pp 33–34

[4] Chuprov I V 2006 The relationship of specific energy of rock destruction with single impact energy *News of the universities. Mountain journal* 5 pp 66–69

[5] Al-Saqqaf Khaled Sayed Taha and Zedgenizov V G 2013 The results of the study of the working process of the striker for destrying oversized rocks using the mathematical model *Bulletin of ISTU* 7 (78) pp 18–22

[6] Chernykh I V 2008 *Modeling of electrical devices in MATLAB, SimPowerSystems and Simulink* (Moscow: DMK Press, St. Petersburg: Peter)

Justification of the system parameters for alerting drivers about the emergency situation occurrence

V A Zeer, E V Grazhdantsev, R S Glukhikh, R M Avdeev, A I Elagin

Polytechnic School, Siberian Federal University, 79 Svobodny pr., 660041 Krasnoyarsk, Russia

E-mail: egrazhdantsev@sfu-kras.ru

Abstract. The article analyzes the road traffic accident rate in Russia and the Krasnoyarsk territory. As a way to reduce traffic accident rate on road sections outside built-up areas, a system for alerting drivers in conditions of a lack of visual information about a possible emergency situation is proposed. The results of mathematical modeling and physical experiments are presented, which allow us to justify the values of the system parameters and algorithms for its functioning.

1. Introduction

In the Krasnoyarsk Territory (KT), as in the Russian Federation (RF), over the past 5 years, there has been a decrease in the number of road traffic accidents (RTA) (with a slight increase in 2019), injured and killed in them (Fig. 1) [1,2].

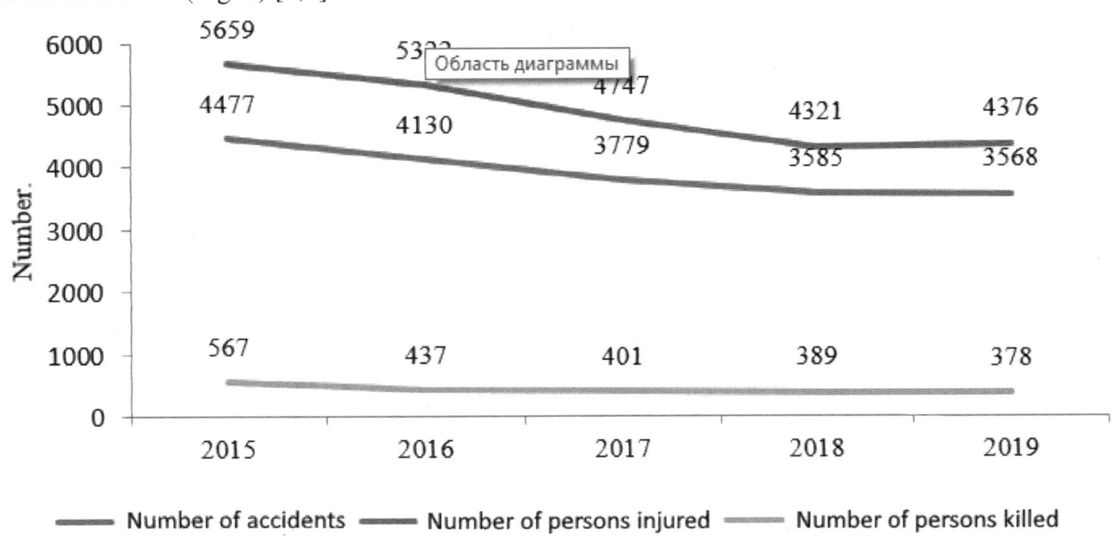

Fig. 1. Number of road traffic accidents, deaths and injuries in the Krasnoyarsk Territory

The percentage of injured and dead in road traffic accidents in the Krasnoyarsk Territory from the total number in the Russian Federation practically does not change from year to year and ranges from 2 to 2.5% (Fig. 2).

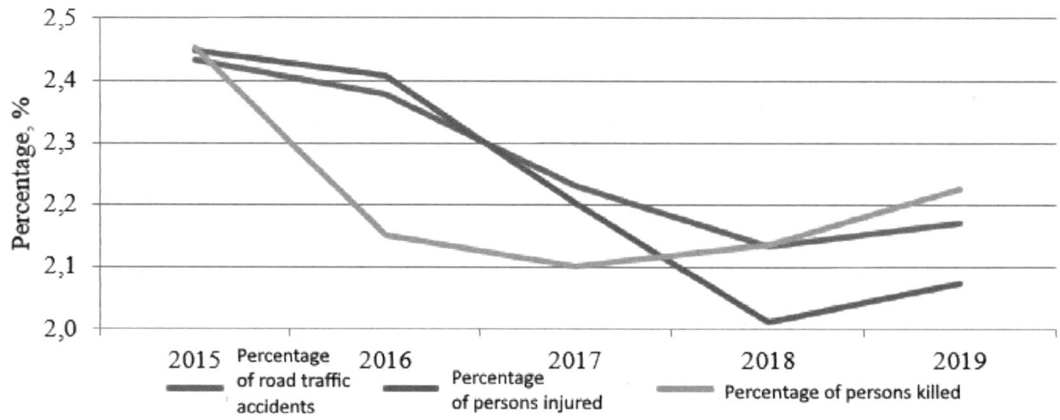

Fig. 2. Percentage of road traffic accidents, deaths and injuries in them in the Krasnoyarsk Territory from the total number in the Russian Federation

The Krasnoyarsk Territory is among the 10 regions of the Russian Federation with the largest number and the highest percentage of road traffic accidents and injuries.

Analysis of the number of accidents, injuries and deaths per 1000 km of roads in the Krasnoyarsk Territory (Fig. 3) shows a significant risk of getting into an accident, which is 110 accidents for every 1000 km of roads. At the same time, the number of injured and dead for every 1000 km is 132 and 12 people, respectively, which is also a significant negative indicator. The value of the number of accidents and fatalities per 1000 km in the Krasnoyarsk Territory corresponds to the average value of these indicators for the country.

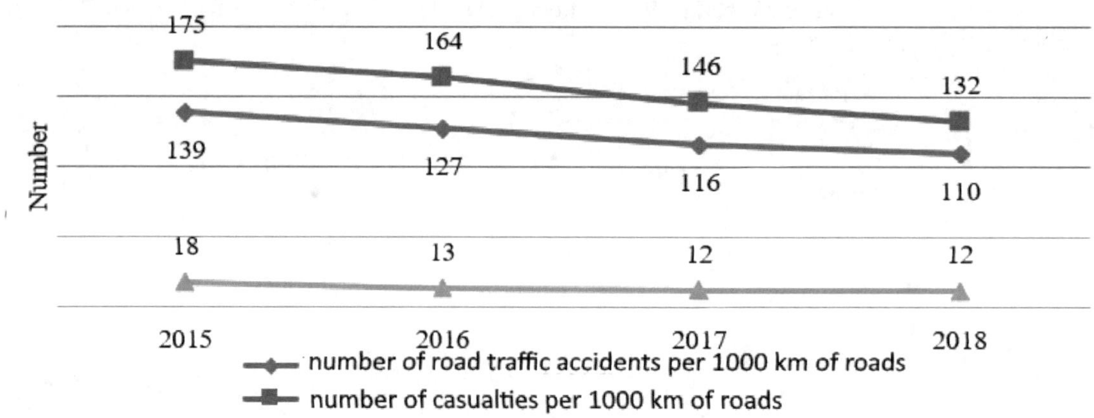

Fig. 3. Dynamics of the number of road traffic accidents, injured and killed persons in accidents per 1000 km of roads in the Krasnoyarsk Territory

The analysis of the number of injured and dead persons per 100 accidents and the number of dead per 100 injured persons (Fig. 4) also indicates a high severity of the consequences of accidents, despite the positive dynamics of their decline. In the Krasnoyarsk Territory, 123 and 11 people are injured and killed in 100 road traffic accidents, respectively, and 9 people are killed per 100 casualties.

The accident rate analysis in the Krasnoyarsk Territory shows an annual decrease in both absolute and relative indicators against the background of a decrease in similar indicators in the Russian Federation as a whole. In addition to the above, the share of accidents, persons injured and killed in them in the Krasnoyarsk Territory remains at a high level year on year, and in recent years it has been increasing, despite the fact that by the number of cars the Krasnoyarsk Territory is not among the 10 regions with the largest vehicle fleet.

The Krasnoyarsk Territory ranks 10th in the number of people killed in collisions (2.17% of the total in the country) and 11th in the number of people killed in collisions with a pedestrian (2.25% of the total in the country).

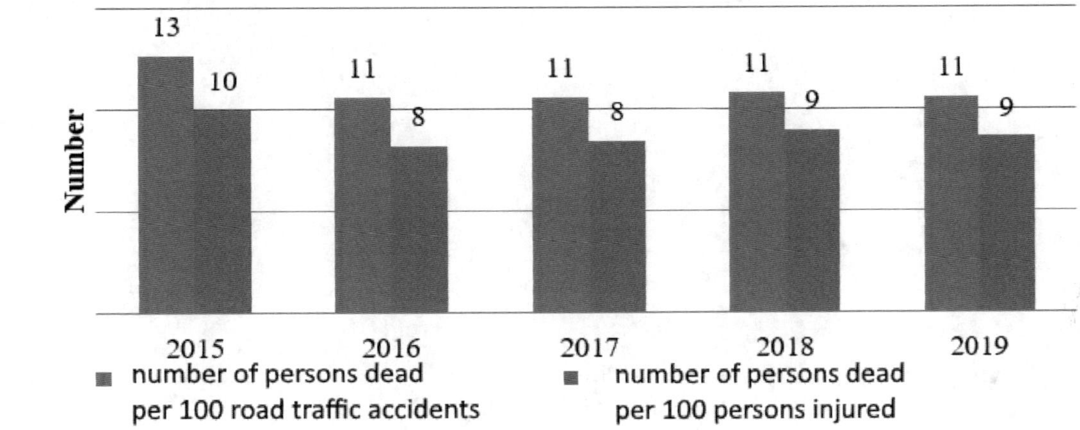

Fig. 4. Severity of the consequences of road traffic accidents in the Krasnoyarsk Territory

The most serious types of road accidents in the Krasnoyarsk Territory are collisions, rollovers, hitting a stationary vehicle, an obstacle and a pedestrian, and over the past 5 years, there has been no trend towards a decrease in the severity of consequences in the region (Fig. 5).

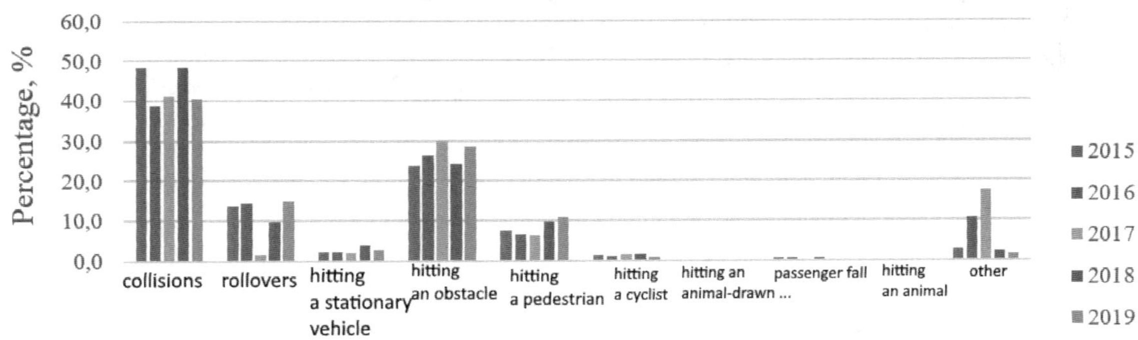

Fig. 5. Percentage of deaths in road accidents by type of the accident in the Krasnoyarsk Territory

The absolute indicators of accidents by type of road accidents show that significant highest values are achieved with such types of road accidents as collisions and collisions with pedestrians. These accidents are usually caused by reduced visibility for various reasons: insufficient (due to changes in weather events) and limited visibility (due to the complexity of the terrain). Such road sections are potentially dangerous.

Collisions with high severity of consequences are typical for roads outside built-up areas, where high vehicle speeds contribute to an increase in the number of victims [3].

Despite the fact that highways and their elements are designed to provide the necessary visibility to enable road users to prevent accidents, in practice these conditions are not always met.

One of the ways to reduce traffic accident rate on roads outside of built-up areas on emergency-dangerous sections can be a system for alerting drivers in conditions of a lack of visual information. The effective result of the system operation is guaranteed and timely notification of vehicle drivers about a possible collision under the following conditions: adverse weather conditions, complex road geometry, steep ascents, descents, dangerous turns, etc. [4, 5, 6].

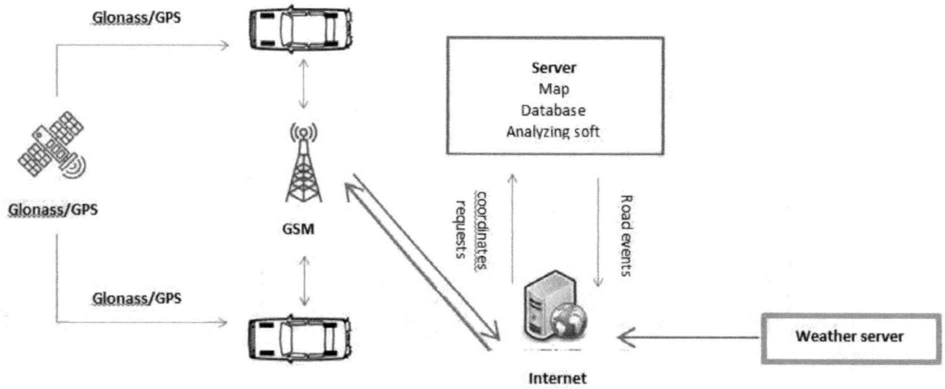

Fig. 6. A system for alerting drivers in conditions of a lack of visual information about a possible emergency situation

2. Methods and Equipment

To provide the mathematical apparatus for the functioning of the algorithms of the designed system, it is necessary to determine its basic parameters. Dangerous situations are those situations that, if no response measures are taken, will lead to an accident in the form of a collision. In fact, the response measures aimed at preventing road accidents are reduced either to a decreasing vehicle speed up to a complete stop, or to a changing the direction of the vehicle movement, or to a combination of these actions [7, 8, 9].

According to regulatory documents, the visibility distance along the entire length of the road must be at least the stopping distance of the vehicle to the obstacle:

$$S_в \geq S_o \quad (1)$$

where $S_в$– the visibility distance;

S_o – stopping distance.

Stopping distance – the distance taken by the vehicle from the time the danger was detected to its complete stop, plus the remaining distance to the object [10]:

$$S_о = S_р + S_{ср} + S_н + S_т + l_0 \quad (2)$$

where $S_р$ – the distance traversed by the car during the reaction time of the driver;

$S_{ср}$ – the distance traversed by the car during the operation of the brake drive;

$S_н$ – the distance traversed by the car during the deceleration increase;

$S_т$, – braking distance;

l_0 – margin of distance equal to 5-10 m.

In expanded form, formula (2) can be written as follows, m:

$$S_о = t_р \times V_а + t_{ср} \times V_а + 0{,}5t_н \times V_а + \frac{V_а^2}{2g\varphi} + l_0 \quad (3)$$

where $t_р$ – driver reaction time, s;

$t_{ср}$ – brake drive reaction time, s;

$t_н$ – deceleration reaction time, s;

$V_а$ – vehicle speed, m/s;

g – acceleration of gravity, m/S[2];

φ – the road adhesion coefficient.

When assigning geometric parameters of roads on sections of ascents and descents, the distance of visibility of an oncoming car is used [11], which is calculated, m:

$$S_{\text{в}} = \frac{V_{85\%}}{3{,}6} t_{\text{p}} + \frac{К_{э} V_{85\%}^2}{254(\varphi \pm i)} + l_0 \tag{4}$$

where $V_{85\%}$ – the speed of the traffic flow, provided 85% km/h;

$К_{э}$ – the coefficient of braking efficiency, assumed to be 1.2 when the road surface is dry;

i – road inclination.

Taking into account the above, we will determine the minimum visibility parameters based on normative documents regulating the geometric parameters of highways for the following typical road traffic situations:

- there is a car moving in the same direction ahead;
- there is a car moving in the opposite direction ahead;

Situation 1. In this situation, the car meets an obstacle in the same lane and requires a complete stop at a safe distance from the obstacle.

The visibility distance without taking into account the road inclination is:

$$S_{\text{в}} = t_{\text{p}} \times V_{\text{a}} + \frac{V_{\text{a}}^2}{2g\varphi} + l_0 \tag{5}$$

In accordance with the recommendations on road safety, the driver reaction timet_{p} is assumed to be 2.5 s. The values of the adhesion coefficient φ for asphalt concrete pavement are set by the methodological manuals [10, 11] and range from 0.8 (dry, clean) to 0.2 (icy).

Situation 2. In this situation, the cars must stop before reaching each other.

The visibility distance is the sum of the distances that cars travel during the reaction time of the drivers $S_{\text{p}1}$ and $S_{\text{p}2}$, the braking distances of the cars S_{t1} и S_{t2}, as well as the distance margin:

$$S_{\text{в}} = S_{\text{p}1} + S_{t1} + S_{\text{p}2} + S_{t2} + l_0 \tag{6}$$

or

$$S_{\text{в}} = \left(t_{\text{p}1} \times V_{\text{a}1} + \frac{V_{\text{a}1}^2}{2g\varphi}\right) + (t_{\text{p}2} \times V_{\text{a}2} + \frac{V_{\text{a}2}^2}{2g\varphi}) + l_0 \tag{7}$$

In the case of the system being developed, as a criterion for deciding on the alarm to a vehicle driver of a possible "obstacle" on the road in the direction of its movement, it is proposed to use the minimum safe distance between road users ($S_{\text{общ}}$), which is determined by calculation. The condition for sending an alert signal is written as follows:

$$S_{\text{дейст}} \leq S_{\text{общ}} \tag{8}$$

where $S_{\text{дейст}}$ – the actual distance between road users, m;

$S_{\text{общ}}$ – minimum safe distance between road users, m.

To determine the distance taken by the i-th car, the formula of the stopping distance, m, is used:

$$S_{i\text{общ}} = S_{\text{p}} + S_{\text{cp}} + S_{\text{т}} + S_{\text{откл}} + l_0 \tag{9}$$

where $S_{\text{откл}}$ – the distance traversed by the vehicle during the system response time, m.

We express $S_{i\text{общ}}$ from (9), assuming that the braking process corresponds to an equidistant movement to a complete stop with the maximum possible deceleration for specific road conditions, m:

$$S_{i\text{общ}} = \vartheta_i * t_{\text{p}} + \vartheta_i * t_{\text{cp}} + \vartheta_i * t_{Ti} - \frac{j * t_{Ti}^2}{2} + \vartheta_i * t_{\text{откл}} + l_0$$

or

$$S_{i\text{общ}} = \vartheta_i * \left(t_{\text{p}} + t_{\text{cp}} + t_{Ti} + t_{\text{откл}}\right) - \frac{j * {t_{Ti}}^2}{2} + l_0$$

or

$$S_{i\text{общ}} = \vartheta_i * \left(t_{\text{p}} + t_{\text{cp}} + t_{\text{откл}}\right) + \frac{{\vartheta_i}^2}{2j} + l_0 \tag{10}$$

where ϑ_i – the initial speed of the i-th car, m/s;

IOP Conf. Series: Materials Science and Engineering 1061 (2021) 012043 doi:10.1088/1757-899X/1061/1/012043

t_{Ti} – braking time of the of the i-th car, s;
$t_{откл}$ – the response time of the system, s;
j – maximum possible deceleration, m/c^2.

In order to account for weather conditions, jfrom the equation for determining the braking force of the car, we define N:

$$P_t = m_a * g * \varphi = m_a * j$$

hence

$$j = g * \varphi$$

where m_a – the vehicle weight, kg.

It should also be noted that the car deceleration during emergency braking can be assumed to be j=7.84 m/c^2 for dry surfaces, 3.92 m/c^2 for wet surfaces, and 1.96 m/c^{2}for icy surfaces. During service braking, the car deceleration is assumed to be 3 times less than during emergency braking, and not exceeding 2.6 m/c^2 for dry surfaces, 1.3 m/c^2 for wet surfaces, and 0.65 m/c^2 for icy surfaces.

The braking time of the i-th car under the condition of equally slow movement to a complete stop can be found:

$$t_{Ti} = \frac{\vartheta_i}{j}$$

Consider the situation 1, when the cars are moving in the same direction, then $S_{общ}$ is determined by the formula (10) as a difference of the ways, passed by cars for the time required to complete stop of the catching-up car (with index 2) taking into account the rest of the way between the objects, m:

$$S_{общ} = S_{2общ} - S_{1общ}$$

then

$$S_{общ} = (\vartheta_2 - \vartheta_1) * (t_р + t_{ср} + t_{T2} + t_{откл}) - \frac{j * t_{T2}{}^2}{2} + l_0$$

or

$$S_{общ} = (\vartheta_2 - \vartheta_1) * (t_р + t_{ср} + t_{откл}) + \frac{(\vartheta_2 - \vartheta_1)^2}{2j} + l_0 \quad (11)$$

For situation 2, when cars are moving in the opposite direction, $S_{общ}$ is defined as the sum of the ways, passed by cars to a complete stop, taking into account the rest of the way between the objects, m:

$$S_{общ} = S_{1общ} + S_{2общ}$$

then

$$S_{общ} = (\vartheta_1 + \vartheta_2) * (t_р + t_{ср} + t_{откл}) + \frac{\vartheta_1{}^2 + \vartheta_2{}^2}{2j} + l_0 \quad (12)$$

3.The research results

Based on the previously proposed mathematical models, we constructed graphs that reflect the effects of the response time of the projected system (Fig. 6), the coupling coefficient (Fig. 7) and the braking mode (Fig. 8) on the stopping way or the minimum safe distance at different vehicle speeds in comparison with the standard values. At the same time, the stopping distance was calculated without taking into account the reserve of the remaining way l_0 The speed range of the vehicle (30-240 km/h) is conditioned by the following data: the minimum value at which in the event of an accident there may be serious consequences for the driver and passengers; the maximum value in the conditions of the speed limit on road sections outside the locality (90 km/h), taking into account the speedometer error (+20 km/h) and oncoming traffic (multiply by 2).

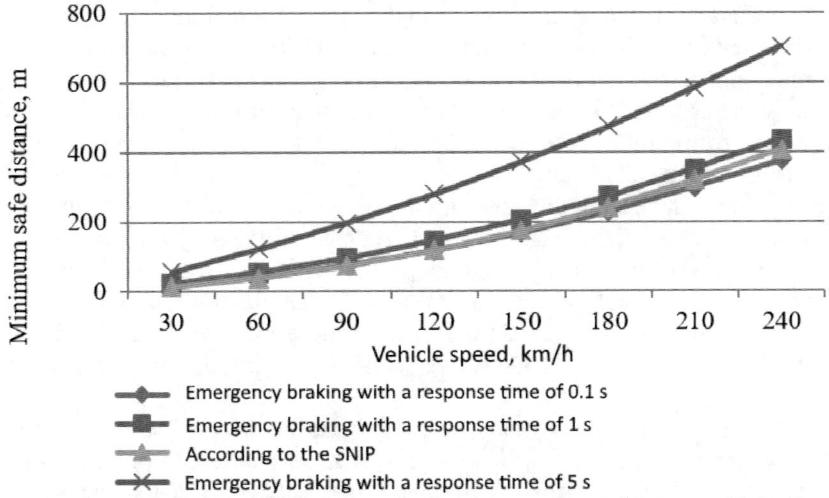

Fig. 7. Dependence of the minimum safe distance on the vehicle speed during emergency braking and the coefficient of adhesion of 0.8

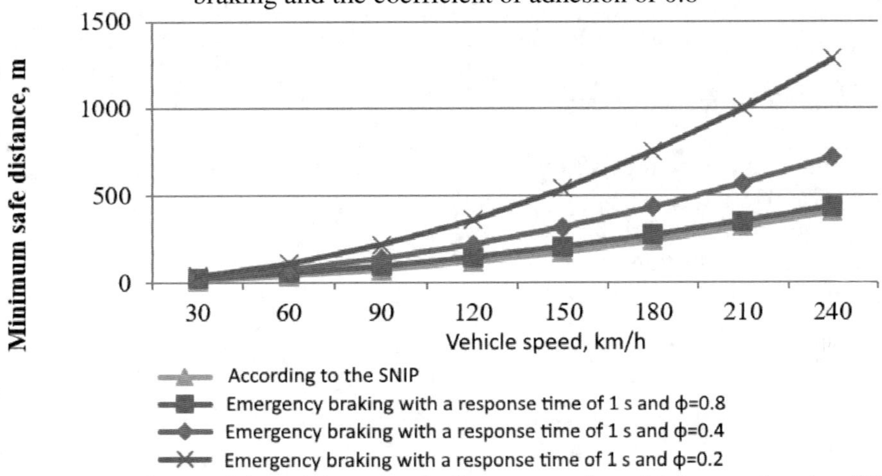

Fig. 8. Dependence of the minimum safe distance on the vehicle speed under different road surface conditions and emergency braking

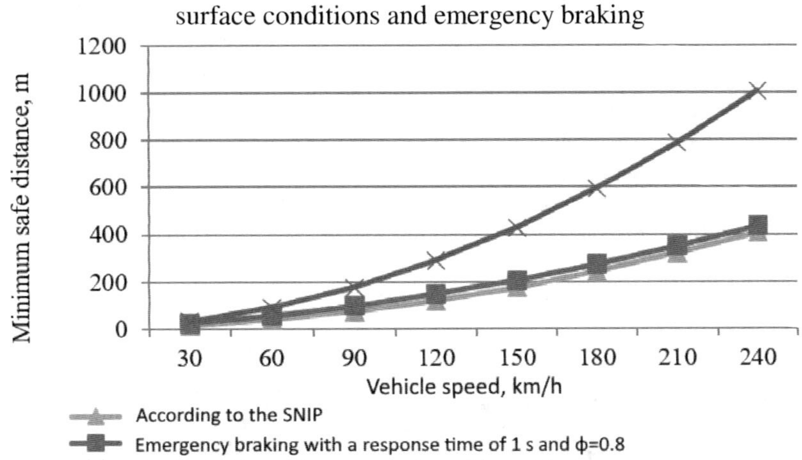

Fig. 9. Dependence of the minimum safe distance on the vehicle speed under different braking modes

Analyzing the obtained dependencies, we can draw the following conclusions: - the proposed mathematical model for determining the minimum safe distance between vehicles does not contradict regulatory documents in the field of road design and safety; - for the correct operation of the system, it is necessary and sufficient that the response time constitutes no more than 1 s; - road conditions significantly affect the minimum safe distance between vehicles, so the projected system must determine and operate in the calculation algorithms with the current values of the tire coupling coefficient with the road surface; - the braking mode (deceleration) also significantly affects the minimum safe distance, especially at high vehicle speeds, so in the projected system it is advisable to use a two-stage notification of drivers, corresponding to service and emergency braking.

To check the technical capabilities of the system, an application for mobile phones has been developed. This application provides transmission of vehicle location coordinates to the server and confirmation of their correct fixation on the server while registering the time of sending and receiving data.

The time of sending service data and receiving confirmation allowed us to determine the response time of the system taking into account various factors, such as the quality of communication in places with different terrain and distance from Krasnoyarsk, as well as to check the accuracy of vehicle positioning. The results of the processed experimental data are presented in table 1.

Table 1. Results of processed experimental data

Place of measurement, factors of influence	Positioning accuracy, m	Average response time, ms	Quality of communi cation
Rain, urban environment, large number of houses and buildings	7-12	135	4G
Rain, urban environment, forest plantations, small dwellings	5-10	165	3G
Rain, urban environment, forest plantations, small dwellings	5-10	513	2G
Rain, track outside the city, flat terrain, buildings	3-4	121	4G
Rain, track outside the city, flat terrain, buildings	3-4	444	3G
Rain, track outside the city, flat terrain, buildings	3-4	503	2G
Rain, track outside the city, flat terrain, forest, developed road infrastructure	4-6	161	4G
Rain, track outside the city, flat terrain, forest, developed road infrastructure	4-6	150	3G
Rain, track outside the city, flat terrain, forest, developed road infrastructure	4-6	419	2G
Rain, country road, mountainous terrain, forested area	5-7	218	4G
Rain, country road, mountainous terrain, forested area	5-7	202	3G
Rain, country road, mountainous terrain, forested area	5-7	653	2G

According to the experimental data obtained, the response time of the system satisfies the conditions of the mathematical model, even if poor GSM communication quality and unfavorable meteorological conditions are taken into account.

The maximum values of the vehicle positioning accuracy can be used as a justification for the remaining path margin $l_0$in the mathematical model of the designed system.

4. Conclusions

Analysis of accidents on road transport in the Russian Federation and the Krasnoyarsk territory shows that, despite the positive dynamics over the last 5 years of reducing the number of accidents, injuries and deaths, the severity of the consequences of accidents remains high.

One of the ways to reduce accidents, especially on road sections outside localities that are characterized by high vehicle speeds, can be a system for notifying drivers in conditions of a lack of visual information about the possible occurrence of an emergency situation.

The results of mathematical modeling and physical experiments allowed us to identify the most significant parameters of the system and justify their numerical values, as well as algorithms for its functioning.

Collectively, the results of the research showed the possibility of technical implementation of the designed system.

References

[1] *Official website of the Federal state statistics service in the information and telecommunications network "Internet"* Available at: http://www.gks.ru

[2] *Road traffic safety indicators. Official website of the state traffic Inspectorate* Available at: http://stat/gibdd.ru

[3] *Information and analytical center for coordinate-time and navigation support. Global navigation satellite system GLONASS* Available at: https://www.glonass-iac.ru/guide/gnss/glonass.php.

[4] Grazhdantsev E V, Zeer V A, Pulianova K V, Voevodin E S, Askhabov A M and Elagin A I 2020 Development of a model of a system for notifying drivers of vehicles about the likely occurrence of an accident. *Truck: transport complex, special equipment* **4** pp 35–42

[5] Glukhikh R S, Grazhdantsev E V, Zeer V A, Rastovtsev P A and Elagin A I 2020 Drivers alert system in the conditions of information shortage *IOP Publishing Ltd Journal of Physics: Conference Series* Vol 1515 *Instrumentation Technologies and Environmental Engineering*

[6] Elistratov V V, Bezrukov S I, Stenin P G and Klimakov V S 2015 Concept of development of collision prevention systems for vehicles *Scientific review. Technical science* **1** pp 178–178

[7] Augustsson N and Wilhelmsson M 2010 *Verification of Collision Avoidance Functionality. A mileage multiplier approach to verify future Collision Avoidance Systems.* Master of Science Thesis in the Master Degree Programme Systems, Control and Mechatronics (Department of Signals and Systems. Division of Automatic Control, Automation and Mechatronics Chalmers University of Technology, Goteborg, Sweden) Report No. EX034/2010

[8] Abboud K and Zhuang W 2009 Modeling and analysis for emergency messaging delay in vehicular ad hoc networks *Proceedings of the IEEE GLOBECOM*

[9] Ho Ting Cheng, Hangguan Shan, Weihua Zhuang 2011 Infotainment and road safety service support in vehicular networking: From a communication perspective *Mechanical Systems and Signal Processing* **25** pp 2020–2038

[10] SP 34.13330.2012. Set of rules. Roadways. Updated version of Building Code 2.05.02-85*, approved by order no. 266 of the Ministry of regional development of Russia dated 30.06.2012

[11] *Recommendations on road safety* 2011 ODM 218.4.005-2010 (Moscow)

AUTHOR INDEX